Layered, Functional Gradient Ceramics, and Thermal Barrier Coatings: Design, Fabrication and Applications

Layered, Functional Gradient Ceramics, and Thermal Barrier Coatings: Design, Fabrication and Applications

Proceedings of the SICMAC Summer School on
Layered, Functional Gradient Ceramics, and Thermal Barrier Coatings
held in Maó, Menorca Island (Spain) on June 11-16, 2006

Edited by

M. Anglada, E. Jiménez-Piqué and P. Hvizdoš

Universitat Politècnica de Catalunya, Barcelona, Spain

ttp TRANS TECH PUBLICATIONS LTD
Switzerland • UK • USA

Trans Tech Publications Ltd
Laubisrutistr. 24
CH-8712 Stafa-Zuerich
Switzerland
http://www.ttp.net

ISBN 0-87849-424-3
ISBN-13 978-087849-424-8

Volume 333 of
Key Engineering Materials
ISSN 1013-9826
Covered by Science Citation Index
Full text available online at *http://www.scientific.net*

Distributed worldwide by

Trans Tech Publications Ltd.
Laubisrutistr. 24
CH-8712 Stafa-Zuerich
Switzerland

Fax: +41 (44) 922 10 33
e-mail: sales@ttp.net

and in the Americas by

Trans Tech Publications Inc.
PO Box 699, May Street
Enfield, NH 03748
USA

Phone: +1 (603) 632-7377
Fax: +1 (603) 632-5611
e-mail: sales-usa@ttp.net

Printed in the Netherlands

Organizing Comittee

M. Anglada	chairman
M. Bartsch	member
C. Baudín	member
R. Danzer	member
G. de Portu	member
J. Dusza	member
O. Van der Biest	member
E. Jiménez-Piqué	secretary
P. Hvizdoš	member

Preface

The Summer School on "Layered, Functional Gradient Ceramics, and Thermal Barrier Coatings: Design, Fabrication and Applications" was organized as a final training activity of the European Research Training Network SICMAC (Structural Integrity of Ceramic Multilayers And Coatings), for which it meant the summit of its last three years work. The Summer School was aimed to disseminate and exchange knowledge on the *design, fabrication and reliability of layered ceramics, ceramic coatings and functionally graded ceramics in order to present an overview on the fundamentals in this area as well as an insight into critical issues and perspectives of these materials.*

The main topics can be divided in five different groups:
- design and processing of multilayers;
- functionally graded materials;
- structural integrity of multilayers;
- thermal barrier coatings: processing and lifetime assessment;
- fatigue, damping and high temperature behaviour of coatings and multilayers.

In the course of the Summer School 18 invited speakers presented 20 comprehensive talks, each focused in a specific area of this timely subject. Other 20 contributions were made by researchers and SICMAC young researchers in form of posters. Moreover, the Summer School provided plenty of space for discussions with an outstanding and diverse group of scientists at the forefront of research and helped to establish new professional contacts between participants.

It is my pleasure and honor to express my gratitude to all speakers, collaborators and participants for making the Summer School possible and successful. I would also like to thank all my colleagues from the RTN SICMAC for their collaboration during the period of its functioning. I am also grateful for the support of the European Community's Human Potential Programme under contract HPRN-CT-2002-00203, [SICMAC] and to the "Institut Menorquí d'Estudis" of Menorca for the support received in the organization of the event.

Barcelona, July 2006

Marc J Anglada
Coordinator of SICMAC

Table of Contents

Key Engineering Materials Vol. 333 (2007) pp 1-16
online at http://www.scientific.net
© 2007 Trans Tech Publications, Switzerland

Stresses and Crack Extension in Multi-Layered Ceramic Composites

F. F. Lange

Materials Department, University of California at Santa Barbara
Santa Barbara, CA 93106 USA

flange@engineering.ucsb.edu

Keywords: laminar, ceramics, composites, compressive stresses, threshold strength, bifurcation, edge-cracking, crack interaction

Abstract. It has been demonstrated, through theory and experiments, that compressive layers arrest large surface and internal cracks to produce a stress below which the material will not fail. This enables the materials to have a Threshold Strength. The stress intensity function, K, was derived for a crack sandwiched between two compressive layers. This function suggests that the threshold strength is proportional to the magnitude of the residual, compressive stress, the thickness of the compressive region, and inversely proportional to the distance between the compressive regions. All of these factors have been experimentally examined for laminar composites containing thin, compressive layers. Cracks that propagate straight though the layer obey the K function used to model this behavior. Crack bifurcation, which occurs at high compressive stresses, produces a larger threshold strength than predicted. Crack bifurcation is not fully understood.

During the initial studies, differential thermal contraction during cooling from the densification temperature was used to develop the compressive stresses. A molar volume change to induce the compressive stress was also used to develop the compressive stresses. In one case, it was shown that the compressive stresses could arise when the compressive layer contained a material that underwent a structural phase transformation during cooling. In another, ion exchanged glass plates that are subsequently bonded together also produce a threshold strength. Factors that affect the threshold strength are reviewed.

Introduction

The strength of a brittle material is not a singular value, but a distributed set of values that reflect the large variety of flaws (types and sizes) that are incorporated during processing. The distribution of strength values obtained during testing, all made at one time, are generally characterized by statistical parameters that vary with the processing method and processing period. That is, most manufactures cannot control the nature of the flaws they inadvertently incorporate during processing. Proof testing, i.e., the application of a specific stress to a component, can be used to truncate the statistical distribution, to define a minimum, or threshold strength, for components that do not fail the proof test. Although there is a significant cost to proof testing, it allows the designer to ensure reliability.

It was recently shown that a threshold strength (i.e., a strength below which the probability of failure is zero) can be obtained in laminar ceramics composed of periodic, alternating layers of one material separated by thinner layers of a second material. [1] The second layer must contain a residual, biaxial compressive stress produced by either differential thermal contraction or a molar volume change, e.g., a phase transformation. A threshold strength has been the 'holy grail' of structural ceramics. It has been demonstrated that large flaws within the thicker layers that extend at a low stress will arrest as they entered the compressive layers. An increasing stress must be applied to 'push' the crack through the compressive layers to cause catastrophic failure. For periodic laminates, failure never occurs below a threshold stress despite large differences in the initial size of the crack present in the thicker layers.

Although this review will concentrate on laminates that can exhibit a threshold strength, the general background concerning stresses and crack extension in laminate composites will be reviewed first.

Stresses in Multi-Layered Composites

Bi-axial Stresses Deep within Layers. Laminates formed at elevated temperatures with two or more materials develop stresses during cooling due to different thermal-elastic properties. For example, during cooling from T_0 to T, the differential thermal contraction of one material sandwiched between two identical layers of a second material will produce a strain given by [2]

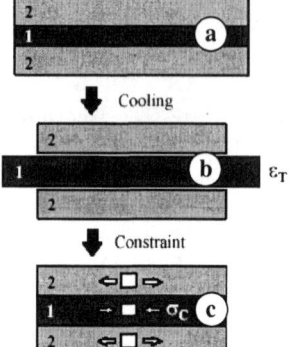

Fig. 1 Symmetric laminate composed of two materials.

$$\varepsilon_r = \int_T^{T_0} (\alpha_2 - \alpha_1) dT, \qquad (1)$$

where α_1 and α_2 are the thermal expansion coefficients of the two materials.

Fig. 1 helps to visualize how the internal residual stresses arise during cooling. Fig. 1a shows the material with the lower thermal expansion coefficient sandwiched between the other material. When the layers are not bonded together, the two layers with the greater thermal expansion coefficient will contract more than the other during cooling as shown in Fig. 1b. If they were bonded together and cooled, both would contain the same strain at the lower temperature given by eq. (1). To determine the biaxial stresses in both materials, one needs to apply biaxial compressive stresses to the layer(s) with the smaller thermal contraction such that the dimensions of all layers match one another. As the applied compressive stress is released after all are bonded together, all three layers will increase their length. During this step, residual stresses will arise in all three layers. These residual stresses will depend on the residual strain, given by Eq. (1), the different elastic properties of the two materials and their respective volume fractions, which for a laminate, is given by the thickness ratio of the two materials.

To understand the effect of elastic properties, one can assume that the elastic modulus of the two sandwiching layers in Fig. 1 are either infinite or zero. When infinite, they will not expand, and the residual compressive stress within the center layer will not relax as the applied compressive stress is removed. When the elastic modulus of the outer layers is zero, the center layer will completely relax to its unconstrained dimensions as the applied compressive stress is removed; namely, no residual stresses will develop in any of the layers. The effect of the volume fraction of the two materials is easily understood by recognizing that the sum of the tensile and compressive forces (F_1 and F_2) acting across the respective cross sections (A_1, A_2) of the two materials must equal zero; namely, no resultant force exists to cause the laminate to move in the space-time coordinates. Thus, for a laminate shown in Fig. 1,

$$F_1 + F_2 = \sigma_1 A_1 + \sigma_2 A_2 = 0 \quad \text{or} \quad \sigma_2 A_2 = -\sigma_1 A_1 \qquad (2)$$

Since all layers are assumed to have the same width (w), and $A_i = t_i w$ (i = 1, 2), then

$$\sigma_2 = -\sigma_1 \frac{t_1}{t_2}. \qquad (3)$$

Along with Eq. (3) it can be shown that the residual compressive (or tensile) stress that arises in this laminate can be expressed by [2]

$$\sigma_1 = \varepsilon_r E_1' \left(1 + \frac{t_1}{t_2} \frac{E_1'}{E_2'} \right)^{-1} \qquad (4)$$

where $E_i' = E_i/(1-v_i)$, E is the Young's modulus, and v is the Poisson's ratio.

Because crack extension initiates via tensile stresses, one generally attempts to fabricate a laminar composite to minimize the tensile stress. Assuming that the tensile stress (σ_t) resides in material 2, Eq. (3) shows that the tensile stress can be minimized by making the tensile layers thick relative to the compressive layers. Namely, the tensile stresses can be minimized when the tensile layers are much thicker than the compressive layers, $t_2 \gg t_1$. Thus, as $t_1/t_2 \Rightarrow 0$, $\sigma_t \Rightarrow 0$. Like wise, the maximum compressive stresses (σ_c) can be developed when $t_1/t_2 \Rightarrow 0$, which from Eq. (4) becomes

$$\sigma_c = \varepsilon_r E_1'. \qquad (5)$$

The above summary assumes that the composite is symmetric, namely, the net stresses on one side of the center line are balanced by the net stresses on the other side, as shown in Fig. 1. Bending stresses would arise if the laminate was non-symmetric. Although the strain used above was assumed to arise due differential thermal expansion, molar volume changes due to a structural phase transformation and/or a chemical reaction (two phases that react to form a third phase) can produce the same state of stress.

Tri-axial Stresses At and Near the Surface. Although the biaxial stresses, described by Eqs. (3) and (4), exist deep within each layer, different stresses exist at and near the external surface. At the surface, the stresses are biaxial, and become tri-axial just below the surface.[2] Of most interest is the case for the layer that contains, biaxial compressive stresses far from the free surface. In this case, tensile stresses exist perpendicular to the center line at and near the surface as shown in Fig. 2 a.

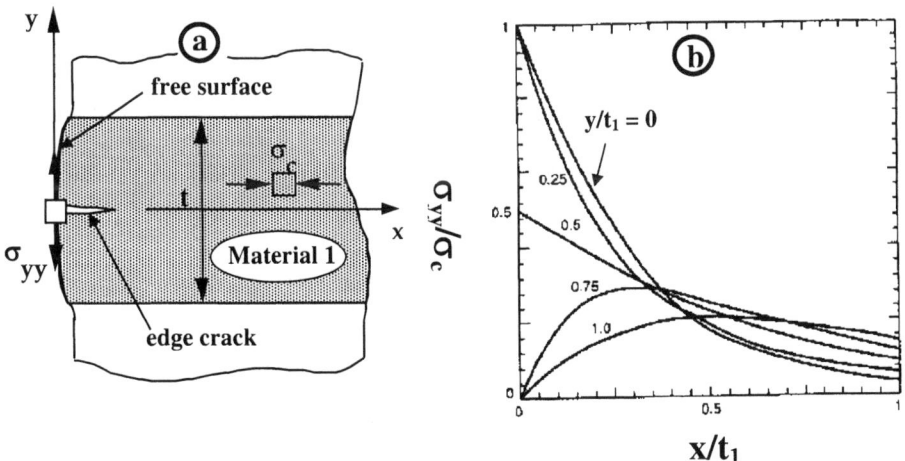

Fig. 2 a) Schematic of the compressive layer terminating at a free surface and the associated tensile stress, σ_{yy} [2]that b) changes as a function of x and y as shown, when normalized by σ_c, the biaxial compressive stress far from the free surface.[3]

The largest tensile stress occurs at y = 0; they are given by [2,3]

$$\sigma_{yy}(x)\big|_{y=0} = \frac{2}{\pi}[\theta - \frac{1}{2}\sin 2\theta]\,\sigma_c \qquad (6)$$

where x is the distance from the free surface, $\tan\theta = t_1/2x$, and σ_c is the absolute value of the residual biaxial compressive stress far from the free surface (x >> 0). As shown by Eq. (6), the tensile stresses have a maximum value at the surface (x = 0) and diminish to a negligible value at a distance from the surface that is approximately equal to the thickness of the compressive layer.

Fig. 2b shows [3] the distribution of the residual stress component σ_{yy} near the surface of a thin layer, assumed to be under biaxial compression far from the free surface. The centerline of the thin layer coincides with the x-axis (y = 0). On the surface (x = 0), σ_{yy} is a step-function, equal to $|\sigma_c|$ in the thin layer, $|\sigma_c/2|$ at the interface between the thin and thick layers, and zero in the two adjacent thick layers. Equation (6) is labeled as y/t = 0 in Fig. 4. Fig. 2b also illustrate σ_{yy} vs. x/t for other specific values of y/t. It can be shown that the other principle stresses close to the surface, namely, σ_{xx} and σ_{zz}, are compressive stresses. It can be shown that $\sigma_{xx}(x) = \sigma_{yy}(x) + \sigma_c$, and $\sigma_{zz} = \sigma_c$, thus, $\sigma_{xx}(0) = 0$ and $\sigma_{xx} \to \sigma_c$ as x → t.

Of less interest are the tri-axial stresses at the free surface that terminates a layer containing biaxial tensile stresses σ_t deep within the layer. For this case, it can be shown [3] that $\sigma_{yy}(x)$ is similar to Eq. (6), except that σ_t is substituted for σ_c; namely, a compressive stress exists where the tensile layers terminates at a free surface.

Crack Extension in Brittle Laminates

Edge-Cracking due to Tri-Axial State of Stress. It has also been shown that because the tensile stresses are highly localized near the surface of the compressive layer, they can give rise to a surface crack, called an edge crack. The edge crack extends along the center line of the compressive layer when either the layer thickness and/or the value of compressive stress exceeds a critical value. The determination of the strain energy release rate function for edge-cracking is very similar to other problems associated with highly localized states of stress. These problems include the formation of microcracks around inclusions, cracking of thin films, crack extension associated with Hertzian contact stresses, and tunnel cracking in the tensile layer of laminar composites that will be discussed below. With the exception of cracking associated with the localized, Hertzian contact stresses, all of these problems are associated with a residual state of stress and the fact that crack extension reduces the stored strain energy within a localized volume of material.

For the current problem, i.e., the formation of an edge crack, the tensile stress is localized within the compressive layer in a region where the layer terminates at a free surface. With the assumption that the compressive layer is very thin compared to the two adjacent tensile layers (Fig. 1), only the stresses associated with compressive layer need be considered, namely, the stresses within the tensile layers are very small. The strain energy release rate function for the edge crack can be derived using a dimensional analysis by further assuming that the laminate is cylindrical, with a radius R. Prior to the formation of an edge crack, the total strain energy within the compressive layer can be defined as U^o_{se}. When a small crack is able to extend to form an edge crack that circumvents the compressive layer, the strain energy associated with the compressive layer containing the edge crack is reduced and can be expressed as

$$U_{se} = U^o_{se} - \left(\frac{F\sigma_c^2}{E*}\right)(2\pi R Z c t_1), \qquad (7)$$

where t is the thickness of the compressive layer, σ_c is the compressive stress deep within the layer, c is the length of the crack that extends from the surface into the compressive layer, $2\pi RZct_1$ is the volume associated with the circumferential surface crack in which the strain energy has been released, and Z is a dimensionless constant that helps defines this volume. The factor $\left(\dfrac{F\sigma_c^2}{E*}\right)$ is the strain energy per unit volume in the compressive layer, close to the free surface, F is a dimensionless function that relates the stresses, $\sigma_{yy}(x)$ and $\sigma_{xx}(x)$ to σ_c.

The energy consumed during crack extension is given by

$$U_s \;=\; 2\pi R\,c\,G_c, \tag{8}$$

where $4\pi R\,c$ is the area of the circumferential surface crack, and G_c is the critical strain energy release rate for the material that forms the compressive layer.

Summing Eqs. (7) and (8), the total free energy as a function of the crack length is given by

$$U_t \;=\; U_{se}^o \;-\; \left(\frac{F\sigma_c^2}{E*}\right)(2\pi R\,Z\,c\,t_1) \;+\; 2\pi R\,c\,G_c. \tag{9}$$

An examination of Eq. (9) shows that the free energy of the system will only decrease when the sum of the second and third terms is negative, namely, when

$$\left(\frac{F\sigma_c^2}{E*}\right)(2\pi R\,Z\,c\,t_1) \;>\; 2\pi R\,c\,G_c \quad \text{or} \quad Z\,t_1\left(\frac{F\sigma_c^2}{E*}\right) \;>\; G_c \tag{10}$$

Thus it can be concluded that for fixed values of σ_c, G_c and E^*, the edge crack will only form when

$$t_1 \;>\; t_c \;=\; \frac{G_c\,E*}{Z\,F\,\sigma_c^2}, \tag{11}$$

where t_c is the critical layer thickness to produce an edge crack. The reason for this is that for given values of σ_c, G_c and E^*, the highly localized strain energy, which depends on the thickness of the compressive layer, is only sufficient to compensate for the work needed to produce a crack when $t_1 \geq t_c$. A rigorous analysis of this problem shows that the numerical constants $ZF = 0.34$. [2]

Tunnel Cracking in the Tensile Layer [4]. In this section, crack extension in the layers containing biaxial tensile stresses will be detailed; these layers will be called 'tensile layers'. If the residual tensile stress is large enough, it will cause pre-existing flaws to "tunnel" through the layer to terminate at the free surface that bound the layer as shown in Fig. 3. Under biaxial tension, multiple tunnel cracks produce a 'mud' crack pattern, similar to those seen in thin films.

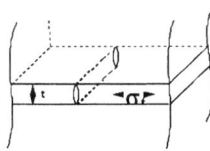

Consider a crack of length 'a' in a tensile layer of thickness t constrained and sandwiched by much thicker layers, such that the biaxial compressive stress in the thicker layers is assumed to be zero. It will be assumed that both materials have identical elastic properties and identical critical strain energy release rates, G_c. We assume that the crack can extend across the interface into the adjacent compressive layers and continue its extension as a tunnel crack shown in Fig 3. The strain energy release rate function for a slit crack, G_s, within a layer subjected to tension and bounded by adjacent layers that contain no stress can be found in Tada, et. al. [5], and given by

Fig. 3 Schematic of a tunnel crack within a tensile layer, terminating at a free surface. [4]

$$G_s = \left(\frac{\pi}{2}\right)\frac{\sigma^2}{E_1^*}\left(\frac{a}{t}\right)t, \qquad a < t \qquad (12)$$

and

$$G_s = \left(\frac{2}{\pi}\right)\frac{\sigma^2}{E_1^*}\left(\frac{a}{t}\right)\left(\sin^{-1}\left(\frac{t}{a}\right)\right)^2 t, \qquad a > t \qquad (13)$$

where $E_1^* = E_1/(1-v_1^2)$, E_1 and v_1 are the Young's modulus and Poisson's ratio of the tensile layer. Fig. 5 plots G normalized by $\dfrac{\sigma^2 t}{E_1^*}$ as a function of the normalized crack length, a/t. As the slit length 'a' increases, the energy release rate increases when the slit is within the tensile layer, a/t < 1, and sharply decays when the slit extends into the adjacent thicker layers, a/t > 1.

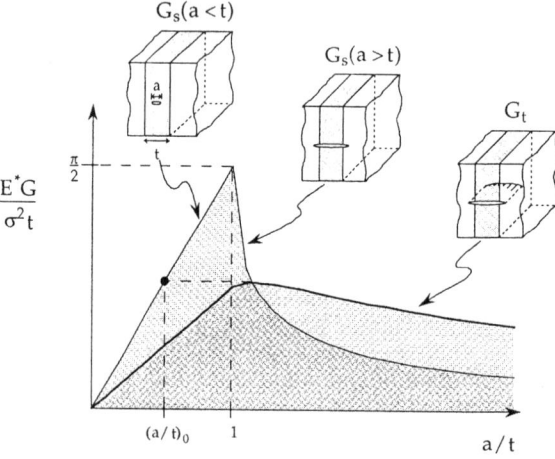

Fig. 4 Plot of the normalized strain energy release rate for crack extension (Gs) and crack tunneling (Gt) versus normalized crack length (a/t). When (a/t) is greater than (a/t)0, crack extension occurs without tunneling. Further cooling (equivalent to increasing the residual stress) is required to initiate crack tunneling. When (a/t) is less than (a/t)0, crack extension occurs with tunneling [4]

G_t was determined numerically [4]; it is plotted in Fig. 4. The strain energy release rate for tunneling, G_t, is small where the width of the crack, 'a' is both small and large; it reaches the maximum value when it slightly extends into the adjacent layers. This maximum point coincides with the intersection of the G_s and G_t functions.

Knowing the critical strain energy release rate, G_c, the elastic modulus E^*, the tensile stress in the layer σ, and the layer thickness t, one can use Fig. 4 as a 'map' to determine the conditions for when the crack will extend across the tensile layer and tunnel along it length.. The value of the normalized fracture energy, $\dfrac{G_c E_1^*}{\sigma^2 t}$, can be represented by a horizontal line in Fig. 4. During cooling, the line moves down because the residual tensile stress increases. Assuming the normalized size of the largest crack in the tensile layer is $(a/t)_s \leq a/t \leq 1$ (see Fig. 4), then, during cooling the horizontal line will intersect the hatched region and the crack will extend without tunneling to a normalized size that is slightly larger that a/t = 1. Further cooling will cause the horizontal line to enter the hatched field for G_t such that the slit crack will tunnel through the tensile layer. On the other hand, if the normalized size of the initial slit crack is $0 < (a/t)_s$, the horizontal line will first enter the hatched area where the initial slit crack will both extend to become a/t > 1 and tunnel at the same time.

Fig. 4 can also be used to interpret crack extension for the case where the laminate contains tensile layers of different thicknesses that has been cooled to room temperature, thus fixing the tensile stress in all of the layers to the same value. By viewing Fig. 4, it can be seen that the initial slit crack cannot tunnel, regardless of its initial size, if the horizontal line of the normalized fracture energy, $\dfrac{G_c E_1^*}{\sigma^2 t}$, is above the maximum point on the G_t curve, which is $\pi/2$. Rearrange it can be shown [4] that one obtains a critical thickness for the tensile layer for which no pre-existing crack, regardless of its size, can tunnel. The critical layer thickness is given by

$$t_c = \left(\frac{2}{\pi}\right)\frac{G_c E_1^*}{\sigma^2}$$

(14)

When $t \geq t_c$ the probability of tunneling will still depend on the size of the flaw that pre-exists within the tensile layer.

The above analysis assumes identical elastic properties and identical fracture energy for both materials. When the adjacent layers are more compliant then the tensile layer (smaller elastic modulus), the normalized energy release rate increases and the slit crack will extend a greater distance into the adjacent layers. The opposite condition occur when the tensile layer is more compliant then the adjacent material.

Crack Bifurcation Under Bending Loads [6,7]. Once it was discovered that edge cracking would occur, as reviewed above, another phenomenon associated with laminar materials was investigated, leading to the discovery that laminates exhibiting edge cracking would also exhibit a

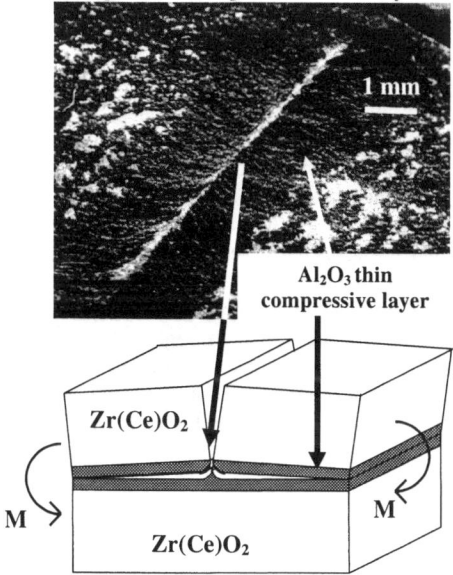

new phenomenon where cracks that would extend into a compressive layer would bifurcate. The observations associated with bifurcation in the compressive layer under bending loads will be reviewed in this section. Bifurcation under tensile loads will be reviewed in a subsequent section.

The thinking that lead to experiments concerning crack bifurcation were based on the following ideas. When a crack propagates through a material, it obviously creates a free surface. When it approaches the compressive layer, the crack will create a new free surface that will terminate the different layers within the composite. A tensile stress similar to that described by Eq. (6) will arise due to the introduction of the free surface. It was reasoned that since the stress in front of the crack would be converted from a compressive stress acting perpendicular to the crack plane to a tensile stress acting parallel to the crack plane, the crack path might be altered to produce a bifurcated crack. It was also expected that the conditions required for bifurcation would be similar to the conditions for edge cracking, namely, bifurcation would be expected to occur when $\sigma_{yy}t_1 >$ critical value, where σ_{yy} is the tensile stress (Eq.6), and t_1 is the thickness of the compressive layer.

Fig. 5 At bottom, illustration of an crack extending into a compressive layer and bifurcating to continue its extension along the center line of the compressive layer under a bending moment, M. Micrograph shows the fracture surface where the white areas are the ZrO_2 and the dark, Al_2O_3. [6]

To test this idea, flexural loading experiments were conducted using specimens fabricated from two different material systems, namely composites formed with $Zr(Ce)O_2$ thick layers and Al_2O_3 thin, compressive layer, [6] and composites formed with Al_2O_3 thick layers and mullite/Al_2O_3 [7] thin, compressive layers. Fig. 5 illustrates the bending moment relative to the layers. For both composite systems it was demonstrated that during fracture, cracks extending perpendicular to the compressive layer would bifurcate and extend along the centerline of the compressive layer. In addition, bifurcation should only occur if the specimen also contained an edge crack prior to fracture. Namely, the conditions for edge cracking and bifurcation were similar, i.e., $\sigma_{yy}t_1 >$ critical value. Fig. 5 illustrates this bifurcation phenomenon. For the specimen shown in Fig. 5, the crack begins to bifurcate prior to entering the compressive layer. Because the two materials have a different atomic number, i.e., the $Zr(Ce)O_2$ scatters more electrons and thus appears white, relative to the darker Al_2O_3 compressive layer, the white ridge on top of the 'mountain' created by the bifurcation clearly shows that bifurcation starts within the $Zr(Ce)O_2$ layer. The micrograph also shows that the crack 'dives' down to a plane near the lower interface between the two materials (white regions) and then rises to propagate close to the center line of the $Zr(Ce)O_2$ compressive layer (not shown).

The experiments also showed that when bifurcation occurred, the extending crack did not continue through the composite. If the loading was stopped, the specimen could be retrieved before it fully fractured. If the loading was continued, crack extension would only occur with the extension of a new crack at some location along the compressive layer. Bifurcation would occur each time the crack entered a new compressive layer, and further crack extension would only continue when a new crack would reinitiate the failure sequence. This failure phenomena produce a much higher strain to failure relative to a composite that did not exhibit crack bifurcation.

Bifurcation during tensile loading, which is related to the threshold strength that special laminar composites can exhibit, will be discussed below.

Architectural Design of Layered Composites that Produce a Threshold Strength

Need for a Threshold Strength. The strength of brittle materials, including ceramics and glasses, must be described by statistical parameters (e.g. Weibull) because they contain an unknown variety of cracks and crack-like flaws inadvertently introduced during processing and surface machining. Proof testing must be used when performance outweighs consumer price sensitivity. The proof test is designed to emulate the thermal/mechanical stresses experienced by the component in severe service. The proof test defines a threshold stress below which components are eliminated by failure prior to service. Eliminating heterogeneities from the ceramic powder that give rise to flaws is another method to ensure reliability. One method to remove inclusions and agglomerates greater than a given size is to disperse the powder in a liquid and pass the slurry through a filter.[8] Providing heterogeneities are not reintroduced in subsequent processing steps and surface cracks introduced during machining are not a critical issue, filtration determines a threshold strength by defining the largest flaw that can be present in the powder, and thus, within the finished ceramic component. Methods for forming engineering shapes with filtered slurries have been developed.[9]

Although others [10-13] have shown that residual, compressive surface stresses will hinder the growth of surface cracks, Green et al. [14] have proposed that the compressive stress should be located a specific distance beneath the surface. By doing so, they suggest the compressive stress will better arrest surface cracks, leading to higher failure stresses and reduced strength variability. However, compressive stresses, either at or just beneath the surface, will not effectively hinder internal cracks and flaws, nor can they produce a threshold strength. As shown below, a threshold strength can only arise when thin compressive layers are placed throughout the body to interact with both surface cracks and internal cracks/flaws.

The hypothesis that multiple, thin compressive layers could lead to a threshold strength had its genesis in an observation made by a co-worker (Sánchez-Herencia) when a crack was observed to initiate and arrest between two compressive layers during experiments to further understand the phenomena associated with crack bifurcation.[6,7] This observation initiated a fracture mechanics analysis to determine the conditions for crack arrest and subsequent failure, and experiments to test the analysis.[1]

The analysis assumed that a pre-existing crack of length 2a spans the thick layer (t_2), sandwiched by the compressive, thin layers of thickness t_1. The biaxial, residual compressive stress within the thin layers is given by σ_c, while the opposing residual tensile stress within the thick layer is given by σ_t. The analysis determines the stress intensity factor for a crack of length 2a when it extends into the compressive layers ($t_2 \le 2a \le t_2 + 2t_1$), under an applied stress, σ_a, parallel to the layers. The stress intensity factor is used to determine the applied stress, σ_{thr}, needed to extend the crack through the compressive layers to produce catastrophic failure.

Stress Intensity Function for Crack Extension Through the Compressive Layer [1]. The stress intensity function, K, is determined by superimposing two stress fields, each applied to the same crack, and each with a known stress intensity factor. Using the rules of super-positioning, the stress intensity factors for each of the stress fields are then summed to determine the stress intensity factor for the crack that extends into the compressive layers. The first stress field is a tensile stress (σ_a – σ_c) applied to the whole specimen; its stress intensity factor is given by the first term in Eq. (2a). The second is a tensile stress of magnitude (σ_c - σ_t), applied only across the portion of the crack that spans the thicker layer, t_2; its stress intensity factor is given by the second term in Eq. (2a).[1]

$$K = (\sigma_a - \sigma_c)\sqrt{\pi a} + (\sigma_c - \sigma_t)\sqrt{\pi a}\left[\frac{2}{\pi}\sin^{-1}\left(\frac{t_2}{2a}\right)\right]. \tag{15}$$

The first term in Eq. (15) is the well know stress intensity factor for a slit crack in an applied tensile field. The second term is always negative and thus reduces the stress intensity factor when the crack enters the compressive layers. Thus, the compressive layers increase the material's resistance to crack extension.

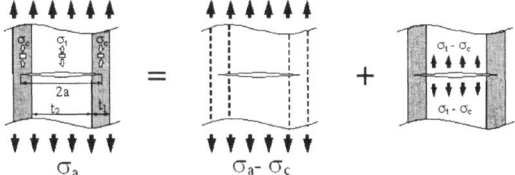

Fig. 6. Crack of length 2a shown penetrating two, bounding compressive layers place under an applied stress σ_a. The stress state on the left can be split into the two superimposed stress states on the right.

Since K decreases as the crack extends into the compressive layers, the maximum stress needed to cause the crack to 'break' though the compressive layers occurs when $2a = t_2 + 2t_1$ and $K = K_c$, the critical stress intensity factor of the thin layer material, a property that describes its intrinsic resistance to crack extension. Substituting these values into Eq. (15) and rearranging, the largest stress needed to extend the crack through the compressive layers is given by

$$\sigma_{thr} = \frac{K_c}{\sqrt{\pi\frac{t_2}{2}\left(1+\frac{2t_1}{t_2}\right)}} + \sigma_c\left[1-\left(1+\frac{t_1}{t_2}\right)\frac{2}{\pi}\sin^{-1}\left(\frac{1}{1+\frac{2t_1}{t_2}}\right)\right]. \tag{16}$$

Equation (16) shows that σ_{thr} increases with the fracture toughness of the thin layer material, K_c, the magnitude of the compressive stress, σ_c, and the thickness of the compressive layer, t_1. One can also show that if the initial crack length in the thick layer is < t_2, and the stress needed to extend it is < σ_{thr}, the crack will be arrested by the compressive layers. On the other hand, if the crack is very small and extends at a stress > σ_{thr}, it will extend though the compressive layers to cause catastrophic failure without being arrested. Thus, Eq. (16) defines a threshold stress, σ_{thr}, below which the laminar body cannot fail when the tensile stress is applied parallel to the layers.

Threshold Strength: Experimental Findings [15,16] Equation (16) was confirmed by using both Finite Element Analysis (FEA) [17] and an experimental method where the stable extension of the crack was sequentially observed to extend through the compressive layer with increasing applied load. The FEA also showed that the threshold strength would increase as the ratio of the elastic

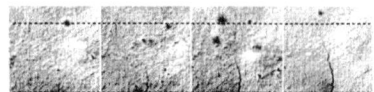

Fig. 7 Micrographs of acetate replicas for a crack extending straight though a compressive layer (defined by broken lines) with increasing applied stress. As shown in Fig. 3, the length of crack as a function of stress was used to confirm the K expression given in eq(15). [15]

modulus of the thin, compressive layer material to the thicker, tensile layer material was decreased. Namely, the strain energy density is smaller in a compressive layer with a smaller elastic modulus.

Because the crack extends in a stable manner, it can be directly observed as it sequentially propagates across the compressive layer with increasing applied stress. Fig. 7 shows a crack that was replicated, in situ, with acetate tape at different stresses as it extended across the compressive layer with increasing applied stress. [15] The shadowed replica was then observed with an optical microscope using Nomarski interference. The stress intensity function can be experimentally determined by plotting the crack length vs. applied stress These measurements confirm the theoretical K function given by Eq. (15). [15]

Equation (15) shows that an increasing stress is needed to extend the crack through the compressive layers, which implies a new toughening mechanism. From an engineering standpoint, Eq. (15) can be rearranged to state a more important concept, i.e., a threshold exists, below which, failure is impossible. The applied stress needed to fully extend the crack through the compressive layer is called the threshold strength, σ_{th}. Failure occurs once the crack fully extends through the compressive layer (i.e. when $2a = t_2 + 2t_1$) and $K \geq K_c$.

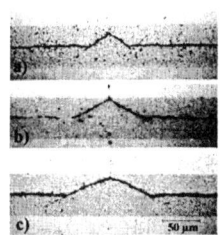

Fig.8 Crack bifurcation in compressive layers with increasing compressive stress (a to c). [16]

Observations show that the crack does not always propagate straight across the compressive layers as implied by Eq. (15). Instead, when either the compressive stress or the thickness of the compressive layer is large, the crack bifurcates as two cracks as it traverses the compressive layer. On the other hand, when the compressive stress and/or compressive layer thickness is small, the crack does not bifurcate. Namely, when a edge crack is observed in the compressive layers before testing, the crack that extends across the interface exhibits bifurcation. This suggests that similar phenomena produce both edge cracking and crack bifurcation both in bending modes of loading as discussed above, and also in tensile modes of loading as discussed here.

Fig. 8 shows that the crack bifurcates either within or as it enters the compressive layer of laminates with the 55μm/550μm architecture. [16] The figure also shows that the angle between the two cracks changes from 115° to 122° to 135° with increasing compressive stress for compressive layers containing 0.40, 0.55 and 0.70 volume fraction mullite, respectively. It should

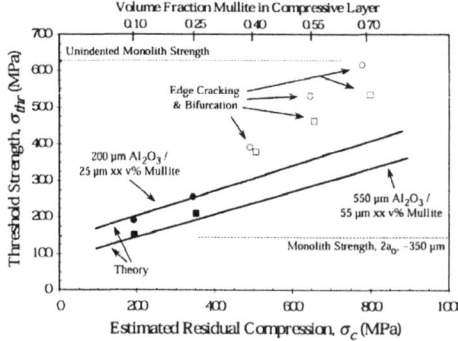

Fig. 9. Effect of compressive stress on threshold strength measure for laminates with two different architectures (laminate dimensions). [16]

be noted that in addition to the bifurcated crack, except for the region between the bifurcated cracks a linear crack is also observed along the centerline of the compressive layer. [16]

Several of the variables can be easily changed to determine their effect on the threshold strength. For these studies, laminar composites composed of thicker alumina layers separated by thinner alumina/mullite layers were fabricated. One of the variables, the compressive stress, was systematically changed by changing the volume fraction of the mullite, to reduce thermal expansion, in the thinner alumina/mullite compressive layers for fixed t_1/t_2 ratios. [16] A second variable, the distance between the compressive layers, t_2, was changed by fabricating different laminates with different values of t_1 at a fixed compressive stress (i.e., fixed mullite content) while also fixing the t_1/t_2 ratio. The residual compressive stresses within the thin compressive layers of some representative architectures were measured using a piezospectroscopic method which determines stress by measuring the stress-induced shift in the fluorescence spectra of trace Cr^{3+} impurities within the alumina. [18]

The results of these studies can be generalized in Fig. 9, which shows the effect of the compressive stress on the threshold strength for two different architectures, one with $t_1/t_2 = 25$ $\mu m/200$ μm and the other with $t_1/t_2 = 55$ $\mu m/550$ μm. The results could be divided into two regimes. [16] In the first regime, in which the compressive stress was < 400 MPa, the threshold strengths for the two architectures agreed well with those predicted by Eq. 16. For this regime, the crack propagated straight through the compressive layer as shown in Fig. 7. On the other hand, for compressive stresses > 400 MPa, the experimental values of the threshold strength were progressively greater than those predicted by Eq. 16. For this second regime, the crack was observed to bifurcate though the compressive layer as shown in Fig. 8.

The bifurcated cracks shown in Fig. 8 were observed after removing the surface by diamond matching to a depth below the penetration depth of the original edge crack. [16] The edge cracking observed in Fig. 8 is an artifact which extended as the new free surface was exposed by grinding, as is evidenced by it's absence between the branches of the bifurcated crack where the tensile surface stresses were relieved by bifurcation.

Hbaieb et al [19] have investigated the bifurcation phenomenon via a finite element analysis based on the material properties for laminar composites studied by Rao and Lange [16] (Fig. 9). The strain energy release rates for the straight and bifurcated cracks were calculated from the results of finite element computations and compared. When the stresses due to edge cracking were ignored, the crack was simulated as a through-thickness crack in an infinite body, and the energy release rate was used to predict crack deviation and bifurcation by comparing the magnitude of the strain energy release rate function for the two cases. Namely, bifurcation would only be favored and expected if the magnitude of the strain energy release rate function, G, was greater than that for the straight through crack. The finite element model successfully predicts bifurcation in only one of the four composites that were experimentally studied. When the stress field, and thus strain energy of the edge cracking was incorporated into the finite element simulations, the strain energy release rate calculations successfully predict the phenomenon of bifurcation in three of the four composites, as observed in the experiments. The presence of edge cracking, was concluded to be important to the occurrence of crack bifurcation in laminar ceramic composites.

Strength of Laminar Composites with Thin, Porous Layers [20]. As discussed above, a finite element analysis showed that a laminar composite could have a much larger threshold strength if the compressive layers had a much smaller elastic modulus relative to the thicker, tensile layers. [19] The reason for this is because the compressive layer would have a smaller strain energy density, reducing the applied stress intensity factor within the compressive layer. This condition would require a higher applied stress to drive the crack through the compressive layer. Thus, for a crack to extend across the compressive layer and cause failure, the applied stress must be sequentially increased to counteract the "shielding" effects of both the residual compressive stress and modulus mismatch. It was estimated that the influence of the elastic modulus mismatch could be significant.

In order to study the effects of elastic modulus mismatch, porosity was introduced into the thin compressive layers via adding pore-forming agents such as starch particles. One composite contained thick Al_2O_3 layers and thin layers composed of a mixture of Al_2O_3 and 0.50 or 0.70 volume fraction of mullite (designated 50M and 70M) used to control the magnitude of the compressive stress, and rice starch particles to introduce the porosity. Rice starch readily disperses in aqueous solutions, and is easily removed from a powder compact by pyrolysis. In a second composite system, the thick layers were also formed with Al_2O_3. The thin layers in the second composite system contained unstabilized zirconia (MZ-ZrO$_2$) mixed with Al$_2$O$_3$ plus the rice starch to produce the porosity. These composites were designated 20MZ and 50MZ. In this case, the ZrO$_2$ would undergo a tetragonal to monoclinic structural phase transformation during cooling and also produce a large density of microcracks as detailed elsewhere.[21] The compressive stresses were measure by the piezospectroscopy technique pioneered by Clarke and his students [18].

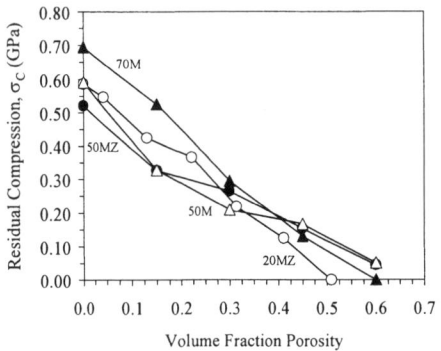

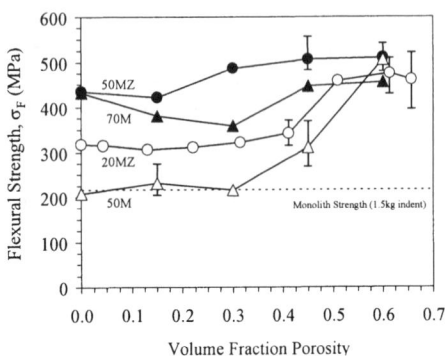

Fig. 10 Values of the compressive stresses produced in the thin (t$_1$) porous layers by laminar constraint, as determined by piezospectroscopic analysis of the thick (t$_2$) layers. Open and closed triangular symbols denote results for the 50M and 70M laminates, while the open and closed circular symbols denote results for the 20MZ and 50MZ laminates, respectively. [20]

Fig. 11 Four-point bending flexural strength versus volume fraction of porosity in the thin compressive layers, for laminate specimens containing a single 1.5 kg Vickers indent. Open and closed triangular symbols denote results for the 50M and 70M laminates, while the open and closed circular symbols denote results for the 20MZ and 50MZ laminates, respectively. The dashed line indicates the strength of an Al$_2$O$_3$ monolith containing the same size indent. [20]

It was shown that as the volume fraction of porosity was increased, the residual compressive stress (σ_c), elastic modulus (E$_1$), and critical stress intensity factor (K$_c$) of the thin layers were reduced, approaching zero at approximately 0.65 volume fraction of porosity. Fig. 10 shows the residual compressive stress as a function of the porosity introduced into the thin layers. Despite the large reduction in σ_c and K$_c$, Fig. 11 shows that the flexural strength did not significantly change with increasing volume fractions of porosity, and actually increased as the volume fraction of porosity in the thin compressive layers was > 0.40. These specimens exhibited a different mode of fracture relative to other laminates studied by the Lange Group at UCSB; namely, a crack propagating across any one of the thick layers was completely arrested by the thin, porous layers, and did not continue into an adjacent thick layer as shown by a number of fractured thick layers in Fig. 12. Micrographs of these highly porous layers showed that they are not able to support a continuous crack front. That is, fracture in these highly porous layers would have to occur by the sequential fracture of one supporting polycrystalline ligament after another. However, it appeared that cracks in adjacent thick layers would initiate, extend, and stop before ligaments failed within the porous layers. This occurrence would not be unreasonable since a stress singularity would be associated with any pre-existing crack in dense, thick layer, whereas the 'cracks' within the porous layer do not have a stress singularity. Alternately, since the elastic modulus of the thin, porous layers → 0, the strain energy that drives the crack through the porous layers also → 0. Thus it appeared that crack extension across each laminate was determined by the statistical distribution of flaws throughout all the dense laminate layers. Eventually, after a number of layers failed, catastrophic failure of the laminate occurs by shear failure (delamination) of the porous layers. Specimens failing in this manor did not exhibit a threshold strength; namely, crack extension in any one of the dense, thick layers was governed by the statistical distribution of cracks in each thick layer.

Both types of mechanical behavior, i.e., the inhibited extension of a crack across a compressive layer and the complete arrest of a crack due to sufficient porosity, are very promising for structural ceramics in general. As shown above, decreasing the elastic modulus of thin compressive layers within a body increases the energy required to drive a crack through the structure, resulting in a greater threshold strength. Additionally, by creating interlayers with volume fractions of porosity within a critical range, the integrity of the ceramic body can be maintained, while cracks within the body are completely arrested. Failure of the body, due to linking of the flaws within separate layers, can be controlled by reducing the number of inherent flaws produced in the body during fabrication or subsequent machining. In addition, during the sequential failure of one thick layer after another, the load-displacement behavior exhibits an extended strain to failure shown in Fig. 13, which in itself is a useful phenomenon.

Interaction of Cracks in Laminates that Exhibit a Threshold Strength [22]

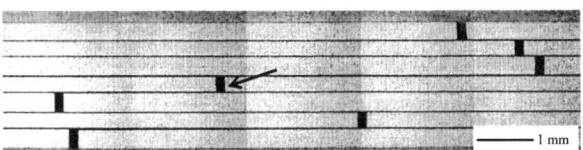

Fig. 12. Optical micrograph showing the fracture pattern on the tensile surface of a 20MZ laminate specimen with thin compressive layers containing 0.61 volume fraction porosity. The arrow denotes the location of a 1.5 kg Vickers indent used to initiate failure during 4-point bending. The thick layers broke randomly and were eventually linked by cracking or delamination within the porous layers. [20]

Although a single crack within a thick, tensile layer can be successfully arrested by the thin compressive layers, there is a finite probability that large cracks can exist close to one another in adjacent thick layers. Several researchers have used different methods to obtain solutions for the stress intensity factor of interacting cracks. It was shown that the two closely spaced cracks can interact to increase the stresses at the outer ends of the two cracks.[23] In photoelastic studies of two closely spaced cracks, it was shown that two cracks tend to act as one large crack once they overlap, and coalescence does not occur until they overlap.[24]

The purpose of this section is to review the experimental findings concerning the interaction of cracks in adjacent thick layers, separated by thin compressive layers. As shown in Fig. 14 cracks were systematically introduced at the surface of the thick layers with an indenter and the strength of these specimens was measured as a function of their off-set, separation distance, x. As discussed below, the crack-crack interaction reduced the threshold strength of the laminar ceramic in a predictable way. It is also shown that the position of the crack affected the threshold strength as well.

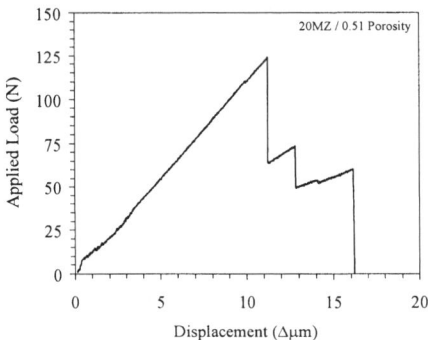

Fig. 13. Load-displacement plot for a 20MZ laminate specimen with thin compressive layers containing 0.51 volume fraction porosity. The specimen, containing a single Vickers indent of 1.5 kg, was tested in 4-poing bending until complete failure. [20]

Four different laminate architectures where fabricated. For each, the threshold strength for non-interactive cracks is (labeled 'Single Crack' in Fig. 14). The off-set distances (x) between two cracks were normalized with respect to t_2, the thickness of the thick, tensile layer. Fig. 14 shows the strength of specimens containing two off-set cracks, normalized by the strength of a single crack versus the normalized off-set separation distance (x/t_2) for the four laminates. Each point represents the strength of one specimen from one of the four different laminates. As shown, the strength of double indented specimens depends on the off-set separation distance. The largest decrease in normalized strength was 0.72, and occurred when the normalized, off-set separation distance was 0, i.e., for two co-planar cracks. The strength of the double indented specimens was < 1 until the normalized off-set separation distance was ≥ 2.

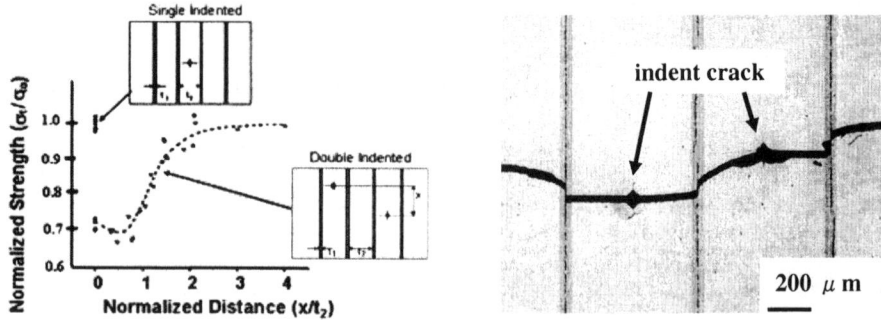

Fig. 14 Reduction of threshold strength when two cracks are introduced into adjacent thick layers, separated by an off-set distance, x. Micrograph show that the two adjacent cracks share a common crack path when the separation distance is small. [22]

Fig. 14 shows the crack paths for interacting of two typical cracks. All evidence suggests that the two indentation cracks extended across their thick layers, and at some load, coalesced after breaking through their separating, compressive layer. The cracks extended though the two outer compressive layers at a prescribed stress that, as discussed below, can be predicted with knowledge of the distance between the outer, compressive layers.

As shown in Fig. 14, the strength decreased with increasing off-set separation distance for normalized, off-set separation distances between 0 and 0.5. Within this range, the two cracks coalesced as shown in Fig. 14. For a normalized off-set separation distance of 1.4 (not shown) the two cracks did not coalescence, yet these specimens did exhibit a reduction in threshold strength as reported in Fig. 14. As the normalized off-set separation distance is increased further, the interaction between the two cracks diminishes such that the normalized strength recovers to 1 when the normalized, off-set separation distance (x/t_2) is between 2 and 3.

Moussa et al. [23] reported the finite element results of two parallel, semi-elliptical, surface cracks with different lateral and off-set separations (x = 0.3, 0.5 and 1.0). They showed that when the off-set separation distance was > 2, each crack behaved as if isolated from one another. As the off-set separation distance decreased to zero, the stress intensity factor of the inter-crack tips increased relative to that for an isolated crack of the same length and applied stress. This increase was greater for smaller values of the off-set separation. When the normalized, lateral separation distance became < 0 (two cracks overlap one another), the stress intensity factor of the inner crack tip was smaller relative to the isolated crack, and larger for the outer crack tips. That is, for substantial overlap, the two smaller cracks would act as one larger crack. The magnitude of the increased stress intensity factor at the inner crack tips before overlap, and the increased stress intensity factor for the outer cracks tips after overlap diminished with increasing normalized, off-set distance (x). In addition, the inner crack tips exhibited an increasing Mode II stress intensity factor, suggesting that the paths of the inner crack tips would change, one, or the other, seeking to extend towards its neighboring crack as suggested by the photoelastic study of Lange [23].

Observations for two cracks, separated by a single compressive layer, are consistent with the finite element (FE) studies summarized above [23]. The FE results suggested that the initial cracks, introduced by indentation, would not interact with one another because their lateral separation distance was too large. If this were the case, they would only begin to interact at the critical applied stress that would cause them to extend across their respective, thicker, tensile layers. Observations, such as the one shown in Fig. 14 indicates that their separating compressive layer did stop the cracks before they further extended to coalesce as predicted by the FE studies.

Equation 16 suggests that when t_1/t_2 is small, as for specimens prepared for the crack interaction study, the threshold strength will be inversely proportional to the square root of the dimension of the thicker layer, t_2. Thus, if one were to double the dimension of the thick layer, the normalized value of the threshold strength should be 0.71 relative to a non-interacting crack within one thick layer. As shown in Fig. 14, for normalized off-set values < 1, the normalized threshold strength is approximately 0.7; this data shows that when the off-set value produces strongly interactive cracks, the threshold strength is reduced by the inverse square root of the number of interacting cracks within adjacent thick layers.

Concluding Remarks

As reviewed above, a threshold strength can be achieved in laminar composites containing thin, compressive layers. But this concept has limitation as implied by the study of interacting cracks. Namely, when the thick layers become approximately the same dimension as the inherent crack size within these layers, there will be an increased probability that cracks within the thick layers will interact such that the concept of a threshold strength will no longer have any meaning. Thus, the concept explored here to produce a threshold strength is only valid when the dimensions of the thick layers are much larger than the size of the inherent cracks within these layer. Further, it was shown that thin layers containing large amounts of porosity could stop cracks, and that the cracks in one thick layer did not progress through to the next thick layer. The implications of this observation requires further study.

Acknowledgements

The author is grateful for the support of the Office of Naval Research via contract Grant: N00014-03-1-0350 and to Steven G. Fishman for his generous support of my research career. The author thanks all the students, Matthias Oechsner, Craig Hillman, Kais Hbaieb, Michael Pontin, Hak-Sung Moon, Geoff Fair and Mark Snyder that took part in this work and Javi Sánchez-Herencia for his initial observations leading to the discovery of the Threshold Strength phenomenon. The author also thanks his UCSB colleagues, Bob McMeeking and Glenn Beltz for their significant contributions to what they called 'baby' fracture mechanics, and to Zhigang Suo, now at Harvard.

References

[1]   M. P. Rao, A. J. Sánchez-Herencia, G. E. Beltz, R. M. McMeeking and F. F. Lange, Science, 286, pp 102-5, Oct. 1 (1999).

[2]   S. Ho, C. Hillman, F.F. Lange and Z. Suo, J. Am. Ceram. Soc. 78 [9] 2353-59 (1995).

[3]   A.J. Monkowski AJ and G. E. Beltz, Inter. J. Solids and Structures, 42 [2] 581 (2005).

[4]   C. Hillman, Z. Suo, and F.F. Lange, J. Am. Ceram. Soc. 79 [8] 2127-2133 (1996).

[5]   H. Tada, P. C. Paris, and G. R. Irwin, The Stress Analysis of Cracks Handbook, 2nd ed., Del Research, St. Louis, MO, 1985.

[6]   M. Oechsner, C. Hillman, and F.F. Lange, J. Am. Ceram. Soc. 79 [7] 1834-38 (1996).

[7]   A. J. Sanchez-Herencia, C. Pascual, J. He, F.F. Lange, J. Am. Ceram. Soc., 82 [6] 1512-1518 (1999).

[8]   F. F. Lange, J. Am. Ceram. Soc. 72 [1] 3-15 (1989).

[9]   B.C. Yu and F.F. Lange, *Advanced Materials*, 13, No. 4, 2001.

[10]   P. Honeyman-Colvin and F. F. Lange, J. Amer. Ceram. Soc. 79 [7], 1810 (1996).

[11]   F. F. Lange and B. I. Davis, J. Am. Ceram. Soc. 62 [11-12], 629 (1979).

[12]   D. J. Green, J. Mater. Sci. 19 [7], 2165 (1984).

[13]   R. Lakshminarayanan, D. K. Shetty, R.A. Cutler, J. Am. Ceram. Soc. **79** [1], 79 (1996).

[14]   D. J. Green, R. Tandon, V. M. Sglavo, Science 283, 1295 (1999).

[15]   M. P. Rao, J. Rödel, and F. F. Lange, J. Am. Ceram. Soc., 84, 2722-24 (2001).

[16]   M. P. Rao and F.F. Lange, J. Am. Ceram. Soc. 85 [5] 1222-8 (2002).

[17]   K. Hbaieb and R. M. McMeeking, Mech. Mat. 34 [12] 755-72 (2002)

[18]   Q. Ma and D. R. Clarke, J. Am. Ceram. Soc. 77 [2] 298-302 (1994).

[19]   K. Hbaieb, R.M. McMeeking and F. F. Lange, to be published.

[20]   M.G. Pontin and F.F. Lange, J. Am. Ceram. Soc., 88 [2] 376-382 (2005).

[21]   M.G. Pontin, M.P. Rao, A.J. Sánchez-Herencia and F.F. Lange, J. Am. Ceram. Soc., 85 (12) 3041-48 (2002).

[22]   H. Moon, MG. Pontin, FF Lange, J Am. Ceram. Soc. 87 [9] 1694-1700 (2004)

[23]   W. A. Moussa, R. Bell, and C. L. Tan, Int. J. Press.Vessels Piping, 76 [3] 135–45 (1999).

[24]   F. F. Lange, Int. J. Fracture Mech. 4 [9] 287-94 (1968).

Key Engineering Materials Vol. 333 (2007) pp 17-26
online at http://www.scientific.net
© 2007 Trans Tech Publications, Switzerland

Flaw Tolerant Ceramic Laminates with Negligible Residual Stresses between Layers

S. Bueno[a], C. Baudín[*b]

Instituto de Cerámica y Vidrio (CSIC). C. Kelsen 5, 28049 Madrid (Spain)

[a]bueno@icv.csic.es, [b]cbaudin@icv.csic.es

Keywords. Laminates, flaw tolerance, strength, interface, microstructure

Abstract. Ceramic laminates can be designed to combine high strength with flaw tolerance. In this paper, the designing approach based on the mechanical response of residual stresses free biological layered structures is revised. The main design tools are analysed and different ceramic-ceramic systems combining stiffness, high strength and flaw tolerance with thermo-mechanical stability are described. Two main approaches have been used depending on the relative toughness of the layers and the interfaces between them. Laminates constituted by layers separated by weak interfaces, to originate crack deflection and delamination along the interface, show high thermal shock resistance but limited resistance to shear stresses and, thus, to wear. Laminates with strong interfaces that combine stiff and high strength external layers with flaw tolerant internal ones are appropriate for wear applications. In this group of materials, the combination of layers with the same phase composition and different microstructures avoids residual stresses due to thermal expansion mismatch, but the attainment of such microstructural differences implies the co-sintering of layers with large differences in the green state. The generation "in situ" during sintering of the desired microstructural differences represents an interesting alternative in terms of processing for this group of materials.

Introduction

Ceramic materials are being proposed and used as substitutes for metals in structural applications that involve high temperature in severe erosive and corrosive environments and/or compressive loads. The major problem for the structural use of ceramics is related with their brittle fracture mode, which implies the variation of strength of different components within the same batch as a function of the distribution of strength limiting flaws. Although particularly weak components can be removed from the batch by proof testing, once a component enters service, subcritical growth of preexisting flaws or the formation of new cracks, for instance by erosion, can lead to unpredicted failure of the components [1].

In the past, much of the effort to improve the mechanical behaviour of ceramics was placed in producing the highest degree of homogeneity in bulk monophase ceramics with very small flaws. However, in the last two decades new strategies fundamentally different from the conventional "flaw elimination" approach of monolithic ceramics have emerged directed to achieve "flaw tolerance". Advances in processing technology have made it possible to produce structures in which the layered architecture combining materials with different properties leads to a laminate with mechanical behaviour superior than those of the individual constituents [2].

There is one large group of ceramic laminates designed on the basis of the development of residual stresses between the layers during cooling from the sintering temperature. When the external layers of the laminate are in compression, large improvements in strength can be achieved specially in systems such as alumina/zirconia in which high residual stresses can be attained [3]. This is the so called strengthening approach, similar to that traditionally used in glass [4]. Moreover, R curve behaviour revealing flaw tolerance is also observed in this kind of laminates for

---

[*] Corresponding author. FAX: 34 91 735 58 43, TEL: 34 91 735 58 40

flaws embedded in the external layer [5]. A challenging approach is that of laminated ceramics with the external layers in tension which present extreme reliability due to threshold strength [6]. Laminated ceramics designed on a residual stresses approach will be extensively analysed in other papers in this book.

Major limit for laminates designed on the basis of residual stresses is the degradation of the mechanical behaviour when the temperature of use approaches that of stress relaxation. In this sense, nature offers a number of residual stresses free simple layered structures, such as shells or teeth, which present improved failure behaviour as compared to that of the individual components. For example, layers of stiff, hard and brittle aragonite platelets, held together by an easily to deform and tough proteinaceous matrix, make nacre a rigid material in which both toughness and strength are significantly higher than those of aragonite, which constitutes the 95vol.% of nacre. Several mechanisms leading to energy dissipation have been identified to occur during the fracture of nacre [7-13]: sliding of the aragonite layers, stretching of the filaments in the proteinaceous matrix and crack deflection around the aragonite plates.

There has been a great effort to design and fabricate ceramic based laminates following the so called biomimetic approach, which consists in fabricating hierarchical structures through artificial methods mimicking the natural structures [14,15]. On the basis of the toughening mechanisms proposed for nacre, two groups of materials have been developed. One group of materials combine relatively thick rigid external layers with thin internal layers capable of deformation and energy absorption during fracture. A number of ceramic-metal and ceramic-polymer laminates have been developed on this basis [14-18]. In these laminates, adhesion at the interface between the layers has to be optimised and their width minimised in order to allow the interaction between the deformation mechanisms of the layers. Main drawback of this family of laminates is the lack of stability at high temperature due to the characteristics of the metal and polymeric layers. As a consequence, different ceramic-ceramic systems have been used as the basis to design residual stresses free layered materials combining stiffness, high strength and flaw tolerance with thermo-mechanical stability. Two main approaches have been used: laminates constituted by layers separated by weak interfaces between them and laminates that combine stiff and high strength external layers with flaw tolerant internal layers. The characteristics of both kinds of laminates will be described in this paper.

Ceramic laminates with weak interfaces

Different ceramic-ceramic layered composites have been designed and processed on the basis of weak interfaces between layers to originate crack deflection and delamination along the interface before penetrating the contiguous layer [2,19-25]. High thermal shock resistance materials can be developed on this basis [26]. Main drawback of this approach is that crack deflection limits the resistance of the materials to shear stresses and, thus, their wear resistance [26].

Crack deflection along interfaces. He and Hutchinson [27] established the requirements for crack deflection along an interface between two layers of different materials: 1, 2. Results of these authors for a stress free laminate, derived from an energy criterion, are plotted in Figure 1 for a crack that propagates through a layer of material 2, with elastic modulus, E_2 and Poisson coefficient v_2, and reaches the interface with layer 1, with elastic properties E_1 and v_1. This crack will penetrate into layer 1 or deflect along the interface depending on the ratio between the energy release rate of the deflecting crack, G_d, and that of the penetrating crack, G_p, and the relationship between elastic properties of the layers given by α (Eq. 1):

$$\alpha = \frac{E_1' - E_2'}{E_1' + E_2'}$$

(1)

where $E_{1,2}'$ are the reduced Young modulus for material 1 and 2 (Eq. 2):

$$E'_{1,2} = \frac{E_{1,2}}{(1-v^2_{1,2})} \qquad (2)$$

From Figure 1, the capability for crack deflection would increase for interfaces between materials with low Young modulus values and large differences between them (i.e: large values of α). Also, "weak" interfaces, with low values of critical energy release rate (G_d) as compared with that of material 1 (G_p) will be required. Further development of this model to include the effect of residual stresses was also published by the same authors [28].

One of the fundamental works on layered materials with weak interfaces was published in 1990 by Clegg et al. [29], that suggested to adapt the technology used to fabricate multilayer capacitors to produce ceramic multilayers. They proposed to form ceramic powders into sheets and to coat them to give suitable interfacial properties. In particular, these authors dealt with a system made of stiff (Young's modulus $\cong$ 450 GPa) and brittle SiC sheets separated by soft (Young's modulus $\cong$ 15-20 GPa) graphite thin layers. In addition to the large differences between the Young's modulus of these compounds (i.e: $\alpha\cong$-1 in Eq. 1), a main point for the performance of this system according to the previously described model (Fig. 1) is that SiC and graphite do not present any reactivity and, thus, the weakness of the interfaces is assured even at high temperature. Moreover, due to its special structure, graphite presents high fracture resistance in the direction perpendicular to the basal planes whereas it is easily fractured in shear, for a crack traversing from a SiC layer towards a graphite layer the deflection conditions according to He and Hutchinson [27] will be fulfilled.

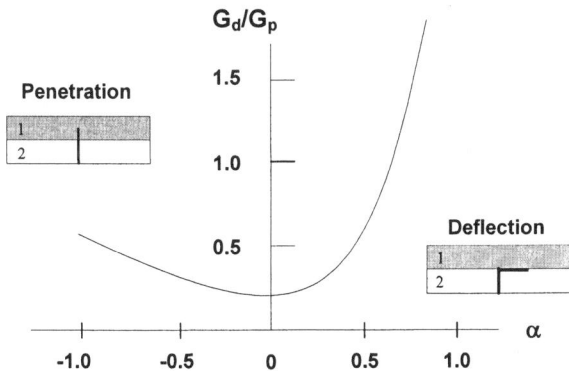

Fig. 1. Graphical representation of the requirements for crack deflection along an interface between two layers of different materials 1, 2, according to He and Hutchinson [27]. A crack that propagates through a layer of material 2 and reaches the interface with layer 1, will penetrate into layer 1 or deflect along the interface depending on the ratio between the energy release rate of the deflecting crack, G_d, and that of the penetrating crack, G_p, and the relationship between elastic properties of the layers given by α in equation (1).

The mechanical behaviour of laminates constituted by SiC/C-SiC was also investigated to determine the influence on the mechanical properties of the thickness of the layers and the possibility of chemical diffusion from the SiC layers to the carbon based interfaces [21].

Other laminated systems that promote delamination along the interfaces as the mechanism to prevent brittle behavior combine silicon nitride layers (Si_3N_4) with weak interfaces of Ti_3SiC_2 [22] or boron nitride (BN) [23,30,31]. In particular, Park et al. [23] studied the effect of incorporating different proportions of silicon nitride or alumina to the BN interfaces and found a direct relation between the proportion of second phase added and the energy release rate of the interface determined during bending tests. There is a critical amount of silicon nitride and alumina that produces the strengthening of the interfaces and for which crack deflection could not occur.

Most of laminated materials with weak interfaces are based on oxides / non-oxides systems and, thus, have limited applications at high temperature due to the low chemical stability (i.e.: oxidation of graphite). Kriven et al. [32,33] and Morgan and Marshall [34] developed laminates based on oxide ceramics (alumina, zirconia, mullite) reinforced with weak interfaces of yttrium phosphate (YPO$_4$), lanthanum phosphate (LaPO$_4$) and aluminum phosphate (AlPO$_4$). These materials had high chemical stability and showed delamination along the interfaces during fracture, giving rise to work of fracture values similar to those obtained in oxide / non-oxide systems [32].

The design of ceramic laminates with porous interlayers [24,25] is other approach to obtain weak interfaces. For a given composition, there is a critical value of porosity that determines the onset of crack deflection. This value matches with an energy release rate of the interlayer low enough to promote crack deflection according to He and Hutchinson [24,25,27].

Crack growth process associated with graceful fracture. The model system proposed by Clegg can be used to describe the general behaviour of laminates with weak interfaces. For a laminate consisting of 150 μm thick layers of SiC separated by graphite layers up to 25 μm thick, graceful fracture such as that plotted in Figure 2 was found [19-21,25,30-33]. Unnotched samples tested in three points bending deformed linearly until brittle fracture of the first SiC layer occurred at load P$_1$. Then, the growing crack was deflected at the first SiC-graphite interface which led to the load increase for increasing imposed deformation, even though stiffness of the sample was lower. For a sufficient load P$_2$, the crack was initiated again at a defect at the next SiC layer. This process was repeated until all SiC layers were eventually cracked. Such failure behaviour increased about 1000 times the work of fracture of the laminate as compared to that of monolithic SiC samples and gave apparent fracture toughness values of about 16 MPa·m$^{1/2}$.

The fracture behaviour of laminated materials, has been modelled by numerous authors [35-41] for different loading configurations and different characteristics of the interfaces between layers, and has been extensively reviewed by Hutchinson and Suo [42]. In particular, for ceramic laminates with weak interfaces tested in bending, the model proposed by Clegg et al. [37,38] accounts for the effect of different parameters on the total energy to cause complete failure. The main conclusions of this model regarding the energy required to cause failure per unit volume are the following:
i) It is constant for different loading spans.
ii) It is little affected by laminate thickness as long as laminates have more than 10 layers.
iii) It has an inverse dependence with Young´s modulus.
iv) It can increase dramatically with the strength of the layers.
v) It increases for increasing toughness of the interfaces as long as the conditions for deflection are respected.
vi) For constant laminate thickness, decreasing the thickness of the stiff layers has the same effect on the fracture energy as increasing the toughness of the interlayers.

Ceramic laminates with strong interfaces and flaw tolerant internal layers.

As described above, main drawback for laminates in which crack deflection and delamination is the main toughening mechanism is the lack of wear resistance. Ceramic laminates with strong interfaces between high strength, stiff and hard external layers and flaw tolerant internal ones is another possibility for materials combining high strength and toughness and, in this case, appropriate for wear applications. The basic requirements for this family of laminates are that the microstructures of the layers have to provide fundamentally different modes of crack control while the possibility of tensile residual stresses in the external layers is minimised. In order to fulfil this latter requirement, elastic and thermal strain mismatches have to be avoided.

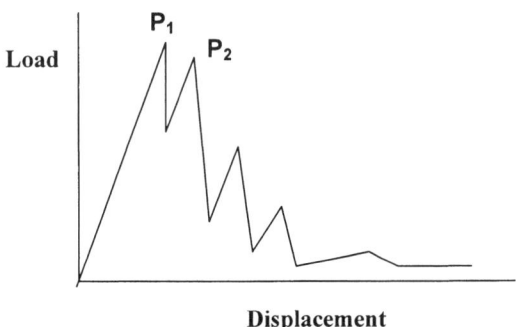

Load

Displacement

Fig. 2. Schematic representation of the load-displacement behaviour of laminates with weak interfaces showing graceful fracture, according to Clegg et al. [14].

Laminates fabricated by the co-sintering of layers with dissimilar microstructures. The first work on the development of laminates constituted of high strength external layers with flaw tolerant internal ones was by Russo et al. [43]. These authors proposed a trilaminated structure that combined surface layers consisting of homogeneous mixtures of alumina with low aluminium titanate contents with an internal layer of the same composition with heterogeneous microstructure. The homogeneous layers presented high strength whereas the heterogeneous one provided high fracture energy. As the thermal expansion behaviours of layers with the same composition are similar, such laminates were free of residual stresses. For optimum thickness of the external layers, the material presented high strength for natural defects, comparable to that of the fine microstructure external layers, and the flaw tolerance of the internal layer for large defects, as shown in Fig. 3.

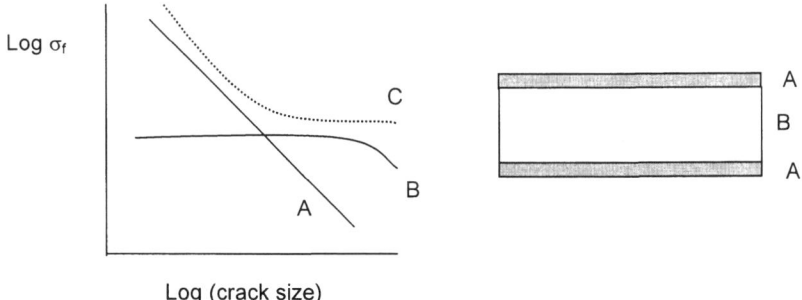

Log σ_f

Log (crack size)

Fig. 3. Trilaminated structure, C, proposed by Russo et al. [43] and the corresponding mechanical response of the layers. The surface layers, A, presents homogeneous microstructure whereas the internal layer, B, of the same composition, has a heterogeneous one. For optimum thickness of the external layers, the material showed high strength (σ_f), comparable to that of the surface layers for natural defects and the flaw tolerance of the internal layer for large defects.

The same approach has been followed for the design of laminated structures in different systems [44-46]. Padture et al. [44] proposed a glass / glass-ceramic bilayer as a model homogeneous / heterogeneous laminate where mica flakes in the glass-ceramic layer inhibited the propagation of indentation cracks. Chan et al. [45] developed trilaminates in the alumina / calcium hexaluminate system (Al_2O_3 / $CaAl_{12}O_{19}$), where alumina external layers with homogeneous microstructure consisting of equiaxed grains were combined with a composite internal layer where the second phase possessed a platelike morphology. The flaw tolerant behaviour of both kind of laminates [44,45] was determined by the action of the grains in the heterogeneous layer as crack bridges.

The response of this kind of laminates [44,46-48] under Hertzian contact tests showed the capability of these structures for dispersion of potentially dangerous surface cracks as the generation of a diffuse shear-fault damage zone in the heterogeneous layer greatly inhibits the extent of cone cracking in the homogeneous outer layer (Fig. 4). Moreover, a synergism effect is produced by the interaction between the fracture and deformation modes in the layers as revealed by shallower cone cracking in the homogeneous layers than in the corresponding bulk materials and the independence of crack penetration on applied load. This shielding of the surface cone cracks is optimised by control of the layer thickness [44,45]. Therefore, for designs in particular applications it is necessary to custom-tailor layer thicknesses to the scale of prospective surface damage events.

In situ formed heterogeneous layers. The main advantage of the above described approach of combining layers with the same phase composition and different microstructures is that residual stresses due to thermal expansion mismatch are avoided. The limit of this approach is the difficulty to obtain constituent layers of the same composition with large microstructural differences, and, therefore, with significant differences in the mechanical behaviour, because it implies the co-sintering of layers with large differences in the green state. One solution is the fabrication of graded materials in which transitional microstructures are tailored between both surfaces of the samples through a green processing in several steps, as it allows reaching specific surface properties different than those of the bulk [49,50].

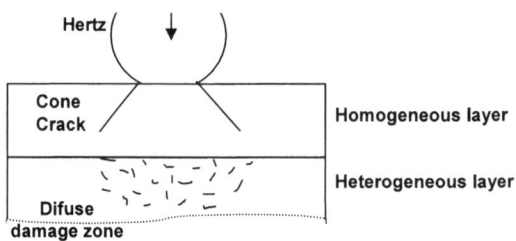

Fig. 4. Schematic representation of a laminate subjected to Hertzian contact in the surface homogeneous layer, so that cone crack extends downward towards the heterogeneous internal layer.

A much more interesting approach in terms of processing is the generation "in situ" during sintering of the desired different microstructures for each layer [51,52]. As the microstructural heterogeneity is developed during sintering, no decohesion of the layers due to differential sintering occurs. Bueno at al. [53,54] described a new method to obtain laminates constituted by layers with large differences in terms of grain size starting from green bodies with similar microstructures in the alumina - titania system. The approach was based on the effect of small amounts of titania as agents for alumina grain growth enhancement. Starting from a fine grained green body that combined alumina layers with composite layers made of mixtures of alumina and titania, the sintering schedule led to a layered structure with external layers of small grain sized alumina combined with additional "in situ" formed layers contiguous to the composite layers constituted by large alumina grains due to the diffusion of titanium (Fig. 5a). The "in situ" formed large grain sized alumina layers reached a thickness up to 200 µm, depending on the thermal treatment [54] and conferred flaw tolerant behaviour to the laminate due to crack branching and bridging (Fig. 5b).

These laminated structures developed low residual stresses during cooling from sintering due to the similarity between the thermal expansions of the layers [55]. Maximum apparent toughness values of about 12 MPa m$^{1/2}$ were reached using this approach.

As discussed above, in laminates for wear resistance applications that combine homogeneous / heterogeneous layers delamination cracks are avoided by ensuring a strong bond between the alternating ceramic layers. In these materials, weakness is confined to the grain scale within the heterogeneous layers, absorbing energy in these layers by toughening mechanisms at the microstructural level rather than a well-defined interfacial crack.

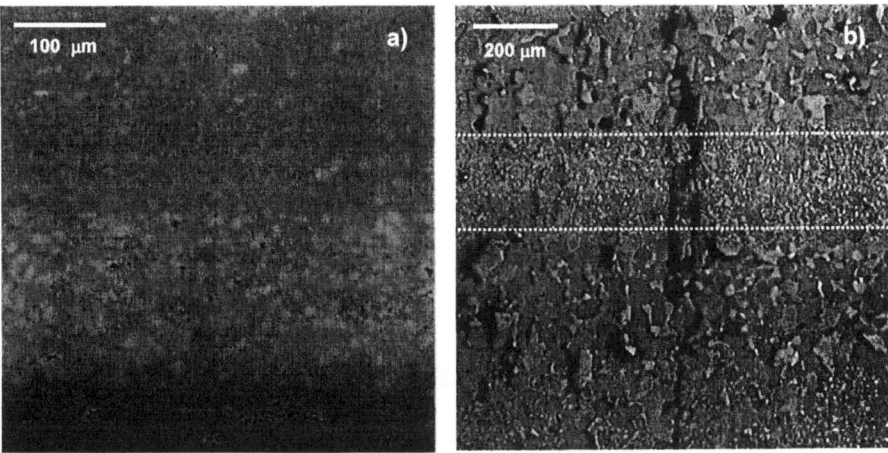

Fig. 5. Characteristic features of the alumina – aluminium titanate laminates proposed by Bueno et al. [53,54].
a) Characteristic microstructure. Low magnification field emission scanning electron micrograph of polished and chemically (HF 10 vol% -1min) etched surfaces. The effect of titania led to the layers of large grain sized alumina (clearest grey) to be formed "in situ" between the original high strength fine grained surface alumina layer (dark grey) at the bottom of the image and the composite layer (intermediate grey) in the top of the micrograph, made of mixtures of alumina and titania in the green state.
b) Optical microscopy micrograph of polished lateral faces of fractured laminated samples showing characteristic crack paths. Crack propagation is from the bottom to the top. Dashed lines mark the composite internal layers. The large alumina grains of the in situ formed large grained alumina layer act as sites for crack branching and as bridges in the wake of the propagating crack.

Summary

Much research is being devoted to the development of residual stresses free laminated structures in order to improve the performance of brittle ceramic materials even at high temperature. Through an adequate design of these structures, and taking into account the mechanisms responsible of the improved failure behaviour as compared to those of the individual components, it is possible to tailor the materials to the requirements of a particular application.

Acknowledgements

Work supported in part by the European Community's Human Potential Programme under contract HPRN-CT-2002-00203 [SICMAC] and by the Project MAT2003-00836 and the grant CSIC I3P-BPD2001-1 (Spain).

References

[1]   W.J. Clegg: Mater. Sci. Technol. Vol. 14 (1998), p. 483.

[2]   H.M. Chan: Annu. Rev. mater. Sci. Vol. 27 (1997), p. 249.

[3]   G. de Portu, J. Gurauskis, L. Micele, A.J. Sánchez-Herencia, C. Baudín, G. Pezzotti: J. Mater. Sci., in press. Published on-line 10th April 2006.

[4]   R.F. Cook: J. Am. Ceram. Soc. Vol. 88, (2005), p. 2798.

[5]   J. Gurauskis, A.J. Sánchez-Herencia, C. Baudín: J. Eur. Ceram. Soc. (2006), in press.

[6]   M.P. Rao, A.J. Sánchez-Herencia, G.E. Beltz, R.M. McMeeking, F.F. Lange: Science Vol. 286 (1999), p. 102.

[7]   K.E. Gunnison, M. Sarikaya, J. Liu, I.A. Aksay, in *Hierarchically Structured Materials*, edited by I.A. Aksay, E. Baer, M. Sarikaya, D.A. Tirrel, Materials Research Society, Pittsburgh, Penn. USA (1992), p. 171.

[8]   J.D. Currey, P. Zioupos, P. Davies, A. Casinos: Proc. R. Soc. London Ser. B. Biological Sciences Vol. 268 (2001), p. 107.

[9]   J.D. Currey: Science Vol. 309 (2005), p. 253.

[10] A.G. Evans, Z. Suo, R.Z. Wang, I.A. Aksay, M.Y. He, J.W. Hutchinson: J. Mat. Res. Vol. 16 (2001), p. 2475.

[11] R.Z. Wang, Z. Suo, A.G. Evans, N. Yao, I.A. Aksay: J. Mat. Res. Vol. 16 (2001), p. 2485.

[12] D.R. Katti, K.S. Katti: J. Mat. Sci. Vol. 36 (2001), p. 1411.

[13] K. Katti, D.R. Katti, J. Tang, S. Pradhan, M. Sarikaya: J. Mat. Sci. Vol. 40 (2005), p. 1749.

[14] I.A. Aksay, M. Sarikaya, in *Ceramics toward the 21st century*, edited by N. Soga, A. Kato, The Ceramic Society of Japan, Tokyo (1991), p. 136.

[15] I.A. Aksay, D.M. Dabbs, J.T. Staley, M. Sarikaya, in *Ceramics toward the 21st century*, edited by N. Soga, A. Kato, The Ceramic Society of Japan, Tokio (1991), p. 1.

[16] E.H. Lutz, S.E. Brunings, R.W. Steinbrech: Ceram. Eng. Sci. Proc. Vol. 19 (1998), p. 457.

[17] A. Hiltner, K. Sung, E. Shin, S. Bazhenov, J. Im, E. Baer, in *Hierarchically Structured Materials*, edited by I.A. Aksay, E. Baer, M. Sarikaya, D.A. Tirrel, Materials Research Society, Pittsburgh, Penn. USA (1992), p. 141.

[18] A.A. Abdala, D.L. Milius, D.H. Adamson, I.A. Aksay: Abstracts of Papers of the American Chemical Society Vol. 227 (2004), p. U525.

[19] W.J. Clegg: Acta Metall. Mater. Vol. 40 (1992), p. 3085.

[20] A.J. Philipps, S.J. Howard, W.J. Clegg, T.W. Clyne: Composites Vol. 7 (1994), p. E24.

[21] J.X. Zhang, D.L. Jiang, Sh.Y. Qin, Zh.R. Huang: Ceramics International Vol. 30 (2004), p. 697.

[22] L.Y. Ming, P. Wei, L. ShuQuin, C. Jian, W. RuiGang, L. JianQuiang: Ceramics International Vol. 28 (2002), p. 223.

[23] L. Zou, Y. Huang, R. Chen, C. An Wang, D. Park: J. Eur. Ceram. Soc. Vol. 23 (2003), p. 1987.

[24] J.B. Davis, A. Kristoffersson, E. Carlstrom, W.J. Clegg: J. Am. Ceram. Soc. Vol. 83 (2000), p. 2369.

[25] J. Ma, H. Wang, L. Weng, G.E.B. Tan: J. Eur. Ceram. Soc. Vol. 24 (2004), p. 825.

[26] W.J. Clegg: Science Vol. 286 (1999), p. 1097.

[27] M.Y. He, J.W. Hutchinson: Int. J. Solids Structures Vol. 25 (1989), p. 1053.

[28] M.Y. He, A.G. Evans, J.W. Hutchinson: Int. J. Solids Structures Vol. 31 (1994), p. 3443.

[29] W.J. Clegg, K. Kendall, N.McN. Alford, J.D. Birchall, T.W. Button: Nature Vol. 347 (1990), p. 45.

[30] H. Liu, S. M. Hsu: J. Am. Ceram. Soc. Vol. 79 (1996), p. 2452.

[31] Q. Zan, C. Wang, Y. Huang, S. Zhao, C. Li: Ceramics International Vol. 30 (2004), p. 441.

[32] D.H. Kuo, W.M. Kriven: Materials Science and Engineering Vol. A241 (1998), p. 241.

[33] D. Kim, W.M. Kriven: J. Am. Ceram. Soc. Vol. 86 (2003), p. 1962.

[34] P.E.D. Morgan, D.B. Marshall: J. Am. Ceram. Soc. Vol. 78 (1995), p. 1553.

[35] J.R. Rice: J. Appl. Mech. Vol. 55 (1988), p. 98.

[36] K.S. Chan, M.Y. He, J.W. Hutchinson: Materials Science and Engineering Vol. A167 (1993), p. 57.

[37] A.J. Philipps, S.J. Howard, W.J. Clegg, T.W. Clyne: Acta Metall. Mater. Vol. 41 (1993), p. 805.

[38] A.J. Philipps, S.J. Howard, W.J. Clegg, T.W. Clyne: Acta Metall. Mater. Vol. 41 (1993), p. 819.

[39] C.A. Folsom, F.W. Zok, F.F. Lange: J. Am. Ceram. Soc. Vol. 77 (1994), p. 689.

[40] M.Y. He, C.H. Hsuch, P.F. Becher: Composites: Part B Vol. 31 (2000), p. 299.

[41] S. Roham, K. Hardikar, P. Woytowitz: J. Mater. Res. Vol. 19 (2004), p. 3019-3027 (2004).

[42] J.W. Hutchinson, Z. Suo: Advances in Applied Mechanics Vol. 29 (1992), p. 63.

[43] C.J. Russo, M.P. Harmer, H.M. Chan, G.A. Miller: J. Am. Ceram. Soc. Vol. 75 (1992), p. 3396.

[44] S. Wuttiphan, B.R. Lawn, N.P. Padture: J. Am. Ceram. Soc. Vol. 79 (1996), p. 634.

[45] L. An, H.M. Chan, N.P. Padture, B.R. Lawn: J. Mater. Res. Vol. 11 (1996), p. 204.

[46] H. Liu, B.R. Lawn, S.M. Hsu: J. Am. Ceram. Soc. Vol. 79 (1996), p. 1009.

[47] L. An, H.M. Chan, N.P. Padture, B.R. Lawn: J. Mater. Res. Vol. 11 (1996), p. 204.

[48] L. An, H. Ha, H.M. Chan: J. Am. Ceram. Soc. Vol. 81 (1998), p. 3321.

[49] K. Morsi, H. Keshavan, S. Bal: Materials Science and Engineering Vol. 386 (2004), p. 384.

[50] I.M. Low: Materials Research Bulletin Vol. 33 (1998), p. 1475.

[51] L. An, H.M. Chan: J. Am. Ceram. Soc. Vol. 79 (1996), p. 3142.

[52] D.G. Brandon, O. Glozman, L. Baum, R. Grylls, in *Proceedings of the Third Euro-Ceramics*, edited by P. Durán, J.F. Fernández, Faenza Editrice Ibérica S. L., Madrid (1993), p. 725.

[53] S. Bueno, C. Baudín: J. Mat. Sci., in press. Published on-line 5th April 2006.

[54] S. Bueno, C. Baudín: J. Eur. Ceram. Soc., in press. Accepted January 2006.

[55] G. de Portu, S. Bueno, L. Micele, C. Baudín, G. Pezzotti: J. Eur. Ceram. Soc., in press. Published on-line 13 September 2005.

Key Engineering Materials Vol. 333 (2007) pp 27-38
online at http://www.scientific.net
© 2007 Trans Tech Publications, Switzerland

Lamination Process to Obtain Structure with Tailored Residual Stress Distribution

Goffredo de Portu[1, a], Lorenzo Micele[1, b]

[1]Institute of Science and Technology for Ceramics, via Granarolo, 64 48018 Faenza, Italy
and
Research Institute for Nanoscience (RIN), Kyoto, Japan

[a]deportu@istec.cnr.it, [b]micele@istec.cnr.it

Keywords: Laminates, multilayers, Ceramic Composites, Tape.Casting

Abstract. In this paper a method to produce laminated ceramic composites containing residual stresses is described. The method consist in superimposing thin layers obtained by tape casting, their worm-pressing and sintering. Detailed information on the process and on the slurry compositions are reported.
The reasons why laminated structure can exhibit improved performances are also illustrated.
The model on which a multilayer composite, containing residual stresses, can be designed is briefly illustrated. The relationship among the physical, chemical and microstructural properties of the different layers, necessary to stimulate the residual stresses outlined.

Introduction

The main limit for an extensive application of ceramics as structural materials is their inherent brittleness and, as a consequence, their poor reliability. In order to overcome this problem the design of new materials and structures with improved flaw tolerance has been considered.

The development of ceramic composites [1, 2], in general, and functional graded materials (FGM) [3, 4] or laminated structures [5 -13], in particular, are different ways to tackle the question.

The motivation for the use of graded materials and laminated composites (considered as a peculiar case of FGM) can be traced back to the observation of biological structures. In those structures the most performing parts of the material are located in regions that experience the highest stresses. See for example bamboo plants, sea shells or bones.

Similarly, considering the tribological aspect, it has been recognized that the performances of wear-resistant materials are mainly related to the properties of thin surface layers [14, 15].

The development of laminated structures is based on the assumption that it is possible to design a material containing controlled residual stresses that can be used to increase the mechanical [16-26] and tribological performances of the system [27, 28]. This goal can be achieved exploiting the differences in thermal-physical properties (i.e. different sintering rates or Coefficients of Thermal Expansion-CTE) among the laminae of dissimilar materials utilized in the process.

The increase in mechanical properties of layered architectures can be also achieved through different strategies such as introduction of weak interfaces [29, 30], containment of martensitic transformation [31] or existence of porous layers [32]. However the approach which involves the stimulation of compressive stresses [13, 33-35] is one of the most promising one, if also contact damage phenomena have to be considered. Such structures are normally obtained by stacking alternating layers of materials with different CTE that will translate in residual stresses during cooling in the sintering stage. Such residual stresses yield strength values, apparent toughness, wear and impact damage resistance higher than that found in monolithic materials of the same composition.

In this work we describe in detail a method, based on tape-casting, suitable to produce laminated ceramic oxide structures, with the required stress field distribution. Before this we briefly evidence the requirements emerged from the theory to obtain such composites.

Modeling macro-stress in laminates

In the case of rigidly bonded layers of two different materials the overall residual stress field is due to different sintering rates, thermal expansion and elastic mismatch between the constituent ceramic phases of laminated composite. In our case as we used Al_2O_3 (A) and Al_2O_3+3Y-TZP composite (AZ) it consists of two separate components: i) a microscopic stress field deriving from the microstructural scale from grain-to-grain thermal and elastic mismatches between Al_2O_3 and 3Y-TZP phases; and ii) a macroscopic stress field, which is established to fulfil equilibrium conditions between adjacent layers.

Several models have been proposed [9,10,13] to predict the latter stress amount and distribution in laminated structures. In the case of rigidly bonded layers of two different materials (in our case A and AZ), the laminated structure suffer a mismatch strain represented by [9]:

$$\varepsilon_M = \int_T^{T_0} (\alpha_{AZ} - \alpha_A) dT, \qquad (1)$$

where α_{AZ} and α_A are the thermal expansion coefficient of the two materials and T_0 is the temperature at which elastic stress develops due to thermal strain mismatch and T is the room temperature.

Considering a perfectly symmetrical architecture, far away from the free surface, the residual stress σ_{res} depends on the ratio between the thickness t_A and t_{AZ} of Al_2O_3 and Al_2O_3+3Y-TZP respectively. The layer with the lower thermal expansion coefficient (Al_2O_3 in our system), undergoes residual biaxial compressive stress given by:

$$\sigma_A^{res} = -\frac{\varepsilon_M E_A'}{1 + \dfrac{t_A E_A'}{t_{AZ} E_{AZ}'}}, \qquad (2)$$

while in the layer with greater thermal expansion coefficient (Al_2O_3+3Y-TZP composite, in our case) experiences a biaxial tensile stress given by:

$$\sigma_{AZ}^{res} = -\sigma_A^{res} \frac{t_A}{t_{AZ}}. \qquad (3)$$

In Equation (2) E' = E/(1-v), where E is the Young's modulus and v is the Poisson's ratio of the relative materials.

In the model the role of thermal-physical properties in developing residual stresses is taken into account, but Equation (2) and (2) are valid as first approximation: they gives the residual stresses induced by the lamination of layers having infinite extent. Near edges, however, the residual stress is not biaxial and the actual stress state can be obtained superimposing onto the bulk solution of Equation (2) and (2) the distribution of line forces necessary to render the edges traction-free. This "edge effect", modelled according Boussinesq stress function [36], has been already employed by previous authors to compare experimental data and theoretical prediction [13, 35, 37]. For brevity, with reference to this model, we address the reader to the mentioned publications.

Design and processing to obtain laminated structures.

Lamination of different thin ceramic layers to form thick specimens is a relatively simple and inexpensive process, which has shown interesting results and can be considered a valid alternative to more sophisticated processes [1, 2]. Several processing routes have been explored for the preparation of these composites including electrophoretic deposition [38 - 41], sequential slip casting [31, 42] and tape casting [12, 43 - 46].

Among the ceramic laminated composites that can be produced, one of the most studied system is the alumina-zirconia one. To stimulate residual stresses in a material several techniques can be used: bonding of layers with different CTE, inducing phase transformation in the constituent materials (i.e. tetragonal to monoclinic phase transformation in zirconia based materials) [47] or promoting phase reactions such as in mullite based materials [48]. According to the model outlined before, in this paper we limited our analysis to the effect produced by rigidly bonded layers of two different materials with different coefficients of thermal expansion and shrinkage. In the range 25-1400°C alumina has a coefficient of thermal expansion (CTE) $\alpha \approx 9 \times 10^{-6}$ °C^{-1}, while zirconia has $\alpha \approx 11 \times 10^{-6}$ °C^{-1}. In addition alumina has different sintering rate and lower shrinkage than zirconia.

On the other hand to control the residual stresses in the structure and avoid defects during sintering, such as edge cracking [9, 11], delamination [49] or tunnelling cracks [50, 51], an appropriate design and controlled mismatch in CTE among different laminae is necessary. Our previous experience showed that the difference in CTE between Al_2O_3 and ZrO_2 is too high and leads to the formation of cracks in the laminated structure. To reduce the mismatch between the two types of layers, one of them was made of an alumina-zirconia composite. The reason for choosing alumina and zirconia as the constituent materials of ceramic laminates can be generally traced back to the excellent bonding between the layers in the absence of excessive diffusion between components, their good thermo-mechanical properties and their relatively ease of processing. These characteristics make the two materials interesting candidates for the production of ceramic multilayers.

In this paper we address our attention to the preparation of hybrid laminated structures constituted by alternate layers of pure alumina and a composite formed by 60 vol% Al_2O_3 + 40 vol% tetragonal ZrO_2 stabilized with 3 mol% Y_2O_3 (3Y-TZP) fabricated by warm pressing and sintering of layers produced by tape casting.

These materials have appropriate thermal expansion coefficients ($\alpha \approx 9.0 \times 10^{-6}$ °C^{-1} and 10.0×10^{-6} °C^{-1} from 25 °C to 1400 °C, respectively). These differences are sufficient to induce residual stresses in the laminated structures that do not exceed the threshold values for spontaneous cracking under the experimental conditions described below.

Preparation of thin ceramic layers by tape-casting. This study entailed the use of a high purity (99.7%) alumina powder (Alcoa A16 -SG, Alcoa Aluminium Co., New York, USA) and a zirconia powder (TZ3Y-S, Tosoh Corp. Japan) doped with 3 mol% of Y_2O_3 (usually referred to as 3Y-TZP) both with an average particle size of 0.3 μm.

On the basis of previous experiences [52, 53], the different powders were mixed with organic binders, dispersant, plasticizers and solvents to obtain suitable slips for tape casting. Slurry compositions were the same for both Al_2O_3 and Al_2O_3-ZrO_2 composite powders.

Because of the high number of components involved in the process, tape-casting slurries are complex and a very precise procedure is recommended. Their preparations involves the use of:

- an azeotropic mixture of methyl ethyl ketone (MEK)[1] and ethanol[2] as SOLVENT
- a phosphate ester as DISPERSANT (often called GTO or TRIOLEIN)[3]
- a polyvinylbutyral as BINDER (referred to as B98)[4]
- a butyl-benzyl phtalate as PLASTICIZER (referred to as S160)[4]

1. Sigma-Aldrich, Germany
2. Merk, Germany
3. Fluka Chemie, Switzerland
4. Monsanto Company, St. Louis, MO, USA

The different slip formulations were calculated considering the density of tetragonal zirconia $\rho_{ZrO2} = 6.13$ g/cm^3 and of alumina $\rho_{Al2O3} = 3.99$ g/cm^3.

Ceramic powders, solvent (ethanol and methyl-ethylketone), surfactant (triolein, to enhance the powder dispersion characteristics and to facilitate removal of the tape from the mylar substrate) and 1/3 of the total binder volume (polyvinyl-butyral, PVB), were first ball-milled with Al_2O_3 or ZrO_2 balls for 24 h. Then, the remaining part of the binder and a plasticizer (dibuthyl phthalate, DBP) were added and the mixtures ball-milled again for 24 h.

Details of slip formulations for Al_2O_3 and Al_2O_3-ZrO_2 composite are reported in Table 1 and Table 2 respectively.

Table 1. Al_2O_3 slip formulation

		wt (g)	wt (%)	Vol (cm^3)	Vol (%)	V s.s. (%)
Powder	Al_2O_3	100	60.02	25.06	24.63	58.25
Dispersant	GTO	1.6	0.96	1.74	1.71	4.05
Solvent	MEK/etOH	47	28.21	58.75	57.73	#
Binder	B98	9	5.40	8.18	8.04	19.02
Plasticizer	S160	9	5.40	8.04	7.90	18.68
		166.6	100	101.77	100	100

Table 2. Al_2O_3-ZrO_2 slip formulation

		wt (g)	wt (%)	Vol (cm^3)	Vol (%)	V s.s. (%)
Powders	ZrO_2	50	30.01	8.16	8.37	21.10
	Al_2O_3	50	30.01	12.53	12.87	32.42
Dispersant	GTO	1.6	0.96	1.74	1.79	4.51
Solvent	MEK/etOH	47	28.21	58.75	60.32	#
Binder	B98	9	5.40	8.18	8.40	21.17
Plasticizer	S160	9	5.40	8.04	8.25	20.79
		166.6	100	97.40	100	100

Where wt (g) is the different weight of the components used, wt (%) is the weigh percentage, Vol (cm^3) is the volume, Vol (%) is the volume percentage and V s.s. (%) is the percentage of the solid solution of the different components in the green tapes.

To prepare a good slurry several steps are necessary and an appropriate sequence of addition of the different organic and inorganic components must be followed.

The sequence of different additions with the time required for the different steps are reported in Table 3.

Table 3. Organic binders and milling time used for tape-casting slurries

		Components	Milling time
Day 1	Step I	Solvent + Dispersant + 1/3 B98	5 min
	Step II	+ Powder (s)*	20h
Day 2	Step III	+ 2/3 B98	30 min Turbula** then 20h milling
Day 3	Step IV	+ S160	20h milling

* In the AZ case, Al_2O_3 powder was put in the suspension first and milled for 4 h, then ZrO_2 powder was added.
** "Turbula" is a stirring machine.

In this way sheets of pure alumina (hereinafter designated as A) and of alumina/zirconia composite with a volume ratio of 60/40, respectively, (hereinafter designated as AZ) were produced by tape-casting. This process was carried out with a laboratory tape-casting bench (Fig. 1) with a stationary double-blade system, called "Doctor-Blade", (Fig. 2) and on a carrier band made of mylar.

Fig. 1. Laboratory tape-casting bench

Before casting, the bench and the "Doctor Blade" were accurately prepared. The mylar band was selected properly to avoid the presence of defects and the atmosphere inside the casting chamber was saturated by solvent (MEK/etOH).

The double blades height was chosen as a function of the powder quantity present into the suspension and the thickness required after sintering. This value was determined assuming a densification shrinkage of about 22% for pure alumina and 24% for the alumina-zirconia composite.

The combinations selected, in general, for our compositions are reported in Table 4.

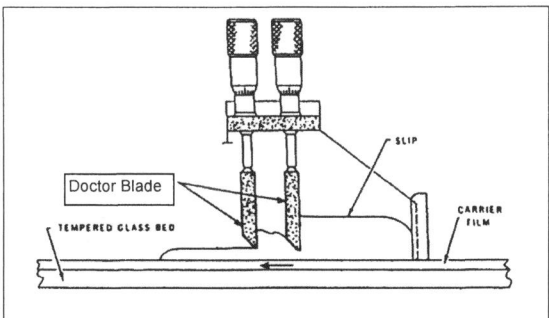

Fig. 2. Schematic of "Doctor-Blade" system

Table 4. Shrinkage evaluation for different blades height adjustment

Material	Blades height (mm)	Green tape thickness (μm)	Sintered tape thickness (μm)
Al_2O_3	0.8	220-240	~180
Al_2O_3-ZrO_2	1.1	320-340	~250

After milling, before casting, the slurries were filtered on a 100 μm mesh (thin tulle) for Al_2O_3 and on a 150 μm mesh (big tulle) for Al_2O_3-ZrO_2. Then suspensions were degassed under vacuum to eliminate the air entrapped during the process. This procedure allows the removal of a large quantity of small bubbles which would be deleterious for the quality of the tapes.

Tape-casting was performed on the carrier band of mylar moving at a speed of 65 cm/min. Tapes were dried in the tape-casting chamber in solvent saturated atmosphere during 20 hours. Then the ceramic sheets were cut in various strips and piled off from the carrier band (see Fig. 3). After that the tape thickness was measured.

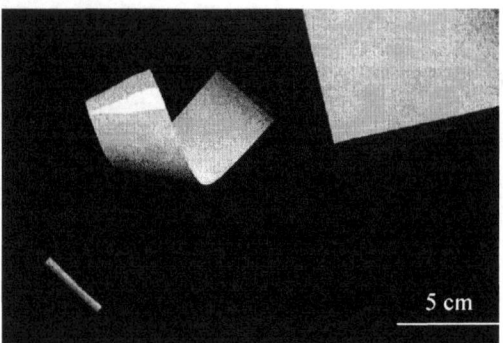

Fig. 3. Ceramic sheet peeled off from the carrier band

Lamination Procedures. Dried tapes were punched into rectangular (34 × 50 mm^2) or circular (Ø = 40 mm) moulds. Different batches of hybrid laminates were prepared changing the numbers of layers and/or A/AZ thickness ratio.

Typical samples consist of 11-13 layers formed with alternated sequence of A and AZ layers (Fig. 4). In general we produced batches according to the sequence A/AZ/.../AZ/A (hereinafter referred as A/AZ) or, in order to change thickness ratio between A and AZ, according to the sequence A/2AZ/.../2AZ/A (hereinafter referred as A/2AZ).

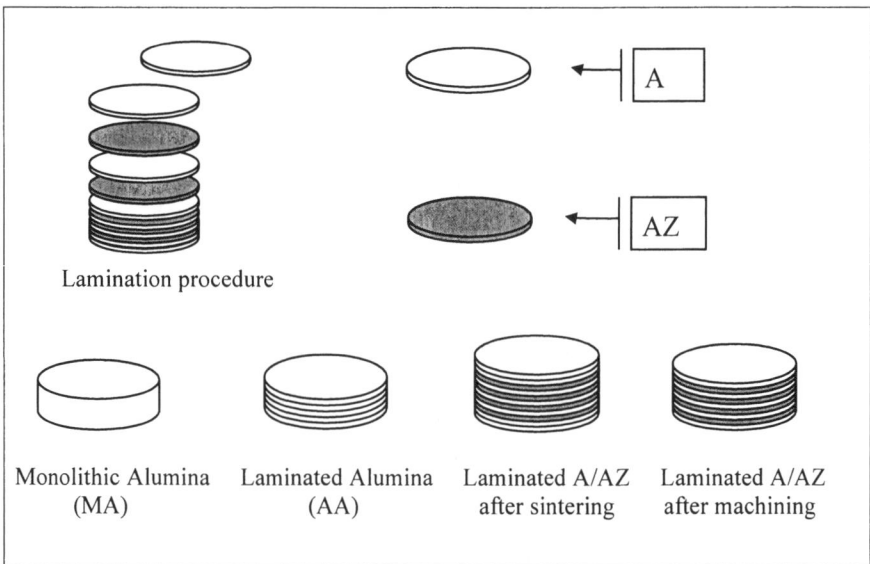

Fig. 4. Typical sequence for laminated hybrid composites and reference materials

To ensure bonding among the stacked green-sheet, warm pressing was carried out for 30 min at 80°C, under a pressure of 30 MPa.

Following the same procedures also laminated structures containing only layers of pure alumina (hereinafter referred to as AA) were obtained. This latter batch of samples was prepared in order to obtain a material with the same surface porosity as the A/AZ material, but with zero (or very low) residual stresses, to be used as reference material.

In the hybrid samples, the structure was designed to leave the layers of alumina (A) on the two surfaces (Fig. 4). Due to lower thermal expansion coefficient and shrinkage during sintering, the external alumina layers underwent residual compressive stresses.

To obtain a perfectly symmetrical structure, two A layers were generally used on each side. This allowed one layer to be removed from each side by grinding for a proper machining of the surface after sintering.

As heating and cooling rates are crucial parameters determining residual stresses in the structure [54, 55], the sintering cycle was carefully controlled. Thermal analysis carried out on green tapes revealed that the burn out of the organic fraction start at about 200°C and finish at about 600 °C.

For this reason to avoid defects in the laminated composites the removal of binders and plasticizers was carried out at 3 °C/h from 200 °C to 600 °C. Finally, the laminates were sintered at 1550°C for 1 h (heating and cooling rates set at 30°C/h). The profile of the thermal cycle is reported in Fig. 5.

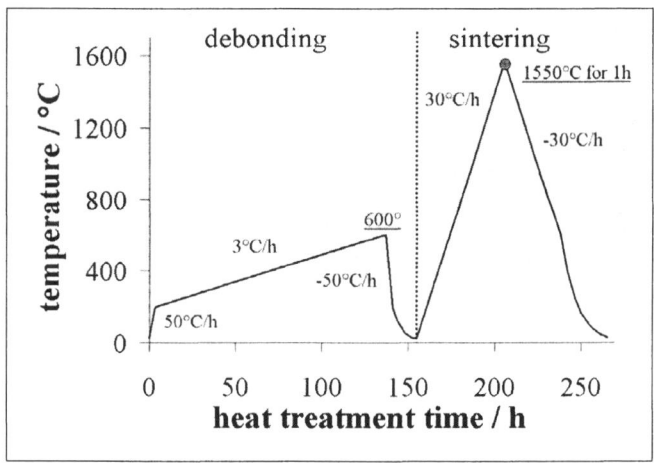

Figure 4. Thermal cycle for densification laminated composites

After sintering, samples dimensions were measured and the final density was determined with the Archimedes method. The density of all composite specimens was comprised between 96.5% and 98% of the theoretical values. For the hybrid composites these values were calculated by the rule of mixture, considering the theoretical density of the different layers and their number. Porosity value was then deduced from the relative density values. As expected [52, 53], a small amount of porosity was observed in sintered components due to the high content of organic substances used in the tape casting process. This porosity was almost equally distributed throughout the layers.

The final thickness for A, AZ and 2AZ laminae after sintering was about 180, 250 and 500 μm, respectively. Due to the tape casting processing, the variation of the thickness of single layer was approximately 10 μm. This variability has an influence also on the precision of the thickness measurement of the overall samples and of the thickness ratio among A and AZ layers. Generally speaking, an error of 3-5% on the value of the A/AZ ratio should be taken in account. However a recent refinement of the process allows us to reduce the variation of the layer thickness below 5 μm.

The photographs of the cross section of the laminated structures mainly used in this work are reported in Fig. 6.

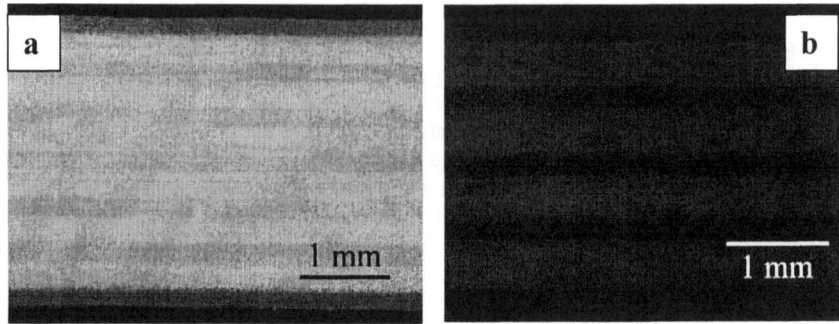

Fig. 6. Sections of different laminated structures. Note the different ratio among the layers of dissimilar materials

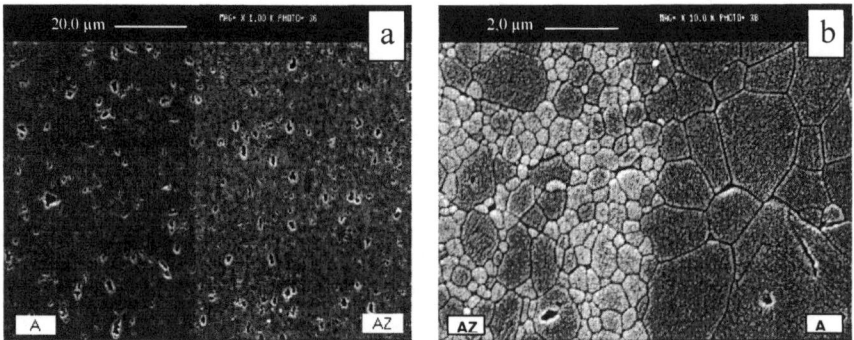

Fig. 7. Cross section of laminated composites. In (a) the interface A/AZ at low magnification is shown and the porosity is clearly visible. In (b) higher magnification showing the interface between AZ composite (left) and A layer (right). Darker grains are alumina grains.

Closer observations (Fig. 7) carried out on SEM micrographs of the surfaces and cross sections of the specimens evidenced, in general (i.e. for a proper design and processing), the absence of both tunneling and delamination cracks. Fig. 7b shows a micrograph of the interface between A and AZ layers. Alumina grains are the darker ones. A very sharp interface can be observed along with a tendency of zirconia particles to segregate at the boundaries with the alumina layers. The grains of Al_2O_3 in the composite are smaller than in the A layer as a consequence grain growth hindering exerted by ZrO_2.

In Fig. 8 is reported a qualitative evidence that the stress state in the different layers respects the distribution foreseen by the model used to design the composite. The compressive stress, parallel to the layer, in the alumina layer hinder the crack propagation perpendicularly to the interface (Fig. 8 a) while in the AZ layer the tensile stress facilitate the propagation of the cracks perpendicularly to the interface (Fig. 8 b). For a detailed analysis of the stress amount and distribution in such laminated structures please refer to the appropriate references reported in this paper.

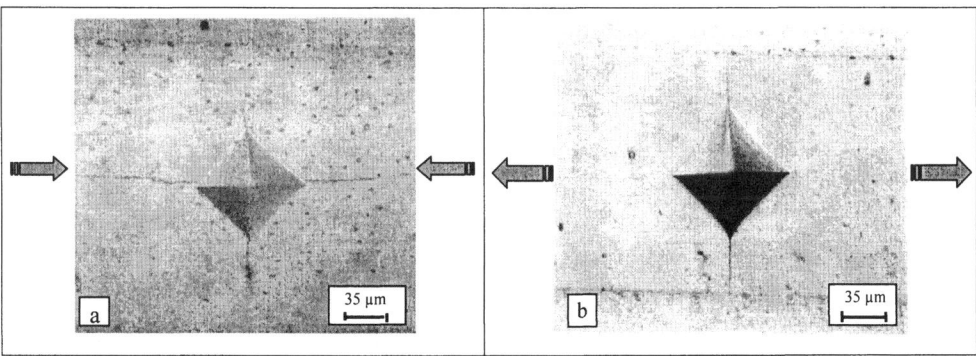

Fig. 8. Effect of residual stresses on crack propagation in laminated structure. a) layers containing compressive stress. Note that crack propagation perpendicular to the stress is hindered.
b) layer containing tensile residual stress. In this case only cracks perpendicular to the interface (i.e. to the direction of the stress) are visible.

Conclusions

The model that evidence the parameters that must be controlled in design a laminated ceramic structure have been briefly illustrated.

A process based to the production of thin laminae obtained by tape casting have been illustrated in detail. The organic and inorganic component suitable for producing good ceramic tapes have been listed. Processing parameters to be used have been proposed.

Good layered ceramic composites with desired residual stress distribution have been obtained.

Acknowledgments

This work has been partially supported by European Community's Human Potential Programme under contract HPRN-CT-2002-00203, [SICMAC].

References

[1] A.G. Evans, , J. Am. Ceram. Soc., **73** [2] (1990) 187-206.

[2] W.C. Tu, F.F. Lange, A.G. Evans, J. Am. Ceram. Soc., **79** [2] (1996) 417-24.

[3] S. Suresh, A. Mortensen, Fundamentals of functionally graded materials. Institute of Materials, London, 1998.

[4] S. Suresh, Science **292** (2001) 2447-2451.

[5] J.S. Moya, Adv. Mater, 7 [2] (1995) 185-189.

[6] M.P. Harner, H.M. Chan, G.A. Miller, J. Am. Ceram. Soc., **75** [7] (1992) 1715-1728.

[7] H.M. Chan, Annul Rev. Mater. Sci., **27** (1997) 249-282.

[8] J.W. Hutchinson, Z. Suo, Adv. Appl. Mech., **29** (1992) 63-191.

[9] S. Ho, C. Hillman, F.F. Lange, Z. Suo, J. Am. Ceram. Soc., **78** [9] (1995) 2353-2357.

[10] T. Chartier, D. Merle, J.L. Besson, J. Eur. Ceram. Soc., **15** (1995) 101-107.

[11] C. Hillman, Z. Suo, F.F. Lange, J. Am. Ceram. Soc.,79 [8] (1996) 2127-2133.

[12] T. Chartier, T. Rouxel, J. Eur. Ceram. Soc., 17 (1997) 299-308.

[13] V. Sergo, D.M. Lipkin, G. de Portu, D.R. Clarke, J. Am. Ceram. Soc., 80 [7] (1997) 1633-1638.

[14] A.J. Perez-Unzueta, J.H. Beynon, M.G. Gee, Wear, 140 (1991) 179-196.

[15] I.M. Hutchings, Tribology: Friction and Wear of Engineering Materials. Edward Arnold: London, 1992.

[16] R. Lakshminarayanan, D.K. Shetty, R.A. Cutler, J. Am. Ceram. Soc., 79 [1] (1996) 79-87.

[17] D.H. Kuo, W.M. Kriven, J. Am. Ceram. Soc., 80 [9] (1997) 2421-24.

[18] P.Z. Cai, D.J. Green, G.L. Messing, J. Eur. Ceram. Soc., 5 (1998) 2025-34.

[19] J. She, S. Scheppokat, R. Janssen, N. Claussen, J. Am. Ceram. Soc., 81 [5] (1998) 1374-76.

[20] H. Tomaszewski, J. Strzeszewski, W. Gebicki, J. Eur. Ceram. Soc., 19 (1999) 255-262.

[21] G. de Portu, L. Micele, S. Guicciardi, S. Fujimura, Y. Sekiguchi, G. Pezzotti, Composite Science and Technology, 65 (2005) 1501-1506

[22] J. Pascual, F. Chalvet, T. Lube, G. de Portu, Materials Science Forum, Vols. 492-493 (2005) pp. 581-586.

[23] E. Jiménez-Piqué, L. Ceseracciu, M. Anglada, F. Chalvet, G. de Portu, J. Eur. Ceram. Soc., 25 (2005) 3393-3401.

[24] A.J. Sánchez-Herencia, C. Pascual, J. He, F.F.Lange, J. Am. Ceram. Soc., 82 [6] (1999) 1512-1518.

[25] A.J. Sánchez-Herencia, L. James, F.F. Lange, J. Eur. Ceram. Soc., 20 [9] (2000) 1297-1300.

[26] M.P. Rao, A.J. Sánchez-Herencia, G.E. Beltz, R.M. McMeeking, F.F. Lange, Science, 286 (1999) 102-105.

[27] S. Conoci, C. Melandri, G. de Portu, J. Mat. Sci., 34 (1999) 1009-1015.

[28] G. de Portu, L. Micele, D. Prandstraller, G. Palombarini, G. Pezzotti, Wear, 206 (2006) 1104-1111.

[29] W.J. Clegg, K. Kendall, N. Alford, T. Button, J.D. Birchall, Nature, 347 (1990) 455-457.

[30] J.R. Mawdsley, D. Kovar, J.W. Halloran, J. Am. Ceram. Soc., 83 [4] (2000) 802-808.

[31] D.B. Marshall, J.J. Ratto, F.F. Lange, J. Am. Ceram. Soc., 74 [12] (1991) 2979-87.

[32] B.F. Sorensen, A. Horsewell, J. Am. Ceram. Soc., 84 [9] (2001) 2051-2059.

[33] J. Requena, R. Moreno, J.S Moya. J. Am. Ceram. Soc., 72 [8] (1989) 1511-13.

[34] F. Toschi, C. Melandri, P. Pinasco, E. Roncari, S. Guicciardi, G. de Portu, J. Am. Ceram. Soc., 86 [9] (2003) 1547-53.

[35] G. De Portu, L. Micele, Y. Sekiguchi, G. Pezzotti, Acta Mat., 53 (2005) 1511-1520.

[36] S. P. Timoshenko, J. N. Goodier, Theory of Elasticity, 3rd ed., 104-109, McGraw-Hill, new York, 1987.

[37] W. Zhu, G. Pezzotti, Proceedings of PSEA '04, edited by Masaaki Naka and Toshimi Yamane, November 2004, pp. 397-403.

[38] P. Sarkar, X. Haung, P.S Nicholson, J. Am. Ceram. Soc., 75 [10] (1992) 2907-909.

[39] R. Fischer, E. Fischer, G. de Portu, E. Roncari, J. Mat. Sci. Letters, **141** (995) 25-27.

[40] B. Hatton, P.S. Nicholson, J. Am. Ceram. Soc., **84** [3] (2001) 571-76.

[41] O.O. Van der Biest, L.J. Vandeperre, Annual Review of Material Science, **29** (1999) 327–352.

[42] J. Requena, R. Moreno, J.S. Moya, J. Am. Ceram. Soc., **72** [8] (1989) 1511-13.

[43] K.P. Plucknett, C.H. Caceres, C. Hughes, D.S. Wilkinson, J. Am. Ceram. Soc., **77** [8] (1994) 2145-53.

[44] G. de Portu, E. Roncari, L. Agostini, P. Pinasco, In: Advanced Ceramics 96, Proceedings of International Workshop, Science and Technology Agency (STA), Inuyama, Japan, 12-14 March, 1996. p.175-180.

[45] M. Jiménez-Melendo, C. Clauss, A. Dominguez-Rodriguez, G. de Portu, E. Roncari, P. Pinasco, Acta Mater., **46** [11] (1998) 3995-4004.

[46] K.P. Plucknett, C.H. Caceres, F. Fremont, D.S. Wilkinson, In: Ceramic Engineering and Science Proceedings, September-October 1992, 16th Annual Conference on Composites and Advanced Ceramic Materials, Part 2, The Am. Ceram. Soc., pp. 873.

[47] G. de Portu, S. Conoci, J. Am. Ceram. Soc., 80 [12] 3242-44 (1997).

[48] L. Micele, M. Brach, F. Chalvet, G. De Portu, G. Pezzotti, Proceedings of The Ninth Japan International SAMPE Symposium & Exhibition, Tokyo, Japan, November 29 – December 2, 2005, Edited by K Kageyama and J Takahashi, The Japan Chapter of the Socyety for the Advancement of the Material and process Engineering.

[49] G. de Portu, L. Micele, G. Pezzotti, J. Mat. Sci. Lett., **40** (2005) 1505-1508.

[50] S. Ho, Z .Suo, J. Appl. Mech., **60** (4) (1993) 890-894.

[51] G. de Portu, L. Micele, G. Pezzotti, Appl. Spectr., **59** [2] (2005) 537-541.

[52] C. Fiori, G. de Portu, In: British Ceramic Proceedings n° 38, Novel Ceramic Fabrication Processes and Applications, edited by R.W. Davidge, Shelton, Stoke-on-Trent, UK, Dec. 1986, 213-225.

[53] G. de Portu, C. Fiori, O. Sbaizero, In: Advances in Ceramics, vol. 24: Science and Technology of Zirconia III, edited by S. Somia, N. Yamamoto, H. Yanagida, The Am. Ceram Soc. Inc., Columbus OH, USA, (1988) pp. 1063-1073.

[54] P.Z. Cai, D.J. Green, G.L. Messing, J. Am. Ceram. Soc., **80** [8] (1997) 1929-1939.

[55] P.Z. Cai, D.J. Green, G.L. Messing, J. Am. Ceram. Soc., **80** [8] (1997) 1940-1948.

Key Engineering Materials Vol. 333 (2007) pp 39-48
online at http://www.scientific.net
© 2007 Trans Tech Publications, Switzerland

Water Based Colloidal Processing of Ceramic Laminates

A. J. Sánchez-Herencia

Instituto de Cerámica y Vidrio, C/ Kelsen 5, 28049 Madrid, Spain

ajsanchez@icv.csic.es

Keywords: Slurries, multilayers, shaping, coatings, water

Abstract. Multilayered materials and coating are complex structures proposed among others to face the structural requirements of ceramics. The development of reinforcement mechanism by laminated structures can be due to deflection criteria or to the presence of residual stresses and requires of tailored laminates. These designs are characterized by the phases, thickness and distribution of the layers as well as the joining strength between them. In this sense water based colloidal processing techniques are used to fabricate layered structures by consolidating the layers from fluid dispersions of the powders in water. In these processing methods the phases presented in the final laminate are mainly given by the composition of the starting slurries while the changes in thickness and sharpness of the layers are controlled by acting on the processing parameters. The achievement of stable slurries is a shared step for all the colloidal processing techniques. In the water based slurries the stability will be dominated by the polar media, the surface behavior of the particles and the presence of dispersant additives to increase the repulsion between particles. The stable slurry ensures an effective milling and dispersion of the phases as well as high solid loadings, if required. Further processes associated to shaping and consolidation of the layers requires the incorporation of additives and-or water removal. The shaping methods based on aqueous slurries can be classified taking into account the process of solid-water separation. For each of those shaping methods, the nature and amount of the additives is different in order to get the optimum rheological behavior and green strength after drying. Depending on the thickness of layers and coatings as well as the shape and dimensions of the samples, the shaping method can be selected alone and combined with others.

Introduction

Laminar ceramic composites have received a great deal of attention due to their potential for use in emerging structural applications in which higher mechanical performance is demanded. The development of laminated ceramic composites allow the fabrication of new materials with properties superior to those of monolithic ceramics [1]. In general, the enhanced mechanical behavior exhibited by these composites can not be attributed to a simple rule of mixtures of the properties of the constituent materials but also the laminar structure itself [2-4]. Notable examples of some of these reinforcement methods include the broadening of transformation zones ahead of a crack [5], the crack deflection by weak interfaces or porous layers [2,6,7] or the reinforcement mechanism associated to residual stresses. If residual stresses are present, they can be controlled by designing the proper thickness and distribution of the layers in order to optimize the mechanical properties. They can be generated at the surface [8,9] in order to increase the fracture strength as a result of the stresses superposition acting on the intrinsic critical flaw. On the other hand, if the compressive residual stresses are induced at the internal layers [10-12], damage tolerance, and consequently reliability is the property primarily favoured. In ceramic laminates with strong interfaces, the differential strain that develops the residual stresses came from the differences in the thermal expansion coefficients (α) [12,13] or from the phase transformation of one of the components [8,14,15].

Compressive stresses usually are beneficial for the mechanical response as they oppose to the crack growth [16,17] and/or may develop a threshold strength [12]. On the other hand, tensile stresses should be subtracted to the strength of the material, and if they overpass a critical value tunnelling cracks will appear and consequently the mechanical response will degrade [14,18]. For these reasons thin compressive layers are desirable, as they will create an additional reinforcement as well as diminish the tension associated. Moreover, the thickness of the layers are referred to other observations related to the residual stresses such as edge crack and crack bifurcation for what a critical thickness "t_c" has to be achieved [11,13,19].

All this considerations clearly show that for a laminate composite the composition, thickness, number and position of the layers are very important design parameters that must be controlled in order to get the desired properties. The fabrication of multilayered ceramics by colloidal processing techniques has been widely used due its versatility and reliability. These methods have the advantages that allow to precise control in the composition and, depending on the technique, the thickness of the layers. The colloidal processing techniques described for the fabrication of laminated ceramic include tape-casting [6,15,20,21], centrifugal casting [5,22], sequential slip-casting [12,23-25], electrophoretic deposition (EPD) [26,27], and others [28-30]. All of them are based on the preparation of stable slurries with specific compositions that are piled up by sequentially adding a layer to a previously formed one. Stable slurries that ensure a homogenous and well dispersed composition are obtained by controlling the interparticle potentials developed within the liquid media [31-33]. The thickness is controlled by controlling the processing parameter associated to the technique (casting time [25,27], blades gap [6], amount of slurry [5], etc...).

Preparation of slurries in water.

Environmental, health and economic requirements made to look for aqueous based suspension systems that provide processing capabilities as effective as the organic based systems. As an alternative to the established processes, the use of water as the dispersion media to fabricate the laminates has both advantages and disadvantages. Advantageous is the know-how derived from the traditional water-based processing techniques such as slip casting , enamel dip coating and others that require highly stable slurries [32,34,35]. On the other hand, the slow drying rate of the layers and tapes, the higher crack sensitivity and the reactivity of the powders with water are causes of concern [36-38]. To overcome these problems, a high solid loading is required to increase the drying rate and simultaneously minimize drying stresses in the tape.

The main problem in a colloidal suspension, which is drastic at high solid loadings, is to maintain the particles dispersed throughout the medium since an attractive force exists as particles approach each other [39]. The DLVO theory (Derjaguin, Verwey, Landau and Overbeek) proposed that the stability of a particle in a liquid is dependent upon its total potential energy function V_T. The general equation that define the overall energy of interaction between particles is

$$V_T = \sum V_A + \sum V_B + \sum V_S \tag{1}$$

The term V_S is the potential energy due to the solvent, and usually only makes a marginal contribution to the total potential energy over the last few nanometers of separation. Much more important is the balance between V_A and V_R, these are attractive and repulsive contributions respectively. Attraction between particles is due to the London-Van der Waals dispersion forces. Those are the result of the interaction between permanent, induced or no permanent dipoles and its effect extends to distances of several nanometers from the particle surface. Repulsive interactions has to be developed with a value high enough to make the particle pair potentials to be dominated by the repulsive terms and make the particles to stay one apart from the others.

This repulsion can be achieved by using two different ways denominated electrostatic and steric mechanisms. The electrostatic mechanism consists in provide the surface of the particles, suspended in a polar media (e.g. water), with a high enough electric charge to develop repulsive electrical potential. The steric mechanism involves polymers added to the system that adsorb on the surface and prevent the particles coming onto close contact. When polar media are used, electrostatic and steric mechanism can act, but if a non-polar media is employed, only steric mechanism are employed to disperse the ceramic powders [31].

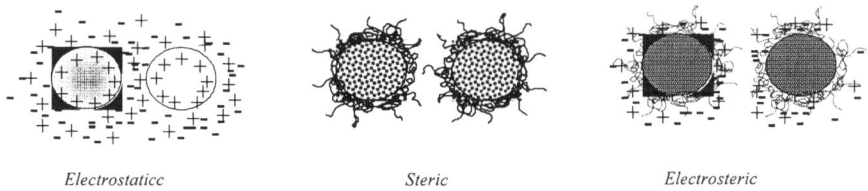

| Electrostaticc | Steric | Electrosteric |

Figure 1. Scheme of the different repulsion mechanism between particles dispersed in a polar media

A third mechanism that involves the sum of those two described previously is the electrosteric mechanism. This one is the most extensively used for water based slurries and consists on the absorption on the particle surface of polymeric chains with functional groups that provide an electric charge to the surface. In this case the repulsive term is due to the sum of both the electrostatic and the steric effect. Fig. 1 show a scheme of those three mechanisms

The achievement of high and stable repulsion forces between the suspended particles in the slurry has several effects.

Effectiveness of milling. As the particles break in two parts, the new particles became on the conditions of repulsion avoiding the agglomeration, very common when new surfaces are generated. This effectiveness is also achieved for deagglomeration processes without milling (High shear or ultrasounds).

Dispersion of different components. The particles in the fluid media have a high mobility. If particles of different composition develop a high repulsive charge between them, a high dispersion will be achieved. Even more, by changing the conditions of the media different microstructures can be achieved and designed [40]

Low viscosity of the slurries. High repulsion between particles makes the slurries to show a lower viscosity than in the case of lower repulsions. For the shaping ceramics that requires a mold, it is desirable a low viscosity that facilitate the mold filling and pouring processes.

High solid content in the slurries. Usually for shaping methods, a high solid content in the slurry is required in order to achieve a high green density, decrease the shrinkage and the drying rate.

High packing density. For a given processing technique and solid content, the slurry with a high repulsion between particles, allows higher green densities. This is because of the mobility of the particles can arrange a packing distribution with the lower energy, that is, the closer packing.

Shaping from slurries.

Fabrication of a ceramic green samples from stable slurries requires the consolidation of the dispersed powders with the desired shape (and microstructure). All the shaping methods are based on the increment of the viscosity of the slurry that ends the viscous behaviour and become a solid. The mechanism employed to increase the viscosity can be used to classify the processing method as it is indicated in Table 1. After shaping a compact, a final drying process until complete water removal is required.

Filtration	Evaporation Deposition	Floculation Coagulation	Gelification
Slip casting Pressure casting Vacuum casting	Tape Casting Electrophoretic deposition Screen printing Dip coating Centrifugal casting	Short range forces Polimeric bridging flouclation (TIF) Direct coagulation casting (DCC)	Gel-Casting Chemical polymerization Thermal polymerization Robocasting Freeze casting

Table 1.- Classification of the different processing methods

The study of the processing methods and its associated variables permits to fabricate multilayered materials by using one or combining several of them. Some methods has been reported to fabricate continuous changes in composition [41,42], but the most common strategy used for laminates has been to use sequential processing of slurries. That is, the layer composition changes are created by changing the composition of the starting slurries and stacking them together after or before drying.

Consolidation of the powders from the slurry requires the addition of additives binders and plasticers that keep the desired shape, strengthen the layers and help the further handling processes. For the fabrication of multilayered ceramics, those additives are the also responsible of keeping the layers together in the green state. Traditionally organic solvents have been used as vehicle for dispersion of ceramic particles to fabricate layered materials, due to their compatibility with the most common binders and plasticizers [31,43].The use of water as the dispersion media requires of compatible binders capable to provide consistency and joining capability to the green layer. There are four main types of binders used for shaping from based slurries [44,45]: cellulose ethers, polyvinyl alcohols, latexes and gelling binders. Latexes are polymeric substances with a molecular weight around 10000 Daltons that had been stabilized in water based emulsions. They have sizes ranging from 20 to 300 nm and are prepared in suspensions with a solid content up to 65 vol.%. As they are polymeric particles in suspension, these binders do not have an important influence on the rheology other than the increment in the solid loading. In contrast, cellulose ethers and polyvinyl alcohols do have an effect on the liquid viscosity as they are polymeric chains dissolved in the media and their molecular weight has a strong influence on the rheology. The great thickening of the viscosity caused in the slurries by these binders results in low density green tapes. The fourth types of binders are the chemically induced gelling binders like alginates or synthetic gels. These have inert components in the liquid media that increase the viscosity after a reaction or physical transformation takes place [45-47].

Processing methods to fabricate ceramic thick layers from slurries

Filtration methods have been extensively used and studied to fabricate both traditional and advanced ceramics. In these methods the particles are consolidated by passing the slurry through a filter (plaster of Paris, porous polymers or metals). The solid-liquid separation process conducts to the formation of the consolidated wall with the shape of the membrane. The shape is maintained because of the separation process has an increment in the solid loading and consequently in the viscosity. Requirements for the slip casting process are a high solid loading in the suspension, in order to get a good packing density, as well as low viscosity, to avoid problems with the mould filling and pouring. The wall formation kinetic is governed by the Darcy law and is summarized in Eq. 2

$$e = K \cdot t^{1/2} \qquad\qquad (2)$$

Where "e" is the thickness of the consolidated layer, "t" is the time and K is the kinetic constant. The kinetic constant can be calculated experimentally and depends of the mould and the slurry conditions. Sequential slip casting has been successfully used to fabricate layered ceramics [12,23-25] where the layer composition is the determined by the composition of the slurries and the thickness is controlled by the casting time according to the kinetic equation. The process is represented in fig. 2. It starts by filling the mould with the slurry of the first layer composition. After a time t_1 this slurry is poured out and the second slurry is cast in the mould. The new layer consolidates on the previous one and after a time t_2, the slurry is poured out. This process can be repeated all the times necessaries to achieve the designed layered structure. The limits of this method are related to the layer thickness (difficult to get lower than 50μm) and the total thickness of the sample (difficult to get larger than 1,5 cm).

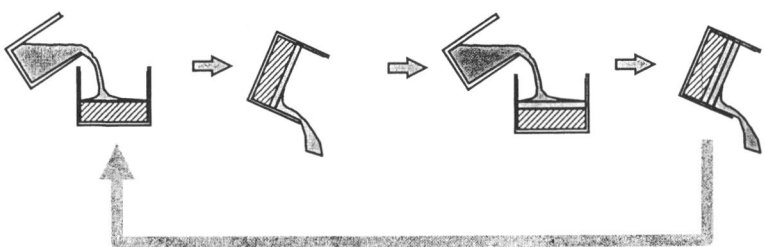

Fig 2. Scheme of the sequential slip casting process

Plot in Fig. 3.a shows an example of casting kinetics measured for slurry with compositions of Al_2O_3-YTZP. This has been used to calculate the casting times employed to fabricate a multilayer with thick and thin layers presented in Fig. 3.b. It has thick layers of alumina with 5vol% of Y-TZP and alumina with 30 vol% of m-ZrO_2 [24]. From this curve two considerations can be extracted. First of all is that the thickness of the sample is limited. Although the equation has not an asymptotic point, thicknesses higher than 1 cm are very difficult to fabricate because of the large time required. The other important consequence is that as the casting time increases the thickness control is more precise.

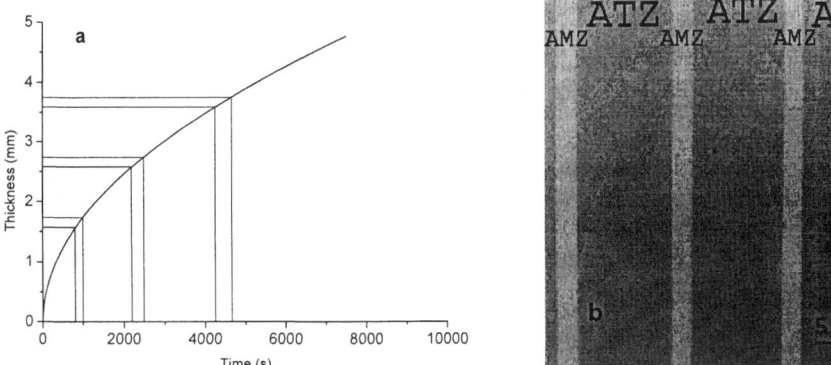

Figure 3. Plot of the casting kinetic for a suspension (a) with the casting time indicated to get the multilayer material shown in (b)

From the **Evaporation/Deposition** methods, the most employed method to develop layered structures has been the tape casting. It consists in the deposition of the slurry on a flat surface by moving a carrier that has one or two blades to determine the height limit (Dr. Blade method). To achieve high green densities in the tapes, slurries with high solid content are required as in this method there are not a driving force for consolidation different that evaporation of water. Both the binders addition and the high solid content of the slurry, confers the plastic behaviour that will make it fluid while passing thought the blades-substrate gap and keeps the tape dimensions after the tape is shaped.

Thermopressure of tapes cast from organic based slurries has been the widest method used to fabricate multilayer. To fabricate ceramic green tapes from water based slurries, latexes are the mostly binders employed nowadays. From the latexes employed for water based tape casting, the acrylic/styrene binders has been the more commonly used as an adhesive to pile up tapes in order to fabricate monoliths [48-50], and multilayers [6,24,51,52]. The formulation of the tapes employed to create those complex structures requires of a relative elevated amount of acrylic/styrene binders (25-30 vol.%) if compared with other casting processes. This high amount of organic leads to low sintered densities and, consequently, to reach a high flaws population as a consequence of the high porosity. To fabricate dense materials for structural applications, it is required to increase the solids content of the slurries and minimize the organics content. Suspensions with high solid loading can be obtained by ensuring the proper stability between ceramic-ceramic, latex-latex and ceramic-latex particles[53].

The use of latexes as an alternative for tape casting has driven new studies in the piling up mechanism. It has been reported that pre-wetting of the tapes and the use of glue products based on the tapes binder conducts to the joining of tapes with different composition at relatively low pressure at room temperature [54]. Excessive amount of glue or small pressures can generate lack of union between layers [50,55]. In this fabrication method for multilayers, the layer thickness is controlled by the substrate-blade gap. The limits in the thickness of the layers are given by the viscosity of the samples and the drying rate. Usually layer thickness range from 200 to 800 µm. Thinner layers present problems of low strength and sticking to the substrate while thick layers usually has associated problems of drying, segregation and rigidity. Figs. 4 a. and b show two different laminates obtained by piling tapes with different composition. To fabricate these laminated the cast tapes were immersed in water to make them plastic, later they were coated with a 5wt% dilution of the binder utilized and pressed at 18 MPa. Fig. 4.c shows a close up to one of the interfaces.

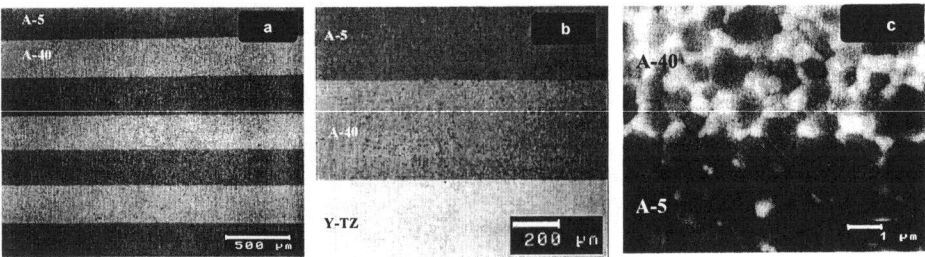

Figure 4. Examples of layered ceramic of in the alumina zirconia system fabricated by piling up the water based tapes.

Tape casting and slip casting are two of the most important methods employed to fabricate laminated materials from slurries with relatively high layer thicknesses. Other techniques have been used to develop multilayers materials from slurries or pastes such us co-extrusion or centrifugal casting [5,56], but they have not achieved the interface definition and layer thickness control as in those above mentioned.

Processing methods to fabricate ceramic thin layers from slurries

The processing of thin ceramic layers usually requires of a substrate (dense or porous) that support the consolidated layer. From the methods described in table 2 screen printing and dip coating has been used from older times to fabricate the glazing of traditional ceramics. Latter the slurry spray coating and the electrophoretic deposition has been used to fabricate thin coatings and multilayers from water based slurries, usually using low solid contents.

In the **screen printing**, the ceramic slurry with binders is deposited onto the substrate surface through a screen using a squeegee. The screen mesh, geometry and quality determines the thickness of the resulting layer [57]. Also the mesh may have mask areas so that a pattern can be deposited on the surface. The thicknesses of the layers usually range from 20µm to 100µm.

The **dip coating** process allows the achievement of layers that range from a couple of microns to a hundred of them. The process can be divided in three parts as it is schemed in Fig. 5.

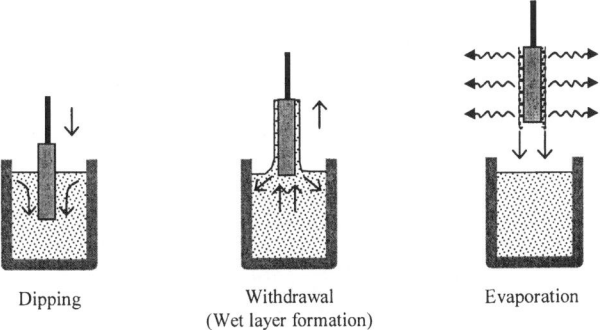

Dipping   Withdrawal   Evaporation
                 (Wet layer formation)

Figure 5. Scheme of the dip coating process.

First the substrate is dipped and soaked in the slurry. Second the withdrawal rate of the substrate at a given speed where the layers is created and finally the solvent evaporation. The layer is generated during the wet layer formation. The thickness depends on the substrate parameters (wettability, roughness and porosity), suspension parameter (viscosity, solid content, binders content and surface tension) and process parameters (withdrawal rate, soaking time, temperature and humidity). If the withdrawal rate is in the Newtonian regime of the slurry, the thickness of the layer (h) can be calculated by the Landau-Levich equation (Eq.3)[58].

$$h = K \frac{(\eta \cdot v)^{2/3}}{\gamma_{LV}^{1/6} (\rho)^{1/2}} \qquad (3)$$

Where k is a constant, η is the viscosity of the slurry, v is the withdrawal rate, γ_{LV} is the liquid-vapor surface tension ρ is the density of the slurry and g is the gravity.

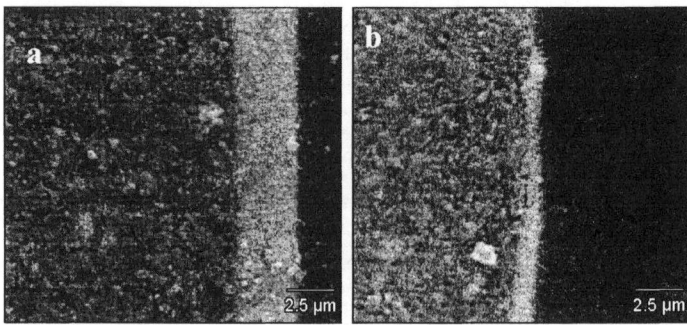

Figure 6. ZrO$_2$ coatings obtained on a green Al$_2$O$_3$ tape by dip coating using slurries with a solid content of 15 (a) and 7.5 (b) wt%.

As an example of the thickness variation, Fig. 6 shows two coating with different thicknesses obtained by dipping a pre-wet green alumina tape (fabricated by tape casting) in zirconia slurries with different solid contents. Coatings obtained from slurry with the 15 wt % (Fig. 6.a) show a higher thickness than the coating obtained with the 7.5 wt% (Fig. 6.b). The thickness vs. solid content curve can be experimentally determined in order to control the thickness. Due to the different porosity between sizes, the thickness is different depending of the tape side. If the coated tapes described above are pressed together (as explained in the tape casting section) and sintered, a multilayered system with different layers thicknesses can be fabricated showing a very high thickness ratio. Fig. 8 shows the cross section one of these multilayers designed with very thin zirconia layers (bright phase) between thick alumina layers.

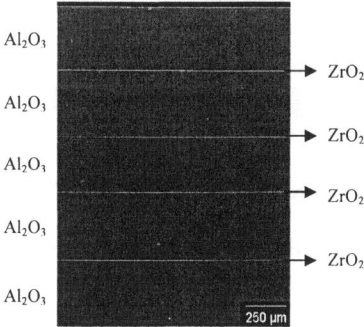

Figure 7. (a) Cross section of a multilayer ceramic obtained by piling and pressing zirconia coated alumina tapes. Thin lines (bright) correspond to zirconia and thick layers (dark) to alumina.

Electrophoretic deposition (EPD) has proved to be also useful to fabricate laminates as coating or as multilayered structures. The method is based on the movement of the charged particles suspended in water when an electric field is applied. It allows the fabrication of layers with different thicknesses over a conductive substrate. Continuous methods have been described to fabricate coatings with changes in compositions, but they still do not achieve a precise control in the composition nor in the layer thickness. The sequential methods have allowed to fabricate multilayers and graded materials. They have been also help by the incorporation of binders and gelling agents to slurry. As in previous methods, a kinetics study is precise. The factors that affect the thickness of the layers fabricated by this processing technique are among others: Electrical conditions (voltage and current), solid contents binders and current time [59,60]. Fig. 8 shows a self-supported multilayered material obtained by EPD changing the deposition time.

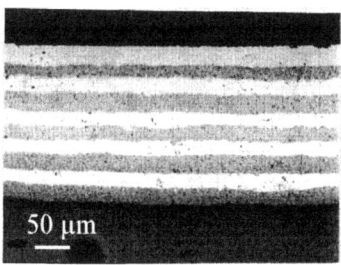

Figure 8. Self supported multilayer ceramic fabricated by sequential EPD

Also the EPD process can assist the packing of the slurries with binders or gelling agents in the composition that will help to fix the coating on the substrate and help the formation of a multilayer [61].

References
[1] H. M. Chan: Annu Rev Mater Sci **27** (1997), 249
[2] M. Y. He and J. W. Hutchinson: Int J Solids Struct **25** (1989), 1053
[3] A. J. Phillipps, W. J. Clegg, and T. W. Clyne: Acta Metall Mater **41** (1993), 805
[4] D. H. Kuo and W. M. Kriven: Mat Sci Eng a-Struct **241** (1998), 241
[5] D. B. Marshall, J. J. Ratto, and F. F. Lange: J Am Ceram Soc **74** (1991), 2979
[6] J. B. Davis, A. Kristoffersson, E. Carlstrom, and W. J. Clegg: J Am Ceram Soc **83** (2000), 2369
[7] W. J. Clegg, K. Kendall, N. M. Alford, T. W. Button, and J. D. Birchall: Nature **347** (1990), 455
[8] A. V. Virkar, J. L. Huang, and R. A. Cutler: J Am Ceram Soc **70** (1987), 164
[9] D. J. Green, R. Tandon, and V. M. Sglavo: Science **283** (1999), 1295
[10] C. Hillman, Z. G. Suo, and F. F. Lange: J Am Ceram Soc **79** (1996), 2127
[11] A. J. Sanchez-Herencia, C. Pascual, J. He, and F. F. Lange: J Am Ceram Soc **82** (1999), 1512
[12] M. P. Rao, A. J. Sanchez-Herencia, G. E. Beltz, R. M. McMeeking, and F. F. Lange: Science **286** (1999), 102
[13] M. Oechsner, C. Hillman, and F. F. Lange: J Am Ceram Soc **79** (1996), 1834
[14] J. S. Moya, J. A. SanchezHerencia, J. F. Bartolome, and T. Tanimoto: Scripta Mater **37** (1997), 1095
[15] N. Claussen and J. Steeb: J Am Ceram Soc **59** (1976), 457
[16] R. A. Cutler, J. D. Bright, A. V. Virkar, and D. K. Shetty: J. Am. Ceram. Soc. **70** (1987), 714
[17] R. Tandon and D. J. Green: J Am Ceram Soc **74** (1991), 1981
[18] S. Ho and Z. Suo: J Appl Mech-T Asme **60** (1993), 890
[19] A. J. Sanchez-Herencia, L. James, and F. F. Lange: J Eur Ceram Soc **20** (2000), 1297
[20] K. Ettre and G. R. Castles: Am Ceram Soc Bull **51** (1972), 482
[21] A. Tarlazzi, E. Roncari, P. Pinasco, S. Guicciardi, C. Melandri, and G. de Portu: Wear Wear **244** (2000), 29
[22] E. Lucchini and O. Sbaizero: J Eur Ceram Soc **15** (1995), 975
[23] L. Zhang and V. D. Krstic: Theor Appl Fract Mec **24** (1995), 13

[24] R. Bermejo, Y. Torres, A. J. Sanchez-Herencia, C. Baudin, M. Anglada, and L. Llanes: Fatigue Fract. Eng. Mater. Struct. **29** (2006), 71
[25] J. Requena, R. Moreno, and J. S. Moya: J Am Ceram Soc **72** (1989), 1511
[26] P. S. Nicholson, P. Sarkar, and X. Huang: J Mater Sci **28** (1993), 6274
[27] B. Ferrari, A. J. Sanchez-Herencia, and R. Moreno: Mater Res Bull **33** (1998), 487
[28] C. E. P. Willoughby and J. R. G. Evans: J Mater Sci **31** (1996), 2333
[29] I. M. Low, R. D. Skala, and D. S. Perera: J Mater Sci Lett **13** (1994), 1334
[30] Y. H. Koh, H. W. Kim, and H. E. Kim: J Am Ceram Soc **85** (2002), 2840
[31] R. Moreno: Am Ceram Soc Bull **71** (1992), 1521
[32] C. A. Gutierrez and R. Moreno: Mater Res Bull **36** (2001), 2059
[33] C. A. Gutierrez and R. Moreno: Br. Ceram. Trans. **102** (2003), 219
[34] F. F. Lange: J. Am. Ceram. Soc. **72** (1989), 3
[35] J. A. Lewis: J. Am. Ceram. Soc. **83** (2000), 2341
[36] B. J. Briscoe, G. Lo Biundo, and N. Ozkan: Ceram. Int. **24** (1998), 347
[37] R. E. Mistler: Am. Ceram. Soc. Bull. **77** (1998), 82
[38] J. Kiennemann, T. Chartier, C. Pagnoux, J. F. Baumard, M. Huger, and J. M. Lamerant: J. Eur. Ceram. Soc. **25** (2005), 1551
[39] R. Moreno and C. A. Gutierrez, in *Proceedings of the 9th CIMTEC*, edited by P. Vincencini (Techna, Florencia, 1999), pp. 611.
[40] A. Bleier and C. G. Westmoreland: J. Am. Ceram. Soc. **74** (1991), 3100
[41] B. R. Marple and J. Boulanger: J. Am. Ceram. Soc. **77** (1994), 2747
[42] A. J. Sanchez-Herencia, K. Morinaga, and J. S. Moya: J. European Ceram. Soc. **17** (1997), 1551
[43] R. Moreno: Am. Ceram. Soc. Bull. **71** (1992), 1647
[44] E. Carlstrom and A. Kristoffersson: Key Eng. Mat. **206-2** (2002), 205
[45] I. Santacruz, C. A. Gutierrez, M. I. Nieto, and R. Moreno: Adv. Eng. Mater. **3** (2001), 906
[46] J. H. Xiang, Y. Huang, and Z. P. Xie: Mat. Sci. Eng. A-Struct **323** (2002), 336
[47] Q. Q. Tan, M. Gao, Z. T. Zhang, and Z. L. Tang: Mater. Sci. Eng. A-Struct. **382** (2004), 1
[48] A. Kristoffersson and E. Carlstrom: J. Eur. Ceram. Soc. **17** (1997), 289
[49] B. Bitterlich and J. G. Heinrich: J. Eur. Ceram. Soc. **22** (2002), 2427
[50] J. Gurauskis, A. J. Sanchez-Herencia, and C. Baudin: J. Eur. Ceram. Soc. **25** (2005), 3403
[51] L. J. Vandeperre, A. Kristofferson, E. Carlstrom, and W. J. Clegg: J. Am. Ceram. Soc. **84** (2001), 104
[52] X. M. Cui, S. Ouyang, Z. Y. Yu, C. G. Wang, and Y. Huang: Mater. Lett. **57** (2003), 1300
[53] J. Gurauskis, C. Baudin, and A. J. Sánchez-Herencia: Ceram. Int. **In press** (2006),
[54] A. J. Sanchez-Herencia, J. Gurauskis, and C. Baudin: Composites B **Available online 4 April 2006** (2006),
[55] J. Gurauskis, A. J. Sanchez-Herencia, and C. Baudin: J Eur Ceram Soc **26** (2006), 1489
[56] Z. Chen, T. Takeda, K. Kikuchi, S.-i. Kikuchi, and K. Ikeda: Journal of the American Ceramic Society **87** (2004), 983
[57] A. Hirt: **81** (2004),
[58] L. D. Landau and B. G. Levich: Acta Physicochim. URS **17** (1942), 42
[59] B. Ferrari, S. Gonzalez, R. Moreno, and C. Baudin: J. European Ceram. Soc. **26** (2006), 27
[60] S. Gonzalez, B. Ferrari, R. Moreno, and C. Baudin: J. Am. Ceram. Soc. **88** (2005), 2645
[61] B. Ferrari, A. J. Sanchez-Herencia, and R. Moreno: J. Eur. Ceram. Soc. **26** (2006), 2205

Key Engineering Materials Vol. 333 (2007) pp 49-58
online at http://www.scientific.net
© 2007 Trans Tech Publications, Switzerland

Laminated and functionally graded ceramics by electrophoretic deposition.

O. Van der Biest [1, a], L. Vandeperre [b], S. Put [c], G. Anné [d], J. Vleugels [e]

[1] Department of Metallurgy and Materials Engineering, K.U. Leuven, Kasteelpark Arenberg, 44, 3001, Heverlee, Belgium

[a]omer.vanderbiest@mtm.kuleuven.be, [b]ljmv2@cam.ac.ukl, [c]stijn.put@umicore.com, [d]guy.anne@bekaert.com, [e]jozef.vleugels@mtm.kuleuven.be

Keywords: Electrophoretic deposition, EPD, functionally graded materials, FGM

Abstract. Electrophoresis is the effect that when an electric field is applied to a suspension of a powder in a liquid, the powder particles move under influence of this field. Frequently the powder particles also deposit at one of the electrodes. The form of the electrode determines the form of the deposit, hence shaping is possible. The current insights into the science and technology of electrophoretic deposition (EPD) will be summarized. EPD is well suited for shaping layered microstructures (laminates), by simply changing repeatedly between two or more suspensions during deposition. Tubular laminates consisting of silicon carbide layers and crack deflecting graphite interlayers have been produced. These tubes demonstrate an enhanced fracture energy and a gradual mode of failure. Another area of advanced ceramics where the use of EPD makes sense are functionally graded materials (FGM) in which one tries to combine in one component high hardness and high toughness. EPD allows the formation of FGM by depositing from a powder suspension to which a second suspension is continuously added during the process. An example will be shown of a graded WC-Co hardmetal.

Introduction

Many ceramics are used in the form of laminates especially functional ceramics in capacitors, sensors, microelectronic devices and others. From the mechanical properties point of view laminates are of interest because one may create ceramics with a threshold bending strength by introducing compressive residual stresses in those surfaces that will be subject to tensile loads in service [1]. Laminates may also be used to increase the toughness by alternating strong and weak layers. The role of the weak layers is to deflect cracks so that sudden catastrophic failure of the ceramic can be avoided [2]. These laminates are cheaper alternatives to fibre toughened ceramic composites.

The stringent requirements of aerospace technologies have given rise to the development of composite structures in which one tries to combine the strong points of two materials and to realize in the same component often irreconcilable material properties such as for instance hardness and toughness. In the simplest of cases, these composite structures are layered systems, e.g. coatings. A well-known example is that of coating a tough substrate by a hard film. The resulting component combines the high hardness of its surface with the high toughness of its core. However, the sharp interface between both is subject to stress concentrations which may lead to delamination or spallation. These problems have inspired the concept of functionally gradient materials (FGM), in which harmful stress concentrations can be reduced by providing a gradual transition in microstructure between the two dissimilar materials. FGM's [3,4] are thus distinguished from isotropic materials by gradients of composition, phase distribution, porosity, texture, and concomitant properties (hardness, density, resistance, thermal conductivity, elastic modulus, etc.). Those gradients are engineered and quantitatively controlled in order to achieve an overall improvement of the final component. In a similar way as in laminates a properly designed gradient may give rise to compressive surface stresses, thus leading to an improvement in strength and wear resistance [5,6]

Laminates can be produced by tape casting or by rolling a polymer based suspension of the powder. FGM's can be produced with a stepwise change in composition for instance by tape casting. This is less desirable because a discontinuity in composition means a discontinuity in elastic properties which can be a source of stress concentration and hence reduces the reliability of these materials. Continuously graded (profile graded) bulk materials are preferred. These have been produced by controlled powder mixing methods, gravitational segregation, centrifugal casting and others.

Electrophoretic deposition (EPD) can handle the processing of both types of composites, laminates and FGM. Moreover, it is a low cost, rapid, near net shaping process capable of producing continuously graded materials and laminates of a fairly complex shape. It also allows the best control of the green shaping process in comparison with other fabrication techniques.

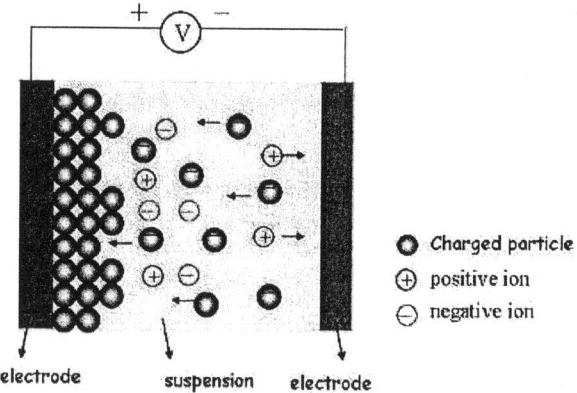

Fig. 1. Schematic of the EPD process

Electrophoretic shaping

Principles of electrophoretic deposition. EPD is a relatively little known ceramic shaping technique where a stable suspension of a powder is used in which the powder particles have acquired an electric charge through interaction with the suspension medium. The electric charge induces the migration of the particles under the influence of an electric field introduced in the suspension (electrophoresis). Under certain circumstances the particles will move to the electrode with opposite charge, lose their surface charge and form a deposit on this electrode (electro-deposition) (Fig. 1). The deposited layer can be several millimetres thick. The *shaping capabilities* are a consequence of the fact that powder can be deposited onto an electrode of almost any shape and that the deposit follows the shape imposed by this electrode. The main limitation is that it should be possible to remove the deposit from the electrode after deposition. Another possible limitation is that the counter electrode should be so designed that a constant electric field is present at the desired area on the deposition electrode (see e.g. [7]). The deposition rate can be in the order of hundreds of microns per minute, making it a rapid forming process. Further, since the apparatus required for electrophoretic deposition is relatively simple (essentially only two electrodes, a bath, and a power source are required), the installation costs are low and it can be scaled up to an industrial size easily. Whereas the successful implementation of EPD is very powder specific, the

EPD-process can in fact be applied to a wide range of materials including metal and organic particles as well as composite particles [8].

Hence, after removal from the electrode and drying, a shaped green ceramic body is obtained. Firing this green body then results in a ceramic component.

Suspension preparation and composition. Suspension parameters determine the structure and cohesion of the green deposit at the electrode. Stable suspensions with well- dispersed particles are necessary to produce densely packed deposits [9]. Because the particles require an electrical charge for electrophoresis, EPD literature considers mostly electrostatic stabilization. There is little systematic information on how suitable suspensions are prepared for electrophoretic deposition. These suspensions are mainly based on organic liquids because electrophoretic deposition from aqueous suspensions has the disadvantage of electrolysis of water at low applied potentials, causing gas formation at the electrodes, giving pores in the deposit and a possible breakdown of the deposits at the electrode. In the case of plate-shaped components, gas bubbles in the deposit can be prevented by using a membrane in front of the deposition electrode [10]. For complex shaped bodies on the other hand, it is not feasible to use a membrane. Another method to avoid gas bubbles is to use palladium electrodes for catholic EPD from aqueous suspensions. Palladium readily absorbs hydrogen [11], but is expensive.

Under wet conditions, the surface charging in protic solvents is governed by a competitive proton transfer between the physisorbed water or the solvent and the particle. The acid sites on the powder particle are able to transfer a proton from the solid to the adsorbed molecule [12]. Proton transfer can also be considered in non-aqueous media, like alcohols, which can also be regarded as neutral amphiprotic solvents. Ketones, ethers and hydrocarbons, on the other hand, are known as aprotic solvents and charging by proton transfer can not take place. However, water can not totally be avoided in these solvents and plays in fact a crucial role in charging particles in non-aqueous suspensions due to the adsorption of water on the particles, loosening surface ions, changing the acid-base character of the surface and the nature of the adsorbed solute- surface interactions [13].

EPD in non-aqueous suspensions is mostly performed from ethanol or ketone or a mixture of these solvents. An advantage of non-aqueous media for EPD is the possibility to use high voltages. Disadvantages are the toxicity of solvents and the higher cost. Recuperation of solvents can lower the costs and reduce the environmental impact.

For colloidal processing like EPD, it is important to have a stable and dispersed suspension to obtain defect free powder compacts. We have used stable non-aqueous suspensions based on alcohols or ketones. Ketone based suspensions with n-butylamine addition were already shown in the past to be suitable suspensions for EPD of silicon carbide, Al_2O_3, ZrO_2 and hardmetals despite their low stability. We have shown recently that specifically for alumina and zirconia a highly stable suspension was formed with the addition of 1 wt% nitrocellulose to methyl ethyl ketone (MEK) with 10-15 vol% n-butylamine. EPD deposits with a high green density, low surface roughness and high deposition yield were obtained. Higher amounts of n-butylamine caused gas bubbles in the deposits, due to water formed by a reaction between the MEK and n-butylamine. Experimental results revealed that the zeta potential is not a straightforward indication of the stability of these suspensions, since the maximum absolute zeta potential did not correspond with a maximum suspension stability, due to the additional electrosteric stabilization of the adsorbed charged nitrocellulose [14].

Controlling deposition rate and composition. It has often been argued that formation of a deposit on the electrode should represent an additional resistance in the electric circuit during the deposition process. The deposit should then have a higher specific resistance than the volume of the suspension that it replaces between the two electrodes. Hence the electric field at the deposition front should gradually decrease as the deposit thickness increases. Most authors mention this additional potential drop over the deposit without giving an explanation or absolute value. Due to the large practical consequences of this issue, a procedure was developed to calculate it from simultaneous measurements of the current flowing through the cell and the conductivity of the suspension during EPD. From these two measured values it is possible to calculate the electric field at the deposition front as a function of time [16]. Results are shown in Table 1 and graphically in Figure 2. It is clear that the suspension composition determines whether an extra potential drop over the deposit is present or not. Alumina suspensions based on MEK and n-butylamine showed no potential drop over the deposit, while suspensions based on ethanol and an acid did cause an extra potential drop over the deposit.

Table. 1. Deposition yield, green density of Al_2O_3 deposits and the E-field strength after 2000 s

Suspension	Yield (g)	Green density (%)	E (after 2000s) (V/cm)
Ethanol + HCl	3.38	53	16.56
Ethanol + HNO$_3$	3.28	51	11.65
Ethanol + PEI	5.76	51	26.97
Ethanol + CH$_3$COOH	6.80	54	25.5
MEK + n-butylamine	9.87	55	48.2

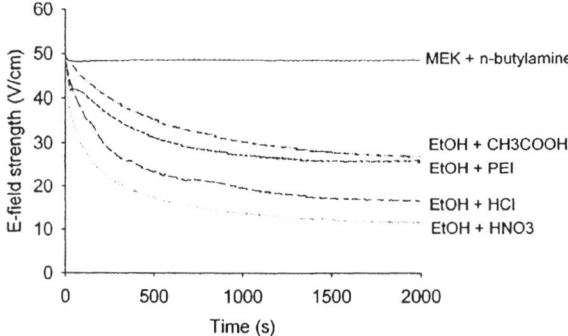

Fig. 2. The E-field strength, calculated from the instantaneous current and suspension conductivity, as a function of time for alumina suspensions based on ethanol with HCl, HNO$_3$, CH$_3$COOH or PEI and on MEK with n-butylamine [16].

In order to understand these differences in behaviour it is useful to think of the deposit as a *membrane* through which ionic transport is controlled by the potential distribution in the pore which is in turn related to the thickness of the electrostatic double layer relative to the pore radius and the magnitude of the surface potential of the powder particles [17].Specifically, for ethanol-based suspensions with HNO$_3$ addition, the specific deposit resistively and the concomitant potential drop over the deposit was found to increase with increasing potential in the centre of the pores of the deposit, as calculated from the zeta potential of the particles in suspension, the Debbie screening length and the measured average pore size. An increased pore potential inhibits ion movement through the pores of the deposit, increasing the resistivity of and the potential drop over the deposit. The absence of an extra potential drop for MEK / n- butylamine suspensions was explained by the very thin double layer relative to the pore diameter.

By means of these findings, it will be possible to develop faster suspensions for specific applications : suspensions with a high potential drop over the deposit are suitable for thin homogeneous coatings on rough surfaces since the electric field drop caused by the deposit can lead to a self-limiting thickness effect. On the other hand suspensions with a low potential drop are suitable for EPD of thick bulk components.

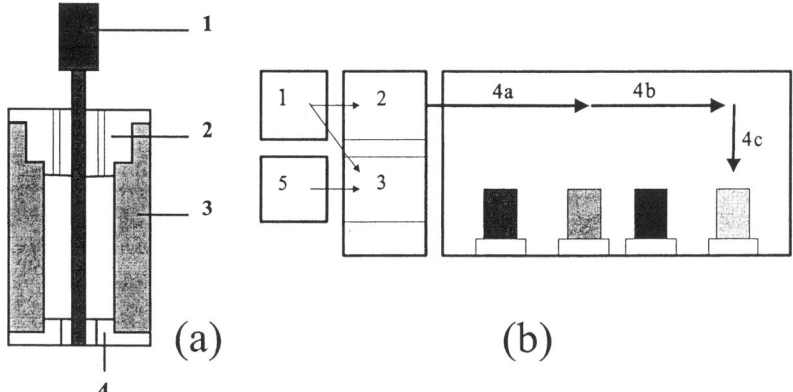

Fig. 3 :(a) the electrode set-up for electrophoretic deposition of tubes: (1) stainless steel counter electrode rod, (2) PTFE top cover, (3) graphite deposition electrode, (4) PTFE bottom cover. (b) schematic overview of the automated cell for electrophoretic deposition of tubes. A personal computer (1) allows to program the unit (2) which controls the movement of a series of pistons (4a,b&c) and the unit (3) which connects or disconnects the voltage from an external power supply (5) to the cell. Two horizontal pistons of unequal length allow to reach 4 positions, the third piston ensures the vertical displacement.

Laminates by EPD

Principles of fabrication An example is shown of the fabrication and properties of laminated silicon carbide tubes. These are intended as prototype combustion tubes. They are build up from silicon carbide laminae separated by thin graphite interlayers which had to be optimized for their ability to deflect the cracks. The electrodes in the EPD process for these laminates naturally are cylindrical in shape. Since the deposit shrinks during drying, it is advantageous to deposit at the inside of a hollow cylindrical electrode (Fig. 3.a). As the deposit dries, it will release from the deposition electrode thus facilitating removal from the electrode. As a result of this choice and considering that symmetry is required for a homogeneous electric field, a rod placed in the centre of the cavity of the deposition electrode is used as a counter electrode (Fig. 3.a). In order to ensure that tubes with well defined dimensions are obtained and in order to prevent that deposition occurs around the edges of the deposition electrode, two electrically insulating PTFE covers are connected to the deposition electrode (Fig 3.a). Holes are drilled in the top cover in order to let the air escape during immersion of the electrode-set in the suspension. The processing was automated by attaching the electrode set-up to an existing apparatus for multi- layered electro-deposition. This device is capable of placing the electrodes in 4 different baths and allows to program the residence time in each bath as well as the applied voltage. Figure 3.b shows the automated cell schematically.

For deposition of the silicon carbide matrix layers, a suspension on the basis of acetone and butyl amine was used with some isopropyl alcohol added. Iso-propyl alcohol was added to match the drying rate of the SiC layers to the drying rate of the graphite layers which are deposited from an iso-propyl alcohol based suspension. Moreover the surface smoothness of the deposit is improved by addition of iso-propyl alcohol. In between the silicon carbide lamella, thin graphite layers were deposited from a commercial colloidal graphite suspension (Superior Graphite #210 in iso-propyl alcohol) diluted further with 70 vol% isopropyl alcohol to obtain good quality layers. At lower concentration no coherent deposit forms, while at higher concentration a highly irregular layer is obtained. Graphite layers obtained by deposition from this suspension were found to have a sufficient shear strength in a laminated plate. In order to ensure a good release of the deposit from the electrode after deposition, the inside of the deposition electrode is covered with a loose dip-coating of a coarse graphite powder. After drying, the tubes were sintered in a graphite furnace under vacuum at 2050°C.

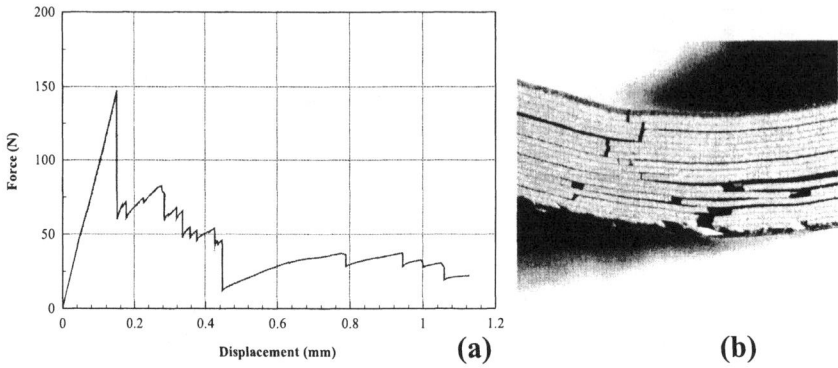

(a)   **(b)**

Fig. 4:(left) load- displacement diagram for a test on a laminated ceramic ring in diametric compression. (right) detail of a laminated ring after testing in diametric compression showing the deflection of cracks into the graphite interlayers

Results. Using this procedure laminated SiC-graphite tubes containing 19 SiC layers (100 ◻m) and 18 graphite layers (12 ◻m) were produced. The tubes have an average diameter of 21 mm and a height of 35 mm. A tough behaviour was observed during testing in diametric compression of rings cut from the tubes: the load-displacement diagram was of the saw-tooth type typical for laminates (Fig. 4 left), and cracks were indeed deflected upon meeting a graphite interlayer (Fig. 4 right). Furthermore, the energy up-take by the sample during failure is indeed higher for a laminate compared to a monolithic sample.

Two observations support this qualitatively. In contrast with tests on monolithic rings, fragments of failed laminae are not ejected from the test set-up at high velocity. Thus, at least less of the elastic energy stored in the sample is converted in kinetic energy of failed parts. Moreover, less elastic energy is converted into sound also compared to tests on monoliths. Both effects can be understood from the laminated architecture: the amount of elastic energy that is released from the sample at any moment is limited, since crack deflection limits the stiffness degradation of the sample and propagation of the crack along the interlayer consumes energy.

FGM by EPD

Principles of fabrication. Since the composition of the deposit during the EPD process is directly related to the composition of the suspension at the moment of deposition, the EPD technique allows processing of functionally graded materials with a continuous gradient in composition by judiciously adjusting the suspension composition as a function of time. Thus by adapting the solids loading, e.g. by injection of a second type of powder, a controlled gradient can be formed in a continuous shaping step. The successful application of EPD for the manufacturing of FGM's has been demonstrated for a variety of systems [18-20]. The EPD process to produce functionally graded materials is schematically illustrated in Fig. 5. The polytetrafluorethylene (PTFE) deposition cell, illustrated in Figure 5b, contains two parallel flat stainless steel electrodes with a surface area of 9 cm². The distance between the electrodes is 3.5 cm.

Complex geometries require an appropriate design of the counter electrode in order to generate a constant electric field at the surface of the deposition electrode. The design of the most suitable counter electrodes needs to be supported by electrical field calculations using finite element analysis [7].

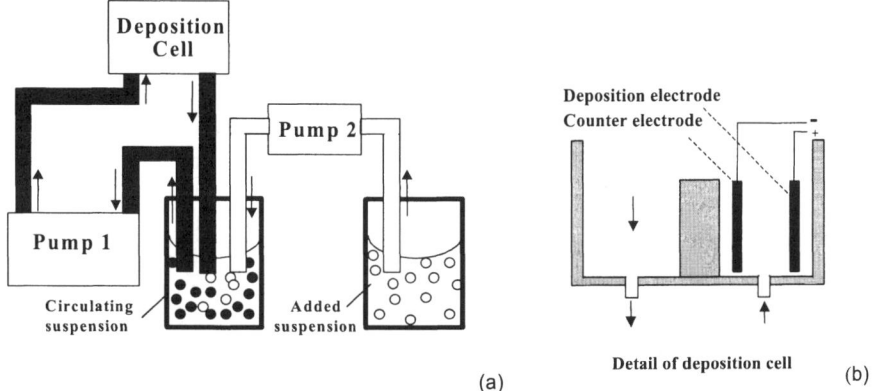

Fig. 5 : Schematic view of the electrophoretic deposition set-up (a) and detail of the deposition cell (b) for functionally graded materials. Both suspensions are stirred continuously to maintain homogeneity.

A process is considered in which first a homogeneous powder suspension is pumped in a circulation system through the deposition cell by a peristaltic pump 1. During time t_{s1}, the powder particles will form a layer with homogeneous composition (Figure 6). In a second step, a second suspension is added at a constant rate to the circulating suspension. The addition of the second powder and thus the creation of a gradient profile is stopped at time t_{f1}. In a third step, electrophoretic deposition is continued until time t_{s2}. during which time the homogeneous core of the plate is being formed. In order to create a plate with a symmetrical structure as shown schematically in Figure 6 (right) , a third suspension is then added to start the second gradient until time t_{f2}. Finally the last homogeneous layer of the plate is formed until time t_e.

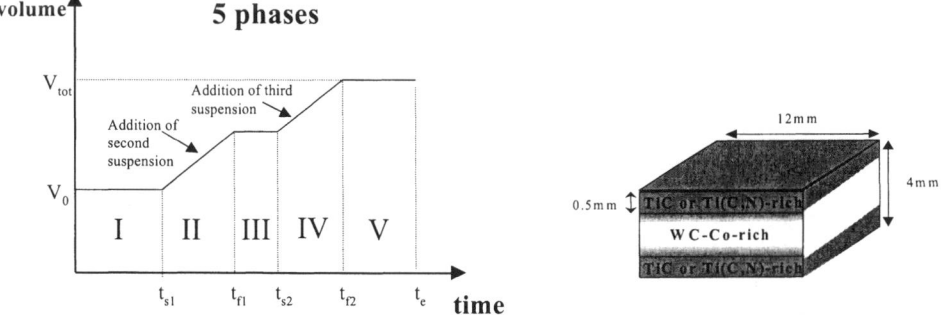

Fig. 6. (left) Volume of the circulating suspension as a function of time during FGM processing. (right) Schematic view of the geometry of a desired WC-Co-Ti(C,N)/WC-Co/WC-Co-Ti(C,N) symmetric FGM.

Example of hardmetal FGM. An example is shown of a plate-shaped FGM, to be used as cutting tool insert, with homogeneous Ti(C,N)-rich surface sides with increased hardness, intermediate gradient layers and a tough core of WC-Co, as schematically illustrated in Figure 8. In this case, a composite WC-Co-5Ti(C,N) powder was used as starting suspension. A suspension of WC-10Co powder was continuously added, decreasing the Ti(C,N) content towards the core of the plate. After deposition of the homogeneous core layer, a Ti(C,N)-rich powder suspension was added in order to create a second gradient from 0 up to 5 wt % on the other side of the plate. Details are given in Table 2. In order to create similar gradients on both sides of the plate, the use of a simulation model is essential [21]. After EPD, the dried deposit was CIPed and sintered at 1450°C for 1 hour. A total deposition time of 25 minutes was needed to deposit a FGM plate with a sintered thickness of about 4.5 mm.

Table 2. Suspension composition and EPD parameters for the symmetric FGM plate.

Suspension	1	2	3
WC-10Co-5Ti(C,N) (g)	13.5	0	0
WC-10Co (g)	0	10	0
WC-10Co-10Ti(C,N) (g)	0	0	6
MEK (ml)	45	22.5	45
n-butylamine (ml)	5	2.5	5
STEP 1 : EPD (sec)	120		
STEP 2 : addition + EPD (sec)		450	
STEP 3 : EPD (sec)		140	
STEP 4 : addition + EPD (sec)			600
STEP 5 : EPD (sec)			190
TOTAL Deposition Time (sec)	1500		

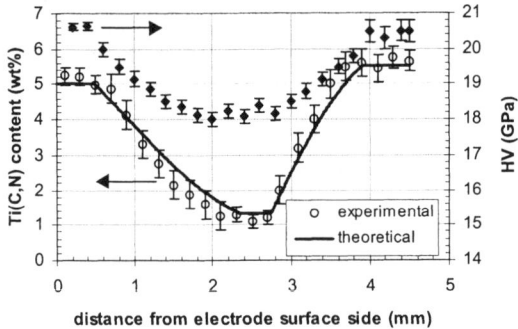

Fig. 7. (a) Ti(C,N) and (b) Vickers hardness profiles along the thickness of the symmetrical WC-Co-Ti(C,N)/WC-Co/WC-Co-Ti(C,N) FGM.

Quantitative electron microprobe measurements clearly revealed the double gradient in Ti(C,N) content (Figure 7). A homogeneous Ti(C,N)-containing layer is obtained on both sides. Towards the core, a continuously decreasing Ti(C,N) concentration is achieved. A good correlation exists between the profile predicted by the model [21] and the experimental results. The hardness and composition profiles along the cross-sectioned plate shows a $HV_{0.5}$ value of 20.5 GPa for the outer

Ti(C,N)-rich layers and 18 GPa for the homogeneous WC-Co core (Figure 7). No warping of the plate was observed after sintering.

These graded structures can achieve high economical and technical competitiveness as tools for applications in highly abrasive as well as in shock loaded conditions, combining a higher level of hardness and wear resistance with adequate toughness.

Final remarks and conclusions

EPD is a versatile powder shaping technique suitable for processing laminated as well as graded composites. In this paper we have tried to highlight some of the improvements in the understanding of the process by recent research. Further progress in understanding is necessary to increase the reproducibility of the process and thus further its acceptance by industry. The added value of laminated and graded structure should justify the expense of designing and implementing a completely new processing route.

References

[1] M. P. Rao, A. J. Sánchez-Herencia, G. E. Beltz, R. M. McMeeking, F. F. Lange, Science. 286. nr 5437, pp. 102 - 105, (1999)

[2] W.J. Clegg, K. Kendall, N. McN. Alford, T.W. Button and J.D. Birchal, Nature, **347**, 455 (1990)

[3] T. Hirai. Functional Gradient Materials, in Processing of Ceramics, part 2; Ed. R.J.Brook, Materials Science and Technology: A Comprehensive Treatment, VCH Verlagsgesellschaft mbH, 17B, 293- 341, (1996)

[4] Y. Miamoto, W.A. Kaysser, B.H. Rabin, A. Kawasaki, R.G. Ford, Functionally Graded Materials : Design, Processing and Applications, Kluwer 1999, p. 296-311.

[5] G. Anné, S. Hecht-Mijic, H. Richter, O. Van der Biest and J. Vleugels, Scripta Materialia, 54, [12], 2053-2056 (2006)

[6] S. Novak, M. Kalin, P. Lukas, G. Anne, J. Vleugels and O. Van Der Biest, J. of the European Ceramic Society, In Press, Available (2006)

[7] G. Anné, K. Vanmeensel, J. Vleugels, O. Van der Biest' Key Engineering Materials ,314, 213-218 (2006).

[8] O. Van der Biest and L. Vandeperre , Annu. Rev. Mater. Sci (1999) vol. 29, 327-52.

[9] F.Bouyer and A.Foissy, J. Am. Ceram. Soc., 1999. 82(8): p. 2001-2010.

[10] R. Clasen and J.Tabellion, Cfi-Ceramic Forum International, 2003. 80(10): p. E40-E45.

[11] T.Uchikoshi, K. Ozawa, B. Hatton, Y.Sakka, J. of Mat. Res.,. 16(2) p. 321-324, (2001).

[12] F. Delannay, *Characterization of heterogeneous catalysts*. eds. Marcel Dekker, Inc. (New York and Basel), 1984.

[13] I.D Morrison., Colloids and Surfaces A-Physicochemical and Engineering Aspects, 71(1): p. 1-37, (1993).

[14] G. Anné, K. Vanmeensel, B. Neirinck, O. Van der Biest, J. Vleugels, J. Europ. Ceramic Soc. in press , available on line (2006)

[15] G. Anné, K. Vanmeensel, J. Vleugels, O. Van der Biest, Colloids and Surfaces A: Physicochemical and Engineering Aspects, Vol. 245, Nr. 1-3, p. 35-39 (2004).

[16] G. Anné, B. Neirinck, K. Vanmeensel, O. Van der Biest, J. Vleugels, Key Engineering Materials Vol. 314 pp 13-18 (2006).

[17] G. Anné, B. Neirinck, K. Vanmeensel, O. Van der Biest, J. Vleugels, J. Am. Ceram. Soc, Volume 89, Issue 3, Page 823-828, (2006)

[18] P. Sarkar, X. Huang, P.S. Nicholson, J. Am. Ceram. Soc., 76, (4), 1055-56 (1993)

[19] P. Sarkar, S. Datta, P.S. Nicholson, Composites part B, 28B, 49- 56, (1997)

[20] L. Vandeperre, O. Van Der Biest, Silicon Carbide laminates with Carbon Interlayers by Electrophoretic Deposition, in Key Engineering Materials, Vols. 127-131, 567-574, (1997)

[21] S. Put, J. Vleugels, O. Van der Biest, Acta Materialia, 2003. 51(20): p. 6303-6317.

Key Engineering Materials Vol. 333 (2007) pp 59-70
online at http://www.scientific.net
© 2007 Trans Tech Publications, Switzerland

Advanced Environmental Barrier Coatings (EBC's)

T. Kulkarni, S.N. Basu, and V.K. Sarin

College of Engineering, Boston University, Boston, MA, USA

Keywords: Mullite coatings; Chemical vapor deposition; Protective coatings; Composition gradation; High-temperature stability.

Abstract. Dense, uniform and crack-free mullite ($3Al_2O_3 \cdot 2SiO_2$) coatings were deposited on Si-based substrates by chemical vapor deposition using the $AlCl_3$–$SiCl_4$–H_2–CO_2 system. The coatings were compositionally graded, with the Al/Si ratio increasing towards the outer surface of the coatings for improved corrosion resistance. Mullite grains nucleated when the surface composition of the growing coating was in a narrow range close to that of stoichiometric mullite. The growth rate and crystal structure of mullite were dependent upon temperature, pressure, reactant concentration, and reactant ratios. The phase transformations occurring in these coatings during high-temperature anneals in the range 1100–1400 °C were studied.

Introduction

Emerging technologies in the 21st century are severely challenging the outer limits of most commercially available structural materials. There is therefore an urgent and ever-increasing need, in all segments of technology and industry, such as energy, defense, and various commercial sectors for new and improved lower density, high temperature materials. The development of such materials is a strategic global concern since these materials contribute to enhanced performance, higher efficiencies, and environmental advantages. To meet the worldwide demand for energy over the next decade may be the most difficult and critical problem facing mankind today. The economic and political consequences of not achieving this goal are unthinkable. For example, increase in the universal demand for electricity is projected to significantly increase from current levels, and can only be met if a major portion is achieved using advanced gas turbine systems [1]. Furthermore, significant improvements in heat engine operating efficiencies can be realized with the use of uncooled ceramic components at temperatures above those attainable with superalloys [2]. Therefore, in order to grow and maintain an environmentally clean energy technology, this next generation of advanced gas turbine systems will have to operate with improved fuel efficiency and reduced emissions, mandating the use of higher operating temperatures.

Currently, superalloys with thermal barrier coatings have been driven close to their limit in high temperature applications. For a significant increase in operating temperatures, a new class of materials will have to be introduced. Silicon-based ceramics such as SiC and Si_3N_4 are leading candidate materials for the next generation of high temperature materials due to their high strength, high thermal conductivity, excellent high temperature stability and oxidation resistance and low coefficient of thermal expansion (CTE).

A large number of investigations have focused on introducing SiC and Si_3N_4 (i.e. gas turbines, stators, etc.) for stringent elevated temperature applications. At moderate temperatures and pressures, the formation of a thin self-healing layer of SiO_2 is effective in preventing catastrophic oxidation by minimizing the diffusion of O_2 to the substrate. The presence of impurities can however: increase the rate of passive oxidation by modifying the transport rate of oxygen through the protective scale, cause active oxidation via formation of SiO which accelerates the degradation process, or produce compounds such as Na_2SiO_3 which chemically attack the ceramic via rapid corrosion [3]. Additionally, susceptibility to contact stress damage has restricted their application due to unpredictable and premature failures. It is now well established that the successful utilization of Si-based ceramic components will require protective coatings

As a consequence, protective environmental barrier coatings (EBCs) have become an essential and critical part of materials development in the 21st century. Such coatings must resist reacting with the aggressive environments to avoid degrading themselves, and must also act as excellent diffusion barriers to effectively separate the substrate from the environment. Surface modification is by far the most versatile and advanced means of enhancing material properties, and the only technique known where materials can be tailored to have combination of surface and bulk characteristics, which are often competing. Mullite ($3Al_2O_3 \cdot 2SiO_2$) has received considerable attention since the late 1980s as a candidate material for EBCs, especially for protecting Si-based ceramics and composites. It has superior corrosion resistance, creep resistance and high temperature strength and toughness, compared to SiC and Si_3N_4. Additionally, mullite has a good CTE match, especially with SiC. Several techniques have been explored to deposit protective coatings of mullite, these include:

- Chemical vapor deposition (CVD)
- Plasma and flame spraying
- Physical vapor Deposition (PVD)
- Deposition from an aqueous slurry
- Self oxidation by oxygen implantation of metal alloys

Of the processes listed above, the CVD and plasma/flame spraying processes have shown the most promise. One of the major drawbacks of the plasma/flame coatings is the inherent existence of pores and cracks, which restricts their utilization in several pertinent applications. There has been an extensive research program at Boston University aimed at developing dense CVD mullite environmental barrier coatings (EBC's) coatings for high temperature applications. Dense, crystalline mullite coatings of uniform thickness have been deposited by chemical vapor deposition (CVD) on Si-based substrates, using the $AlCl_3$-$SiCl_4$-CO_2-H_2 system. Test results have clearly demonstrated that these coatings are highly effective in eliminating the problems of hot corrosion and recession of these ceramics. However, there is some concern that the presence of silica in the mullite structure may lead to long-term recession problems in these coatings. In order to address this potential problem, functionally (i.e. compositionally) graded alumina-rich mullite coatings using the CVD technique are being developed.

This approach has several advantages. The composition grading allows the surface of the coatings to be virtually silica free thus eliminating the problem of recession. Furthermore, the virtually pure alumina coating surface would lead to further improved hot-corrosion and oxidation resistance, while the composition gradation will lead to improved coating adhesion by minimizing the thermal stresses on temperature cycling. These coatings should therefore have the oxidation, corrosion and recession resistance of α-alumina, with the thermal shock resistance of stoichiometric mullite coatings. Needless to say, such EBC coatings should have a very broad impact on not only enabling gas turbines to operate at higher temperatures thus leading to improved fuel efficiency and reduced emissions, but also on several other applications.

CVD Mullite Coatings

Considerable efforts to date have been devoted to the development of CVD mullite coatings on Si-based substrates [4, 5, 6, and 7]. This research has focused upon a thorough thermodynamic analysis, a development of the deposition reactor, and an initial understanding of the structural evolution of the growth of the coatings. Thermodynamic analyses based on the minimization of free energy, considering the mixture of both gaseous and condensed species, were performed on mullite deposition using Al and Si-halides, CO_2, and H_2. Calculations were performed on the $AlCl_3$, $SiCl_4$, CO_2, and H_2 system at different process parameters (specifically temperature, pressure, and gas composition concentrations). The resulting stable species (both gaseous and condensed) were used to generate a series of CVD phase diagrams where mullite was predicted to form as a line compound over a range of temperatures, pressures, and input reaction concentrations.

A CVD reaction is governed by thermodynamics, which basically ascertains the chemical driving force. However, the rate control mechanism is determined by kinetics. Control over a coating's structure and morphology and thus its resultant properties is only possible once a reasonable understanding of the kinetics of deposition has been achieved. The kinetics of CVD mullite deposition was analyzed and established using Al_2O_3 and SiO_2 oxide systems and the much studied water-gas shift reaction (Eq. 1) involving H_2 and CO_2 [8].

$$CO_2+H_2 \rightarrow H_2O+CO \tag{1}$$

Experimental

A CVD reactor (Fig. 1) consisting of a vertical hot-walled chamber with a resistively heated three-zone furnace was used for the deposition of CVD mullite coatings. Reactants entered at the base and the spent gases exited through the exhaust tubes at the top of the chamber. The reactants used in the formation of mullite were $AlCl_3$, $SiCl_4$, CO_2, H_2, and Ar, which was used as a carrier gas. $AlCl_3$ was formed in situ by chlorinating heated Al chips. Excess H_2 was present to ensure complete reduction of the metal chlorides and a uniform temperature of reactant gases. The deposition conditions and reactor have been discussed in detail in the literature [5, 6].

Multi Layered System

The initial approach to grow CVD mullite coating was based on techniques used to form mullite powders for production of bulk material. Diffusion couple experiment of Al_2O_3 and SiO_2 pellets at high temperatures in the range of 1500 °C demonstrated the formation of mullite.[9] It was therefore projected that similar results could be obtained by depositing thin alternating layers of Al_2O_3 and SiO_2 and annealing the coatings and forming mullite via enhanced diffusion.

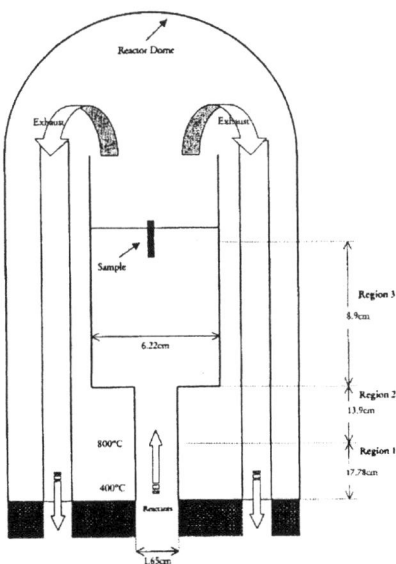

Fig. 1: Schematic representation of the CVD reactor

The widths of the alumina and silica layers were designed to be as thin as possible so as to facilitate complete diffusion across the layers. The thicknesses of the SiO_2 layers were adjusted such that the overall composition of the fully diffused coating would be that of stoichiometric mullite.

Although some conversion to mullite at the Al_2O_3/SiO_2 interfaces was obtained during the growth cycle, post treatment was observed to be necessary in all cases. The initial multilayered coatings grown on SiC and Si_3N_4 substrates, with an overall thickness of 1-1.5 μm (Fig. 2), were annealed in an Ar atmosphere at various temperatures in order to facilitate the formation of mullite via interfacial reactions.

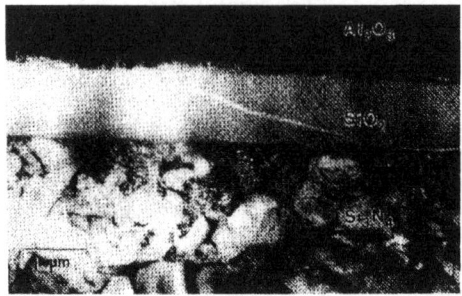

Fig.2: TEM bright-field photomicrograph of a laminated SiO_2/Al_2O_3 coating on a Si_3N_4 substrate

However, this approach for obtaining mullite coatings was abandoned as the vitreous to crystalline transformation in silica (during post treatment) led to coating spallation due to cracking caused by volume changes. Furthermore, mismatch in the thermal expansion coefficient of Al_2O_3 (7.2-8.6 x 10^{-6} K^{-1}) and SiO_2 (0.5 x 10^{-6} K^{-1}) resulted in high residual stresses developing within the coating. Stresses that aided volume contraction in silica enhanced transformation, while those that aided volume expansion inhibited this transformation. So, it was concluded that the multi-layered approach to form mullite coatings was not viable since very high temperatures were needed [9] in order to facilitate reasonable diffusion between Al_2O_3 and SiO_2 layers and this high temperature treatment caused transformations within the coating leading to failure before any considerable diffusion could take place.

Direct Deposition

A multi reactant process was used for the direct deposition of mullite coatings. In this process $AlCl_3$, $SiCl_4$, CO_2, and H_2 were premixed outside the reaction zone to ensure that a homogenous mixture of the chlorides came in contact with water-gas before reaching the substrate surface. The coatings were deposited using the $AlCl_3$–$SiCl_4$–CO_2–H_2 system in a hot-wall CVD reactor [4], with the overall reaction given by (Eq. 2)

$$6AlCl_3 + 2SiCl_4 + 13CO_2 + 13H_2 \leftrightarrow 3Al_2O_3 \cdot 2SiO_2 + 13CO + 26HCl. \tag{2}$$

The coatings were examined by X-ray diffraction (XRD) for phase identification, using monochromatic $Cu_{K\alpha}$ radiation and a 0.02° step size with a 2.0 s dwell time. The surface and cross-section morphologies of the coatings were examined by a scanning electron microscope (SEM), while electron transparent cross-sections were examined in a JEOL 2010FX transmission electron microscope (TEM). The chemical composition of the coatings were analyzed by X-ray energy-dispersive spectrometry (XEDS) using a VG-HB603 dedicated scanning-transmission electron microscope (STEM), using an electron beam focused to a 4 nm diameter.

Fig 3 (a) shows a fracture cross-section of a typical adherent and uniform mullite coating on SiC and Fig 3 (b) shows a cross-sectional TEM micrograph. The pore and crack-free nature of the coating is evident. Fig. 3 (c) shows the X-ray diffraction pattern of the coating, where all non-substrate peaks match with mullite and the substrate (SiC) peaks are marked as S.

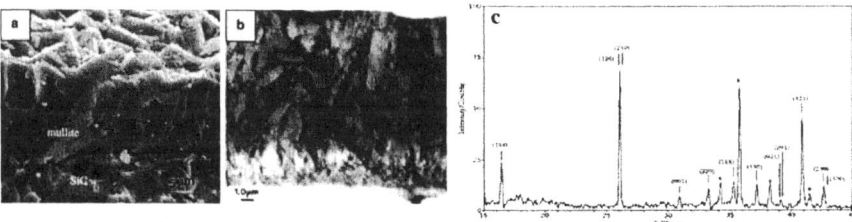

Fig. 3. (a) SEM micrograph of a fracture cross-section of a uniform and adherent CVD mullite coating on SiC. (b) TEM micrograph of the dense mullite coating in cross-section. (c) X-ray diffractogram of a CVD mullite coating.

A TEM bright-field micrograph in Fig. 4 (a) shows a typical CVD mullite coating in cross-section in a region adjacent to the coating/substrate interface. This figure reveals that the coating actually consists of two distinct layers.

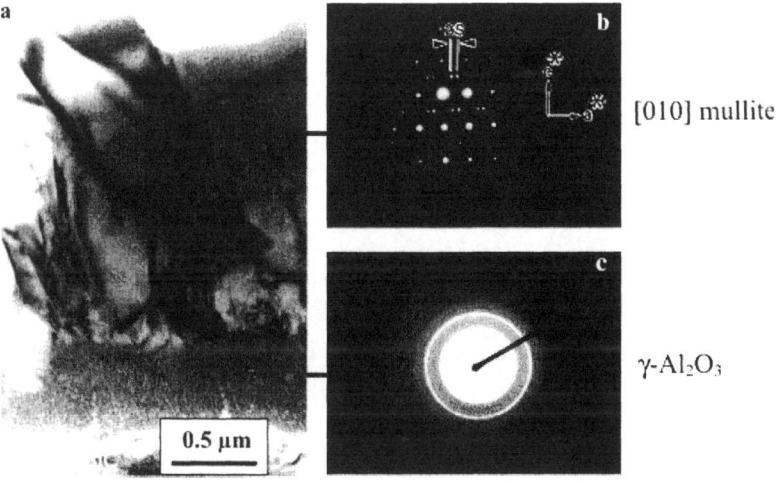

[010] mullite

γ-Al$_2$O$_3$

Fig. 4. (a) Cross-section TEM micrograph close to the coating/substrate interface. (b) [0 1 0] SAED diffraction pattern from a mullite grain in the coating. (c) SAED pattern from the nanocrystalline layer.

The majority of the coating was found to be composed of crystalline grains of columnar morphology. Fig. 4 (b) shows a selected-area electron diffraction (SAED) pattern from a crystalline grain, which is consistent with [0 1 0] mullite. This layer is designated hereon as the 'crystalline layer'. However, a thin non-crystalline layer below the columnar crystalline layer in Fig. 4(a) was always observed to be present in all the coatings. High-resolution TEM showed that the lower layer was composed of very fine ($\sim$5 nm diameter) crystalline particles embedded in a vitreous matrix. These crystallites were initially identified as γ-Al$_2$O$_3$ from the SAED pattern from this region as shown in Fig. 4 (c). Recent transmission electron microscopy based analyses suggest to the presence of a different alumina phase, or a combination of more than one phase, as opposed to previously identified γ-Al$_2$O$_3$. This lower layer was designated as the 'nanocrystalline layer'. The above described microstructure of a nanocrystalline layer at the coating/substrate interface, with a layer of crystalline mullite grains over the nanocrystalline layer was typical for all coatings grown on SiC substrates under the deposition conditions. [5]

Composition gradation

The composition of the mullite coatings was graded from silica-rich close to the coating/substrate interface to alumina-rich towards the outer surface of the coating. Fig. 5 shows the composition profile across a typical coating. The composition is plotted as the Al/Si ratio, whose value is 3 in stoichiometric mullite. Interestingly, the mullite grains nucleated when the Al/Si ratio in the nanocrystalline layer was ∼ 3.2±0.3 [10], which is close to that of stoichiometric mullite. Once nucleated, the mullite grains could be made to grow over a wide range of non-stoichiometric Al-rich compositions. Fig. 5 shows that the surface of the mullite coating was highly alumina-rich (Al/Si∼ 8). By suitable manipulation of the input $AlCl_3/SiCl_4$ ratio, Al/Si ratios as high as 15 have been achieved at the coating surface. These compositions are among the most alumina-rich mullite reported to-date in the literature [11]. Due to long term possible regression of Si from mullite coatings, there is a strong practical motivation to substantially reduce or even virtually eliminate the silica component from the coating surface, which is in direct contact with corrosive atmospheres containing steam.

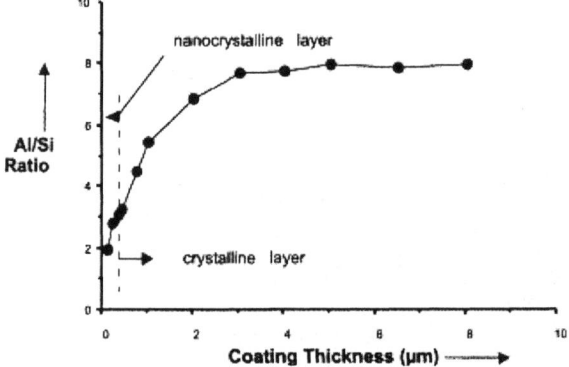

Fig. 5. Gradation of the composition across the coating thickness, expressed as an Al/Si ratio. The interface between the nanocrystalline and crystalline layers occur close to the stoichiometric Al/Si ratio of 3.

Kinetics of the deposition

A kinetic model was designed for a hot-walled vertical reactor. Given that the separate metal chloride deposition systems seemed to be dependent upon the water–gas shift reaction, and that mullite exhibited similar traits to these systems, a starting assumptionfor the kinetics of mullite deposition was that it was also influenced by the water–gas shift reaction.

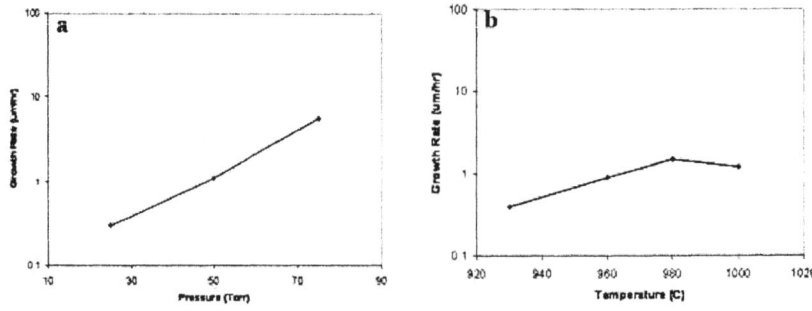

Fig. 6. Growth rate vs. deposition (a) pressure (b) temperature for CVD mullite.

As can be seen from Fig 6, the kinetics of growth of CVD mullite is sensitive to not only deposition temperature but also pressure. The overall model was found to be capable of predicting satisfactorily, both qualitatively and quantitatively, the effects of the various operating parameters on the deposition process.

Structure of Mullite

To understand how mullite can accommodate such a wide range of Al/Si ratio, the crystal structure of mullite needs to be examined. Mullite, with a space group of *Pbam* (space group number 55), is a derivative structure of sillimanite ($Al_2O_3 \bullet SiO_2$) [12]. The unit cell of mullite is orthorhombic with a = 7.5456 Å, b = 7.6895 Å, and c = 2.8842 Å. Mullite is derived from sillimanite by substituting Al^{3+} ions for tetrahedrally coordinated Si^{4+} and removing oxygen atoms to balance the electric charge by the reaction:

$$2Si^{4+} + O^{2} \Leftrightarrow 2Al^{3+} + \square \tag{3}$$

where $\square$ indicates an oxygen vacancy.

As shown in Fig. 7, the formation of an oxygen vacancy causes the two cations (Si, Al) in the tetrahedral positions (T) to shift to the neighboring tetrahedral position (T*). Correspondingly, an oxygen atom adjusts its position from O_c to O_{c*}, leading to a reduction in symmetry [13]. The average crystal structure of stoichiometric mullite is orthorhombic in which edge sharing AlO_6 octahedral chains aligned along the c-axis. These AlO_6 chains form the backbone of the mullite structure, and are cross-linked by the AlO_4 and SiO_4 tetrahedra arranged in a random fashion. Due to the disordering of these (Si, Al) O_4 tetrahedral sites, the c lattice parameter of mullite is half the c lattice parameter of sillimanite [14].

On increasing the Al/Si ratio, the substitution of Si^{4+} by Al^{3+} occurs by replacing the cross-linking SiO_4 tetrahedra by AlO_4. Since this does not disturb the backbone of the mullite structure, i.e., the AlO_6 chains, the mullite structure can be sustained to highly Al-rich compositions.

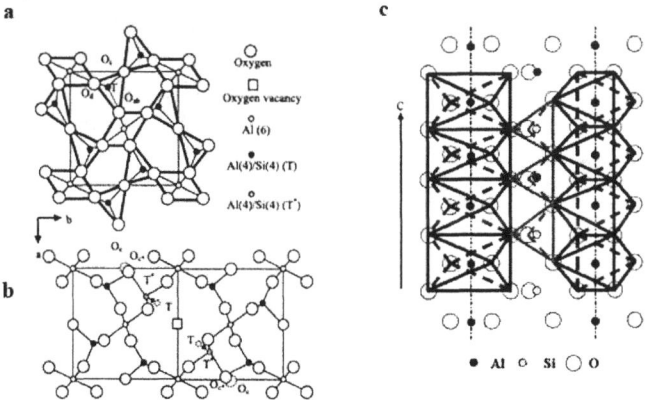

Fig. 7. (a) [0 0 1] projection of the unit cell of sillimanite. (b) Structure of mullite, shown as a derivative structure of sillimanite, where the substitution of Si^{4+} with Al^{3+} leads to the formation of oxygen vacancies and an accompanying movement of cations from the T to the T* sites (c) Structure of mullite along c-axis

Similarly, when decreasing the Al/Si ratio, substitution of cross linking AlO_4 tetrahedra by SiO_4 can occur leading to a Si-rich mullite (Al/Si ~2). However, further reduction of the Al/Si ratio requires the AlO_6 octahedra to be substituted by SiO_6, which is not permissible. Thus, further reduction in the Al content leads to a breakdown of the AlO_6 chains, and the mullite structure cannot be sustained leading to the formation of separate SiO_2 and γ-Al_2O_3 phases.

The number of oxygen vacancies (generated by Eq. 3) in a unit cell, x, increases with increasing Al/Si ratio as [14]:

$$x = \frac{\left(\dfrac{Al}{Si} - 2\right)}{\left(\dfrac{Al}{Si} + 1\right)}. \tag{4}$$

At high Al/Si contents, the oxygen vacancies order to minimize the free energy of the crystal. In addition to the short range ordering of vacancies, long-range 2-D compositions have been reported in mullite [14]. These modulations give rise to domains separated by anti-phase domain boundaries (APB) oriented parallel to (001) planes [13]. These domains produce superlattice spots in the [010] diffraction pattern of mullite and their spacing changes with the coating composition (Al/Si ratio) [7].

Mechanical Properties

Mechanical properties including hardness and elastic modulus of CVD mullite coatings have been evaluated using Nanoindentation. Fig. 8 (a) below shows the variation of elastic modulus as we go from the coating-substrate interface towards the surface of the coating and Fig. 8 (b) shows typical indents on the coating. The increase in elastic modulus of the coating is consistent with previously discussed compositional gradient results.

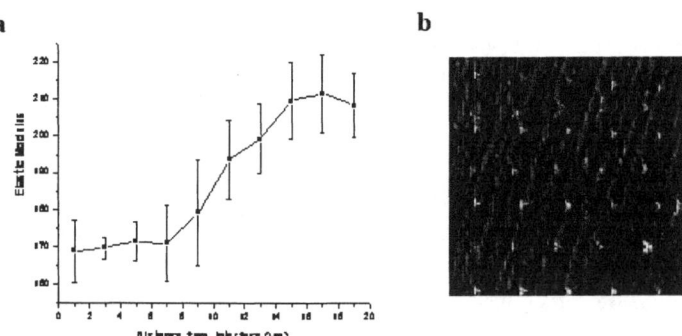

Fig. 8. (a) Variation of elastic modulus across the coating thickness (b) Typical indents on the cross-section of mullite coating

Oxidation and Corrosion Resistance

The functionally graded CVD mullite coatings on SiC are expected to have the hot-corrosion and recession resistance of alumina coatings, while maintaining the thermal shock resistance of stoichiometric mullite coatings. Mullite-coated and uncoated SiC substrates were subjected to hot-corrosion tests by loading the surface with about 5 mg/cm^2 of Na_2SO_4 and subjecting the samples to flowing oxygen at 1100 °C for 300 h [16]. Fig. 9 (a) shows a depth of attack in excess of 20 μm below the uncoated SiC surface. In direct contrast, the mullite-coated SiC sample exhibited no weight loss. Fig. 9 (b) shows the coating in cross-section showing no hot-corrosion attack either within the coating or at the interface, indicating that the mullite coating indeed acted as a very effective hot-corrosion barrier under highly corrosive conditions.

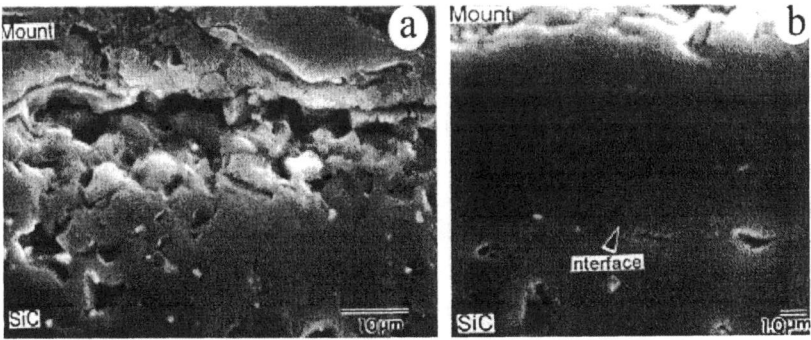

Fig. 9. Hot-corrosion test of: (a) uncoated; (b) CVD mullite-coated SiC substrates.

The adhesion of the mullite coatings was also tested by cycling the temperature between 1250 °C and room temperature. The samples were subjected to substantial thermal shock by rapid insertion and removal of the samples from the hot zone of the furnace and holding the sample for 1 h at each temperature. Fig. 10 (a) shows a fracture cross-section of a coating after 500 cycles. The coating exhibited no signs of cracking and spallation.

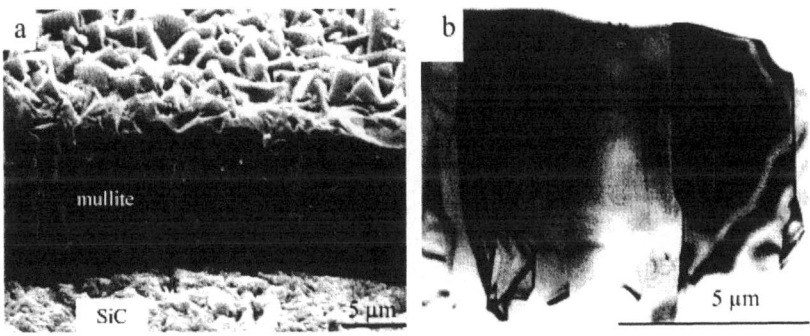

Fig. 10. (a) Fracture cross-section of CVD mullite coating showing excellent adhesion after 500 cycles of cyclic oxidation at 1250 °C. (b) Cross-sectional TEM micrograph showing no phase separation after the 500 h exposure at 1250 °C.

Mullite Phase Transformation

High temperature stability of CVD mullite coatings has been investigated using annealing studies in a temperature range of 1000°C - 1400°C. It has been reported that the composition profile across a graded mullite coating showed little variation after a 100 h anneal at 1250°C [17]. This suggests that long-range-diffusion processes are not occurring in the coating, indicating that mullite is a good diffusion barrier.

Since the a and b lattice parameters of orthorhombic mullite are slightly different, there is typically a splitting of (hkl) and (khl) peak pairs (e.g. 120/210, 230/320, etc) in XRD scans of mullite. However, XRD scans of as-deposited mullite coatings (Fig. 11) almost never exhibit a splitting of such pairs, indicating that the as-deposited mullite coatings have a tetragonal structure. The mullite structure can move towards a tetragonal configuration due to composition effects alone.

In high-Al mullite, increasing the Al content produces oxygen vacancies that increase the relative occupancy of T* sites (Fig. 11). This leads to an increase in the "a" lattice parameter and a decrease in the "b" lattice parameter with increasing Al content. Studies have shown that the mullite structure becomes tetragonal at 78 mole% alumina (Al/Si ~7) [19]. However, in the case of CVD mullite, the as-deposited coatings were tetragonal even when the maximum composition was well below this value, indicating that the tetragonality is not solely composition driven.

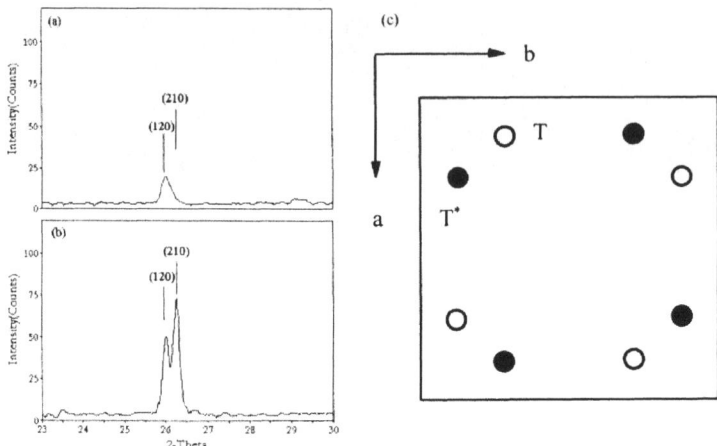

Fig. 11. (a) Lack of splitting of the 120/210 pair of mullite, showing a tetragonal structure. (b) Evidence of tetragonal-to-orthorhombic transformation due to a 100 h anneal at 1250 °C due to the presence of a 120/210 peak split. (c) Schematic of the 4 T and 4 T* sites in sillimanite structural derivatives [18].

Annealing studies of as-deposited coatings for 100 h at temperatures of 1200°C and above led to a tetragonal-to-orthorhombic transformation, with the splitting of the peak pairs becoming more evident at higher temperatures. The lack of compositional changes on annealing suggests that the tetragonal-to-orthorhombic transformation occurs by some short-range structural readjustment. It has been conjectured that in as-deposited CVD mullite, the AlO_4 and SiO_4 tetrahedra populate the T and T* sites randomly, leading to the metastable tetragonal structure in the as-deposited coatings [17]. When annealed at elevated temperatures, the tetrahedra undergo local readjustment to thermodynamically favorable positions, leading to an asymmetric occupation of T and T* sites, causing the tetragonal-to-orthorhombic transformation.

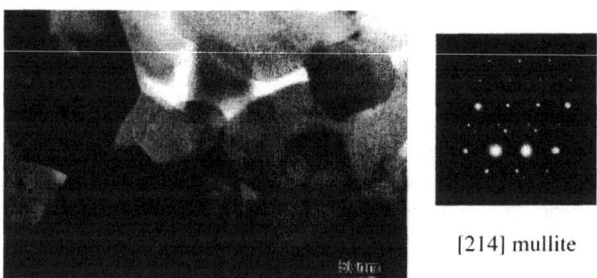

[214] mullite

Fig. 12. (a) Complete mullitization of a nanocrystalline layer as a result of a 100 h anneal at 1300 °C. (b) [2 1 4] SAED pattern of mullite from one of the equiaxed grains.

Annealing mullite coatings for 100 hours at temperatures of 1150°C and above resulted in phase transformations in the nanocrystalline layer. For coatings on SiC substrates, nanocrystalline layers deposited using an $AlCl_3/SiCl_4$ input ratio of 2 and greater transformed completely to equiaxed mullite grains after 100h anneals at temperatures between 1150°C and 1300°C [17]. The TEM micrograph in Fig. 12(a) shows equiaxed mullite grains in the nanocrystalline layer of a coating grown using an input $AlCl_3/SiCl_4$ ratio of 2, with an accompanying [214] mullite SAED pattern (Fig. 12 b) from one such grain formed after a 100 hr anneal at 1300°C . It is clearly evident from the figure that the transformation occurs with no accompanying microcracking or porosity formation.

Conclusions

Uniform, dense and adherent mullite coatings were deposited on Si-based substrates by the CVD process. The coatings were compositionally graded with the Al/Si ratio increasing towards the outer surface of the coating. These compositionally graded coatings exhibited excellent adhesion during annealing at temperatures in excess of 1300 °C. The highly alumina-rich outer surface made these CVD mullite coatings extremely effective corrosion barriers. It is our contention that mullite coatings can be functionally graded to eventually have a silica-free mullite (ι-alumina) on the surface. Such coatings would be unique in that they would have the oxidation and corrosion resistance of pure Al_2O_3 (which is excellent) and the thermal shock resistance of mullite

Acknowledgements

This research has been partially supported by the National Science Foundation under contract No. DMR 0233952. The authors would like to acknowledge contributions by Dr. Michael Auger, Dr. Ping Hou, Dr. Rao Mulpuri, and Dr. Hengzi Wang. Thanks are also due to Dr. Emilio Piqué and Prof. Luis Llanes for their help with evaluating mechanical properties.

References

[1] Vision 21 Program Plan, Clean Energy Plants for the 21st Century, Federal Energy Technology Center, U.S. Department of Energy, 1999.

[2] N.S. Jacobson: J. Am. Ceram. Soc., **76** [1] (1993), p3.

[3] N.S. Jacobson and J.L. Smialek: J. Am. Ceram. Soc. **68** (1985), p. 432.

[4] V.K. Sarin and R.P. Mulpuri: US Patent No. 5,763,008

[5] R.P. Mulpuri and V.K. Sarin: J. Mater. Res. **11** (1996), p. 1315.

[6] R.P. Mulpuri: Ph.D. Dissertation, Department of Manufacturing Engineering, Boston (1996)

[7] D. Doppalapudi and S.N. Basu: Mater. Sci. Eng. **A231** (1997), p. 48.

[8] G.L. Tingey: J. Phys. Chem. **70** (1966), p. 1406

[9] W.L. deKaiser: Science of Ceramics, Vol II, p.243, Academic Press, NY (1963)

[10]P. Hou, S.N. Basu and V.K. Sarin, J Mater Res **14** 7 (1999), p. 2952.

[11]R.X. Fischer, H. Schneider and D. Voll, J. Eur. Ceram. Soc. **16** (1996), p. 109.

[12]C.W. Burnham: Crystal structure of mullite. *Carnegie Institution Washington Year Book* **62** (1963), p. 158.

[13]T. Epicier: J. Am. Ceram. Soc. **74** 10 (1991), p. 2359.

[14]W.E. Cameron: Am. Mineral. **62** (1977), p. 747.

[15] D. Doppalapudi and S.N. Basu: Mater. Sci. Eng. **A231** (1997), p. 48.

[16] A.K. Pattanaik and V.K. Sarin: High temperature oxidation and corrosion of CVD mullite coated SiC. In: Sudarshan TS, Khor TA, Jeandin M, editors. *Surface modification technologies XII.* Materials Park, OH: ASM International; 1998. p. 91–8

[17] P. Hou, S.N. Basu and V.K. Sarin: Int. J. Refr. Metals & Hard Mater **19** (2001), p. 467.

[18] R. Sadanaga, M. Tokonami and Y. Takeuchi: Acta Crystallographica **15** (1962), p. 65.

[19] H. Schneider and T. Rymon-Lipinski: J. Am. Ceram. Soc. **71** 3 (1988), p. C162–C164.

Key Engineering Materials Vol. 333 (2007) pp 71-76
online at http://www.scientific.net
© 2007 Trans Tech Publications, Switzerland

Layered Functional Ceramics via Misted Chemical Solution Deposition

J.F. Scott[1,a], F.D. Morrison[1,b], M. Miyake[1,2c], T Tatsuta[2,d] and O. Tsuji[2,e]

[1]Centre for Ferroics, Earth Sciences Dept., University of Cambridge, Cambridge CB2 3EQ, UK.

[2]Samco Inc., 36 Waraya-cho, Takeda, Fushimi-ku, Kyoto 612-8443, Japan

[a]jsco99@esc.cam.ac.uk, [b]fdm22@cam.ac.uk, [c]miyake@samco.co.jp, [d]tatsuta@samco.co.jp, [e]tsuji@samco.co.jp

Keywords: Chemical solution deposition (CSD); ferroelectrics; thin films; ceramics

Abstract. A review is given of "misted" CSD deposition. This technique uses stoichiometrically correct sol-gel solutions but is not a spin-on process. Instead a monodisperse mist of droplets as large as 3 microns in diameter or as small as 0.3 microns is deposited on a substrate. This technique has the great advantage over sol-gel spin-on processing in that it is suitable for non-planar structures, including nanotubes and nano-wires. One could coat a variety of objects with this technique, including anything from non-planar flash-goggles to a parabolic mirror or focal-plane array of pyroelectric detectors. Yet it is much simpler and less expensive than conventional chemical vapour deposition (CVD). We illustrate its use with functionally graded layers on platinised silicon wafers, on nanotubes of piezoelectrics, and most recently [Pollard, Gregg, et al.] on 100 Gbit/cm2 arrays of Pt nanowires on Si substrates (the latter are 30-nm diameter, spaced 50 nm apart, embedded in porous alumina and capped with lead zirconate titanate capacitors).

Introduction

Our research focuses both on the fundamental physics, chemistry and material science of thin films functional oxides which are deposited here in our lab, and also on extrinsic effects caused by novel geometries and processing required for integration of such oxides into commercially available microelectronic devices. Here we describe recent results from a number of ongoing projects including:

- novel deposition techniques for HfO_2 and ZrO_2 gate oxides and $SrBi_2Ta_2O_9$ (SBT) ferroelectrics (FEs)
- novel nanoscale FE structures in the form of nanotubes
- mechanisms of dielectric leakage in ferroelectric thin films
- geometry effects in FeRAM memory cells and their wider implications for commercial nanoscale high-density memory devices of the future.

Misted Liquid Source Chemical Deposition

The technique involves deposition of a liquid layer of stoichiometrically correct organometallic precursor which is typically converted to the appropriate oxide film by thermal treatment. The liquid precursor is introduced to the substrate as a mist of submicron droplets, Fig 1. Uniform layers as thin as 70 nm have been prepared using this technique.

The first-generation misted deposition machine was the HDF-6000 from Samco Corp.[1,2] Its aerosol was generated with ultrasonic transducers at 1.65 MHz and resulted in 1.0 micron diameter liquid droplets at ca. 10 nm/minute deposition rate with horizontal flow across the substrate (usually a platinized Si wafer). This is the machine in use in our laboratory in Cambridge; it will handle a full 8" wafer. Later a more expensive mist system (Primaxx-2F) was manufactured by SubMicron Systems Co. in the USA, suitable for industrial use.[3] It uses charged droplets and a biasing voltage on the substrate and a vertical flow. The Samco HDF-6000 usually is set up to provide micron-diameter droplets, but both it and the Primaxx-2F can operate with submicron delivery. The latter unit had typically 40 million droplets ccm of mean diameter 0.21 microns and a Gaussian size distribution with standard deviation of ca. 0.10 microns. This is a rather mono-disperse aerosol. Other operating parameters are typically: 1.5-3.0 liters/minute atomizer gas flow; pressure 50 Torr under ambient (n.b., not a high vacuum system); wafer rotation at 15 rpm; dep rate of 10-20 nm/minute. A step coverage of 78% on a 500 nm step height was reported in [3] for thermal ferroelectric SBT on platinized SiO_2.

The misted chemical solution deposition has many advantages as a deposition technique such as:

- excellent step coverage compared with spin-on
- precise composition control
- inexpensive, simple process
- ability to coat 3D substrates

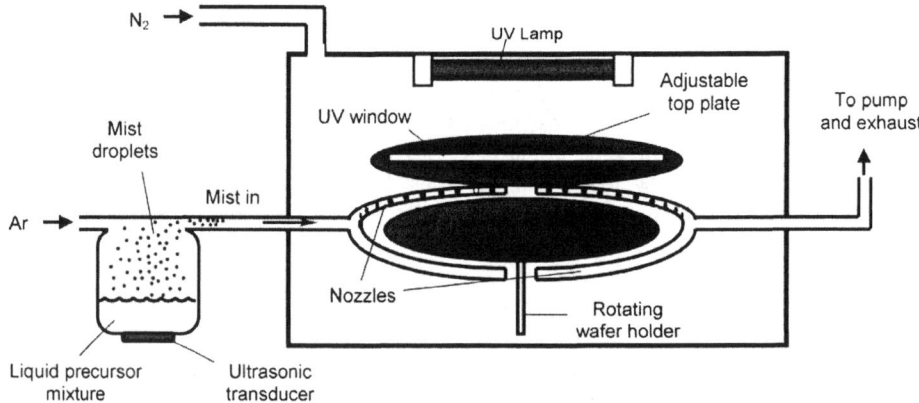

Fig 1 Schematic representation of the misted liquid source chemical deposition apparatus

Deposition of Gate Oxides

Unacceptable leakage characteristics of SiO_2 gate oxides in Field Effect Transistors (FETs) has resulted in the search for more suitable dielectrics. Hafnia and zirconia are believed to be the prime candidates because of their high permittivity (ca. 25) and chemical stability on Si.

Three novel hafnia precursors were investigated for HfO_2 thin films:

- Hafnium tertiary butoxide
- Hafnium (Bis-THD)
- Hafnium (Tris-THD)

The extent of crystallisation of the films was studied by Raman spectroscopy. Hf bis-THD has a significantly higher crystallisation temperature. This may have a potentially large commercial impact in the microelectronics industry, because the gate oxide must withstand a front-end processing step at 800-1000 °C for ca. 7 seconds without crystallisation.

Ferroelectric Films and Nanotubes

Using the misted deposition technique it is possible to produce uniform ferroelectric $SrBi_2Ta_2O_9$ thin films exhibiting excellent FE hysteresis, Fig 2. We have also prepared [3D] FE structures in the form of tubes [4]. These tubes are made from FE (SBT) and have geometries of 100 μm in length, diameters between 400 nm and 2 μm, and wall thickness 40 nm, Fig 3. These have a number of potential applications in both embedded ferroelectric devices (e.g. FeRAMs) and microelectromechanical devices (MEMs):

Data Storage (FE-response)
• High density 3D FeRAMs c.f.existing planar structures
• High aspect ratio coatings (trenching) for DRAMs
MEMs (piezo-response)
• Ink jet printers
• Drug delivery implants
• Micropositioners/movement sensors
• Composites

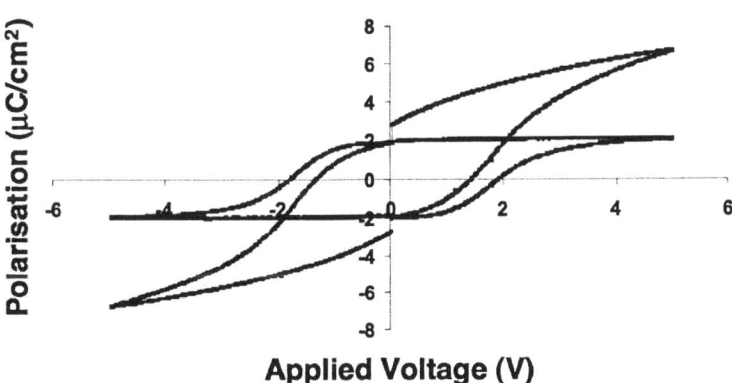

Fig 2 Ferroelectric hysteresis of a 600 nm thick SBT film annealed at 800 °C.

Electroding

With a small modification of the mist system, it has been possible to carry out in situ CVD on the ceramic films in order to put down metal electrodes on them. We have used Ru-DER, and organic ruthenium precursor available from Tosoh Corp. in Japan.[5] Earlier efforts with Pd-acetate failed to achieve good wetted interfaces between dielectric film and electrode.

Nanoscale FE Devices: Geometry

Manufacture of small cell FE capacitors for high density FeRAMs typically involves plasma etching or exposure to chemically reducing forming gases. We have recently shown that this degrades the exposed edge of the FE, leading to increased leakage. The activation energy of the additional leakage response (0.2 eV), figure 4, indicates the presence of H-O dipoles due to hydrogen reduction [6]. This has severe implications for switching of FE devices as the edge/area ratio increases during miniaturisation of FeRAM cells.

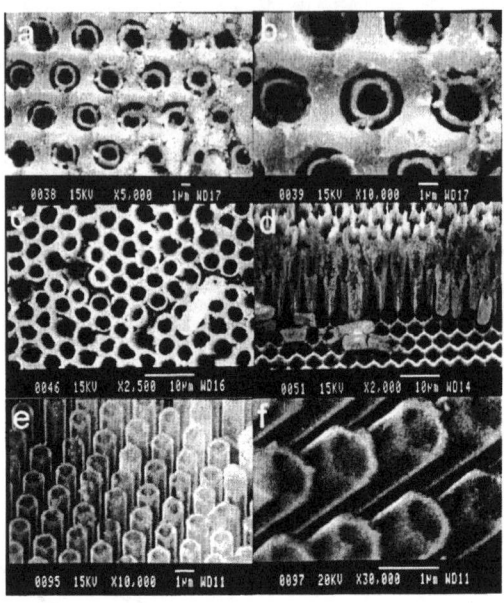

Fig 3 SBT tubes (a) with Ø *ca*. 2 μm and thickness *ca*. 200 nm (b). Larger Ø SBT (c) and cross sectional view (d). Array with Ø = 800 nm (e) and wall thickness < 100 nm (f).

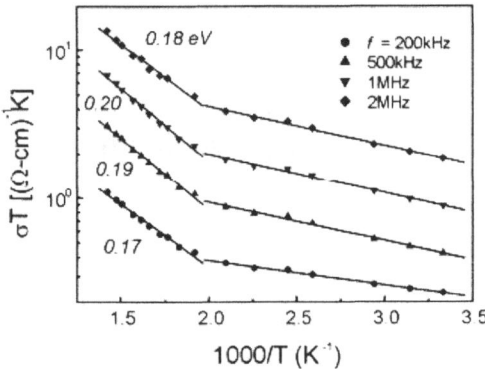

Fig 4 Arrhenius plot of additional leakage response in nanoscale PZT capacitor.

Functionally graded planar structures

LG Electronics Corp. in Korea has utilized the misted system deposition system described above for DRAM capacitor deposition, with $(Ba,Sr)TiO_3$ as the capacitor dielectric. Typically 150 nm of dielectric is deposited in three separate 50-nm layers. In this way the outer layers can be functionally graded to compensate for oxygen vacancy gradients near the electrodes. They found [7] that a carrier gas of N_2O improved performance, compared with a more standard O_2/Ar mix.

Summary

A misted deposition system is described that appears superior to flash MOCVD [8,9]. It provides functional gradients in dielectric films and it gives excellent conformal coverage of non-planar devices. It is an ideal laboratory R&D tool, with one-man operation, low cost, and quick turnaround (several different precursors can be used each week, unlike most CVD systems). It is also scalable to industrial production because dep rates and through-puts are high, and because it takes standard 8" wafers.

References

[1] L. D. McMillan, C. A. Paz de Araujo, T. Roberts, J. Cuchiaro, M. C. Scott, and J. F. Scott, Integrated Ferroelec. **2**, 351 (1992).

[2] M. Huffman, Integrated Ferroelec. **10**, 39 (1995).

[3] N. Solayappan, G. F. Derbenwick, L. D. McMillan, C. A. Araujo, and S. Hayashi, Integrated Ferroelec. **14**, 237 (1997); N. Solyappan, L. D. McMillan, C. A. Paz de Araujo, and B. Grant, Integrated Ferroelec. **18**, 127 (1997).

[4] F. D. Morrison, Laura Ramsay, and J. F. Scott, J. Phys. Condens. Mat. **15**, L527 (2003).

[5] Ru-DER is 2,4-dimethyl penta-dienyl(ethyl cyclopentadienyl) ruthenium.

[6] D. J. Jung, F. D. Morrison, M. Dawber, H. H. Kim, Kinam Kim, and J. F. Scott, J. Appl. Phys. **95**, 4968 (2004).

[7] D. C. Kim, W. Jo, H. M. Lee, and K. Y. Kim, Integrated Ferroelec. **18**, 137 (1997).

[8] T. Ami et al., Integrated Ferroelec. **14**, 95 (1997).

[9] P. C. Van Buskirk, J. F. Roeder, and S. Bilodeau, Integrated Ferroelec. **10**, 9 (1995).

Key Engineering Materials Vol. 333 (2007) pp 77-86
online at http://www.scientific.net
© 2007 Trans Tech Publications, Switzerland

Fracture Mechanics of Ceramics - a Short Introduction

R. Danzer

Institut für Struktur- und Funktionskeramik,

Peter Tunner Straße 5, A-8700, Leoben, Austria

isfk@unileoben.ac.at

Keywords: Linear elastic fracture mechanics, brittle materials, toughening.

Abstract. Failure of ceramics occurs in most cases by fast brittle fracture. This failure mode initiates at flaws in the microstructure, which behave like cracks. In this paper the basic concepts of fracture mechanics are reviewed, which describe the behaviour of cracks in linear elastic materials (LEFM). The strength of this class of materials depends on the toughness of the material and on the flaw size, which acted as fracture origin. In the framework of LEFM, concepts to increase the toughness are discussed, orders of magnitude are exposed and fractography is applied to analyse fracture surfaces.

Introduction

In ceramic materials the compressive strength is around one order of magnitude higher us the tensile strength. Tensile failure is – at least at low and intermediate temperatures – always brittle, since plastic deformation is very limited in ceramics. Fracture initiate at flaws, which acts as local stress concentrations and which can be modelled as cracks. Other damage mechanisms exist – e.g. some sub critical crack growth (some kind of stress corrosion cracking) and fatigue crack growth – but final failure occurs in general due to fast brittle failure. Therefore the review concentrates on fast fracture caused by extension of cracks.

Introduction to fracture mechanics of brittle materials

Instable (fast) crack growth can be modeled by linear elastic fracture mechanics. The behaviour of a crack (length a) in an infinite plate (thickness t) is analysed. The description is strongly influenced by the book "Engineering materials 1" of M.F. Ashby and D.R.H. Jones (see further reading). For simplicity the loading of the crack is made by a homogeneous and uniaxial tensile stress state (as in a tensile test; the stress amplitude is σ) and the crack is perpendicularly oriented to the stress direction. Crack and the loading geometry are indicated in Fig. 1.

The linear elastic fracture mechanics (LEFM) is a mathematical model to describe the behaviour of tensile loaded cracks (see "further reading" for some literature). It is assumed that

(a) the crack is a mathematical section through the material, i.e. the crack tip is infinitely sharp,
(b) the material behaves like a linear elastic continuum (Young's modulus E, Poisson ratio μ) and
(c) no forces can be transferred via the crack borders.

It will be shown later that none of these assumptions is exactly valid in reality, but nevertheless, the LEFM is a model description which can correctly describe many features of brittle fracture in ceramics.

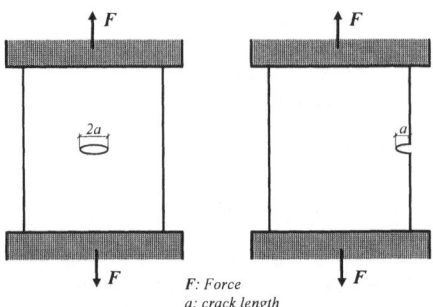

Fig. 1: Internal crack in a uniaxial loaded plate (left) and equivalent situation for an edge crack (right).

F: Force
a: crack length

Energy approach for fast crack propagation. It has been shown in the early work of Griffith [1] that the stability of cracks can be described by an energy criterion: the crack will grow if the decrease in the Gibbs free energy associated with the crack growth is larger as the energy necessary to create the new crack surfaces. The basic concept of the LEFM is also based on this idea. In general two contributions to the changes in the Gibbs free energy have to be considered: the work done by crack growth (δW) and the changes in the elastic energy associated with crack growth (δU_{el}). The crack will growth if

$$\delta W \geq \delta U_{el} + G_c t \delta a .$$ (1)

The energy necessary to create the new crack surfaces (dissipated in the new surfaces) is proportional to the new cracked area ($\delta A = t \delta a$), where δa is the crack extension. The fracture energy (G_c) is the energy necessary to create a unit crack area.

Let us now estimate the energy contributions for the simple example of an edge crack in a strained and fixed infinite plate. Since the plate is fixed, the ends of the plate cannot move and the forces acting on them cannot do any work. Therefore it holds: $\delta W = 0$.

When the crack growths into the plate the material around the crack relaxes and stored elastic energy is released. The exact determination of the released elastic energy needs some effort but a rough estimation is easy. The situation for an edge crack is sketched in Fig. 2.

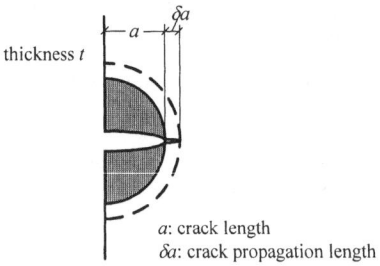

thickness *t*

a: crack length
δa: crack propagation length

Fig. 2:
Relaxed area (shaded) around a strained edge crack. The relaxed area increases (dashed line) by crack growth.

If the plate is homogenously loaded with the strain $\varepsilon = \sigma/E$, the stored elastic energy per volume is $u_{el} = \int \sigma \, d\varepsilon = \sigma^2/2E$. But around the crack the energy is released. For simplicity it is assumed that the stress released volume is a half cylinder: $V = \pi a^2 t/2$ (in reality the volume is a little larger). The elastic energy relaxed by the existence of the crack is therefore: $\Delta u_{el} \approx - (\sigma^2/2E) \cdot (\pi a^2/2) \cdot t$ (the negative sign occurs since the energy is released).

The next question to answer is how that energy changes with crack advance. The released volume increases (see Fig. 2) by a "half onion ring" of width δa, if the crack propagates by the length δa. The volume of the "half onion ring" is: $\delta V_{el} \approx \pi a\, t\, \delta a$. Therefore the increase of released elastic energy due to the crack advance δa is $\delta U_{el} = - (\sigma^2/2E)\cdot \pi a\, t\, \delta a$. An exact calculation (instead of this rough estimation) would yield: $\delta U_{el} = - (\sigma^2/E)\cdot \pi a\, t\, \delta a$. Inserting the energy contributions in Eq. 1 and dividing by the cracked area δA, yields:

$$(\sigma^2 / E)\cdot \pi a \geq G_c \quad , \tag{2}$$

where $(\sigma^2/E)\cdot \pi a$ is the energy released out of the plate per unit area of crack advance (strain energy release rate). The fracture energy, G_c, is often called critical strain energy release rate.

The strength of the plate is the stress to cause fast fracture: $\sigma_f = (G_c E/\pi a)^{1/2}$. The strength scales with the inverse square root of the crack length: $\sigma_f \sim a^{-1/2}$. Note that this is exactly the result of the former Griffith analysis [1, 2].

The power of the above example is that it can be generalized to a large extend. If, for example, the plate is not fixed but loaded with a constant force, the work done by crack growth is not zero but this contribution is balanced by a corresponding change in the released elastic energy: the strain energy release rate remains the same. The result can be generalised for any loading condition, any specimen geometry and any geometry and orientation of the crack giving. The strain energy release rate is than:

$$G = (Y^2 \sigma^2/E)\, \pi a \quad , \tag{3}$$

and the failure criterion reads:

$$G \geq G_c \quad . \tag{4}$$

Now the symbol σ designates the stress at the position of the crack but in the uncracked body.

Y is a geometric factor, which accounts for the geometry of the specimen and the crack and for the slope of the stress field. Data for geometric factors of many different geometries and loading situations can be found in standard handbooks [2, 3].

In ceramics the crack size is very small compared to a typical dimension (D) of the component: $a << D$. In this case (and if the stress in the uncracked body does not change significantly over the length of the crack) Y is around one and does not depend on the length of the crack. To give some examples, for a small penny shaped crack in the interior of the specimen $Y = 2/\pi$ and for a straight through edge crack it is $Y = 1.12$. If the crack length approaches a tenth of the specimen dimension, the geometric factor Y gets dependent on the crack length and the slope of the stress field.

The left hand side of Eq. 4 contains material properties (the Young's modulus E), properties depending on the geometry of the system (crack length a) and on the loading (stress amplitude σ). The fracture energy G_c at the right hand side is a material property. If we rearrange Eq. 4 in a way that all material properties stay at one side of the equality and we extract the square root, the well-known Griffith/Irwin failure criterion arises:

$$K \geq K_c \quad , \tag{4}$$

with $K_c = \sqrt{E G_c}$ being the fracture toughness of the material and $K = \sqrt{E G}$ being the stress intensity factor:

$$K = \sigma Y \sqrt{\pi a} \quad . \tag{5}$$

It should be noted that the fracture toughness depends on the fracture energy and the modulus of the material. The strength of a specimen is

$$\sigma_f = \frac{K_c}{Y\sqrt{\pi a}} \quad , \tag{6}$$

and the size of a critical crack in a (smooth) specimen is inverse proportional to the square of the applied stress:

$$a_c = \frac{1}{\pi} \cdot \left(\frac{K_c}{Y\sigma}\right)^2 \quad . \tag{7}$$

The mean features of brittle fracture in advanced ceramic are described from the Eq. 4 to Eq. 7:

(a) Strength is determined by crack like flaws.
(b) The strength data of nominal identical specimen scatter because the size of the critical cracks is a little different from specimen to specimen.
(c) An increase of strength can be reached by increasing the toughness or by decreasing the frequency and size of cracks. Both aspects will be discussed in the following chapters.

Lastly, because it is more likely to find a large flaw in a large specimen than in a small one, the mean strength of a component depends on its size as demonstrated by Weibull [4, 5]. Therefore the mean strength of large specimens is smaller.

On the basis of these features, the reader could be able to think about 3 different routes to improve the structural integrity of ceramic components:

a) via enhancement of the toughness [6],
b) via improvements in processing [7] or,
c) via reducing the stress state by incorporating shielding stresses in the material [8].

The three alternatives are open fields in research and strong advances have been made the last decades. In the following, detail is given on toughness developments and the influence of processing defects on strength. About the possibility of shielding the material with compressive stresses just outline here that several approaches have been considered, from tempered glasses to multilayers or coatings that contain residual stresses due to thermal mismatch or a transformation phase during sintering[8, 9].

Toughness of Advanced Ceramic Materials

The stress field around the crack tip. In the last paragraph the stress intensity factor was used to describe the energy dissipation by crack growth. In this paragraph it will be shown that it also describes the stress field around the tip of the crack.

The stress to break a bond by stretching the bond length has been estimated a long time ago. The theoretical strength is around $\sigma_{th} \cong E/10$ [10]. This value is in general 2-3 orders of magnitudes higher as the failure stresses observed in technical ceramic materials. What is the reason for that behaviour? There must exist some mechanisms for the local concentration of stresses.

Flaws in the microstructure act as local stress concentration sites. In this respect cracks are extremely effective. Under plain strain conditions the stress field in a surrounding of the crack tip can be approximated by

$$\sigma_{ij}(r,\theta) = \frac{K}{\sqrt{2\pi r}} \cdot f_{ij}(\theta) \quad , \tag{8}$$

where σ_{ij} are the components of the local stress tensor and r and θ are the polar coordinates of a system with the origin in the crack tip. The components of the tensor f_{ij} have values between zero and one. Analytical equations for the components of the tensor can be found in the text books (see "further reading"). It is important to note that σ_{ij} scales with K. The components of the crack tip stress tensor σ_{ij} have a square root singularity at the crack tip. Therefore they become infinite at the tip of a crack (for $r \to 0$) and exceed the theoretical strength in any case, even for very weak solicitations or small cracks.

At a first glance it could be believed that the fracture energy is the energy of the newly created surfaces (surface energy γ), i.e. $G_c = 2\gamma$ (a crack has two borders). In many technical materials the surface energy is in the order of $\gamma = 1$ Jm^{-2}. It can be read off Table 1, which lists some mechanical properties of material classes, that the fracture energy is much higher (some orders of magnitude) for most technical materials. Therefore additional energy dissipating mechanisms must exist and the contribution of the surface energy to the fracture energy is only a lower limit.

Two important classes of "toughening mechanisms" can be identified [11]: process zone mechanisms, where energy dissipation occurs ahead of the crack tip and crack bridging mechanisms, where the energy dissipation occurs behind the crack tip.

Table 1: Typical mechanical properties of material classes

	Young's Modulus [GPa]	Yield strength [MPa]*	Fracture toughness [MPa·m$^{1/2}$]	Fracture Energy [Jm^{-2}]
Polymers	0.1 – 10	1 – 100	0.5 – 10	50 – 50000
Metals	20 – 400	20 – 2000	5 – 200	200 – 500000
Technical Ceramics	100 – 1000	1000 – 10000	2 – 20	5 – 1000

* In tension ceramics fail before they yield. The yield strength in ceramics is therefore determined in compression or by hardness tests. The yield stress of polymers and metals is determined by tensile tests.

Process zone mechanisms. This class of mechanisms occurs in all materials which have a non-linear elastic material behaviour. Examples for the origins of non-linearities are the production and movement of dislocations or twins, phase transformations or the formation of micro cracks. In principle all of those mechanisms may occur in ceramic materials. In this case, there exist a conflict with the presumption assumed at the beginning and linear elastic continuum mechanics seems not longer to be appropriate. This point will be discussed later.

In metals and polymers the high toughness (fracture energy) is caused to a large extend by the plastic deformation of the fractured material. A plastic zone around the crack tip occurs, which – in many cases – can even be detected by the naked eye and can reach – for the toughest metals a diameter of around 1 m.

At the fracture surfaces of ceramics no (macroscopic) traces of any non-elastic deformation can be found. The fracture surfaces fit perfectly together. But even for ceramics small process zones around the crack tip exist, where non-elastic deformations occur. Fig. 3 shows a schematic stress strain curve (the material behaves non-linear for stresses higher as the yield stress, σ_y). Also shown is the slope of the stresses ahead of the crack tip corresponding to Eq. 8. If the stresses reach the yield stress the material reacts by non-elastic deformation. This limits the stress amplitude to a level around the yield stress. The zone where yielding occurs is called process zone (plastic zone in the case of yielding by plastic deformation).

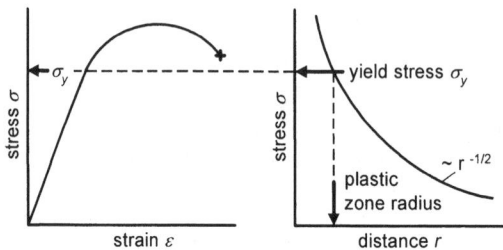

Figure 3:
Left: Stress strain curve indicating non-linear elastic behaviour.
Right: Slope of the stresses ahead of the crack tip. The linear elastic behaviour is indicated by the full curve. Near the crack tip stresses are relaxed by yielding. The size of the process zone is roughly determined by the stress, where the linear elastic solution reaches the yield stress.

The diameter of the process zone, d_p, can be estimated by setting $\sigma_{ij}(r_p) = \sigma_y$ and $K = K_c$. If – for simplicity - we replace the function f_{ij} by one, it yields:

$$d_p \approx \frac{1}{\pi} \cdot \left(\frac{K_c}{\sigma_y}\right)^2 \quad .$$

(9)

Eq. 9 can be evaluated to get an impression for a typical process zone (values in Table 1 can be used in the evaluation). For many advanced ceramics it holds: $K_c / \sigma_y \leq 10^{-3}$ m$^{1/2}$ (see Table 1). Then the diameter of the "plastic" zone (d_{pl}) turns out to be a few micrometers or less: $d_{pl} \leq 3 \cdot 10^{-7}$ m. Therefore it is reasonable that traces of plastic deformations can be hardly found by means of light microscopy.

What is the energy which can be dissipated in such a small zone? The plastic energy density can be estimated to be $u_{pl} = \int \sigma \, d\varepsilon = \sigma_y \, \varepsilon_{pl}$, where ε_{pl} is the "mean plastic strain" occurring in the process zone. The volume of the plastically deformed zone increases with crack extension, the diameter of the plastic zone and – for a straight through crack in a plate – with the thickness of the plate: $\delta V_{pl} \approx d_{pl} \, t \, \delta a$. Therefore the "plastic" energy dissipation caused a propagating crack is: $\delta U_{pl} \approx u_{pl} \, \delta V_{pl} \approx \sigma_y \, \varepsilon_{pl} \, d_{pl} \, t \, \delta a$. As the other contributions to the change in the energy content of the plate the contribution of the plastic deformation is proportional to the new created fracture surface $\delta A = t \, \delta a$ (see Eq. 1). Dividing by the new fracture area, δA, it gives a new contribution ($\delta G_{c,pl} = \sigma_y \, \varepsilon_{pl} \, d_{pl}$) to the fracture energy G_c and the fracture criterion (Eq. 4) keeps valid.

In the following the plastic contribution to the fracture energy is estimated for ceramics and for metals. Typical values for ceramics are (see also Table 1): $\sigma_y \approx 3 \cdot 10^9$ Pa, $d_{pl} \leq 3 \cdot 10^{-7}$ m and $\varepsilon_{pl} \leq 10^{-3}$. and $\delta G_{c,pl} \leq 0.9$ Jm^{-2}. In comparison the contribution of the surface energies is about $2\gamma \approx 1$. It is obvious that plasticity – although it exists in ceramic materials – plays a minor role in toughening of ceramics. For metals some typical values are: $\sigma_y \approx 5 \cdot 10^8$ Pa, $d_{pl} \approx 10^{-3}$ m and $\varepsilon_{pl} \approx 10^{-1}$. Therefore a rough estimation of the contribution is $\delta G_{c,pl} \leq 5 \cdot 10^4$ Jm^{-2}, which is much higher as the contribution of the surface energies. In this case the contribution of plasticity to the toughening is significant. If the Young's modulus of the material is $2 \cdot 10^{11}$ Pa the fracture toughness is $K_c = \sqrt{E \, G_c} \approx 100$ MPa·m$^{1/2}$. This is a typical fracture toughness value of ductile steel.

The above ideas can be transferred to any other mechanism, which produce a non-linear stress strain curve and causes a hysteretic behaviour. The most prominent example in ceramics is the transformation toughening, which occurs in zirconia ceramics [12, 13]. In that case the high tensile stresses at the tip of an approaching crack trigger the transformation of the metastable tetragonal zirconia phase into the monoclinic phase. The monoclinic unit cell has a (5 %) larger volume as the tetragonal unit cell. Therefore the transformation causes non-elastic strains and a hysteretic behaviour. The contribution of that mechanism to the fracture energy may be 1000 Jm^{-2} and more.

Another process zone mechanism sometimes claimed for ceramics is the "micro crack toughening" [14, 15]. It is assumed to occur in multiphase materials, where large internal stresses (remaining from the sintering process) exist on the scale of the micro structure [11]. If a crack propagates in such a microstructure the large tensile stresses ahead of the crack tip add to the internal stresses and may cause the generation of micro cracks. Then the frozen mismatch strains release. This causes a non-linearity in the stress strain curve. Possible contributions to the fracture energy are around 100 Jm^{-2}.

In summary process zone mechanisms are responsible for the high toughness in metals and polymers. Process zone mechanisms also occur in ceramics. But depending of the ceramic system their contribution can be negligible or significant.

Bridging mechanisms. Bridging means some load transfer from on crack border to the other. In that way load is also transferred from the crack tip to the bridging zone behind the tip [16, 17]. This reduces the tensile stresses at the tip. (Nails connecting two planks correspond to crack bridges connecting two crack borders). Bridging is well known in fiber reinforced composites but it also occurs in monolithic ceramic materials. The relatively high toughness of silicon nitride ceramics (up to 10 $MPa \cdot m^{1/2}$) is caused by the bridging action of the needle shaped grains.

If a crack grows, the bridges are pulled out and the crack opening (at the position of the bridge) increases. The work done at a bridge is the force transferred by the bridge times the opening, u, of the crack [16, 17]. This work can be also considered in Eq. 1. As in the case of process zone mechanisms it can be shown that the corresponding energy contribution is proportional to the newly cracked area. Therefore, it can be considered to be a part of the fracture energy G_c and the energy release rate G remains unaffected. The contribution is

$$\delta G_{cb} = \int_0^{u_c} \sigma_b(u)\, du \ , \tag{10}$$

where the bridging stress $\sigma_b(u)$ is the mean force transferred by a typical bridge divided by the newly cracked area. The increase of the bridging stress with the crack opening depends on the type of bridge. The bridges fail at the critical crack opening u_c.

In ceramics the crack path often follows the grain boundaries. The fracture surface is therefore rugged. This may cause a form closure of a cracked body. If such a crack opens elastic deformation of these bridging elements occurs (elastic bridges), some elements may glide on top of each other (frictional bridges) or may wear [18]. All these processes cause some energy dissipation, which can be described by Eq. 10. It is reasonable that some bridging always occurs in ceramics and this class of mechanisms largely contributes to the fracture energy of ceramic materials. Bridging is promoted by elongated grain shapes and coarse grained micro structures. Of course bridging is a significant mechanism in fibre reinforced material [19]. For that class of material the fracture energy can get very large and even reach the level of ductile metals.

Other toughening mechanisms. Other mechanisms, which contribute to the fracture energy, are crack deflection [20, 21] and branching [22]. In the case of both mechanisms the crack area increase and cracks bend out of the plane normal to the most dangerous first principle stress component. In this way the contribution of the fracture energy is increased. Crack deflection and branching are in general prerequisites for the occurrence of bridging processes and should not be discussed separately.

Remarks. The behaviour of materials toughened by a process zone is not linear elastic and in materials toughened with bridging zones, forces are transferred via crack borders. Both do not meet the requirements for the LEFM defined at the beginning of this paper. But the LEFM is still applicable. The reason is that – in the cases analysed – the deviations of the nonlinear behaviour are strongly localized and all energy contributions associated with the growth of cracks were proportional to the newly cracked area. Then the energy dissipation connected with these mechanisms can be treated as a part of the "surface energy". This is not longer possible if process or bridging zones are too large or if the complete specimen is non-linearly deformed. Then other concepts to describe the growth of cracks have to be applied (J-integral).

In most ceramics several mechanisms contribute to the toughness, i.e. the fracture energy can not be ascribed to a single mechanism. The energies of advanced ceramics with optimized microstructure are around 100 Jm^{-2} or more. Due to the high elastic modulus this results in fracture toughness values around 5 $MPa \cdot m^{1/2}$ and more. This fracture energy and fracture toughness should enable the successful application of advanced ceramics for highly loaded components, if the design is properly done.

Influence of flaws on strength

The strength of brittle materials depends on fracture toughness and the size of the critical crack (see Eq. 6). To reduce the size of the critical cracks is an efficient method to improve strength. Indeed the toughening as described above as well as some progress in the processing of ceramic materials resulted in an impressive increase of tensile strength in most classes of ceramic materials.

It is interesting to note that the ratio between fracture toughness and strength is in the range between $1/100$ $m^{1/2}$ and $1/200$ $m^{1/2}$ for most advanced ceramic materials. With the geometry factor of a penny shaped crack the critical (Griffith) crack size (Eq. 7) is between ~ 20 μm and ~ 80 μm. This reflects the state of art of the nowadays processing technology.

Fig. 4 shows the typical appearance of the fracture surface of a tensile tested ceramic specimen. In the overview (Fig. 4a) a smooth area can be clearly recognized, which is called fracture mirror. The mirror is always perpendicularly oriented to the direction of the first principle stress and it contains the fracture origin (Fig. 4b). In this case a large pore that resulted from an inappropriate powder processing

When the crack velocity comes near to the speed of the sound instabilities cause deviations of the crack path out of the original direction. These instabilities cause the so called mist and hackle region [23, 24]. These structures are also indicated in Fig. 4a. Due to this sub critical crack growth the Griffith crack size as defined by Eq. 7 is always a little larger as the size of the fracture origins (microstructural flaws).

It is interesting to note that the Griffith crack size as well as the borderlines between the different regions at the fracture surface scales with the inverse square of the applied stress: $a_i \propto \sigma^{-2}$. This can be used for the determination of the failure stress at hand of the fractured pieces. The proportionality factors of several configurations are published, e.g. in [25].

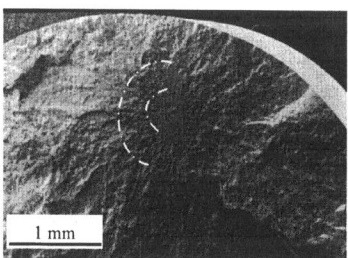
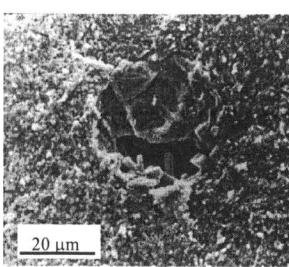

Figure 4. Fracture surface of a tensile tested silicon nitride valve: (left) overview showing the fracture mirror, the mist and hackle region in a fracture surface. (right) close up showing the fracture origin.

Figure 5 shows fracture origins of several technical ceramic materials. It can be recognized, that almost any defect in the microstructure can act as fracture origin, if its size is large enough, i.e. if its size approaches the size that of the corresponding Griffith crack. Therefore the reduction of size and frequency of microstructural defects has an outstanding significance for the development of reliable components made from brittle materials.

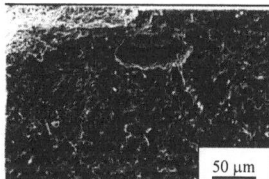
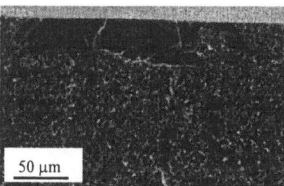
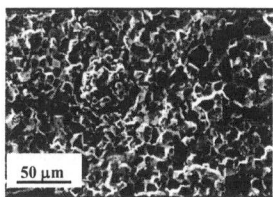

Figure 5. Fracture origins in specimens of several ceramic materials: (a) Pore in a $BaTiO_3$ thermistor ceramic, (b) surface abnormal large grains of Al_3O_2 in Al_3O_2 matrix and (c) agglomerate of alumina leading to porosity.

Concluding Remarks

Fracture of monolithic ceramic materials can be nicely described in the frame work of the linear elastic fracture mechanics. The tensile strength is triggered by flaws, which are distributed in the material and which result from the processing of the material. As larger is the fracture causing flaw as lower is the strength. Therefore reliable processing conditions, which avoid large flaws in the micro structure, are an important prerequisite to have a high strength. The strength also increases with the fracture toughness. In general ceramics are very brittle but in the last twenty years the design of an appropriate microstructure resulted in a significant improvement of the toughness. Nowadays some commercial grades have a fracture toughness of around 10 MPa·√m. Both developments (avoiding large defects and increasing toughness) resulted in an enormous improvement in strength. Commercial materials can have a four point bending strength of more than 1000 MPa. Laboratory materials may have up to 3000 MPa.

A further possible way to strengthen ceramics would be the application of internal stresses as it is, for example, known from quenched glasses. Internal stress fields linearly superimpose with external fields. Compressive internal stresses can, therefore, cause a significant increase in tensile strength. There exist many ways to introduce internal stresses, but the development of layered composite materials seems to be exceptional promising: sheets of different ceramic materials can be stacked over a surface to give a plate like structure. The sequence of materials can be selected in such a way that compressive stresses occur in the surface layer, which is exposed to the highest external stress. The strength and the toughness of such systems will be discussed in some other contributions of that issue.[9, 26]

Acknowledgement

The work was supported in part by the European Community's Human Potential Programme under the contract HPRT-CT-2002-00203, [SICMAC].

Further reading

M.F. Ashby and D.R.H. Jones: *Engineering Materials 1* (Pergamon Press, Oxford 1980)
M.F. Ashby and D.R.H. Jones: *Engineering Materials 2* (Pergamon Press, Oxford 1986)
M.F. Ashby: *Materials Selection in Mechanical Design* (Pergamon Press, Oxford 1992)
J.B. Wachtman: *Mechanical Properties of Ceramics* (John Wiley & Sons, New York 1996)
D.Gross and T. Seelig: *Fracture Mechanics with an Introduction to Micromechanics* (Springer, Berlin 2005)
B. Lawn: *Fracture of brittle solids. Second Edition* (Cambridge University Press, Cambridge 1993)
D. Broek: *Elementary engineering fracture mechanics* (Sijthoff & Noordhoff, Alphen aan den Rijn 1978)
D. Green: An introduction to the mechanical properties of ceramics. (Cambridge University Press, Cambridge 1998)

References

[1] A. Griffith: Phil. Trans. Roy. Soc A221 (1920), p. 163.
[2] H. Tada, P. C. Paris, G. R. Irwin: *The stress analysis of cracks handbook* (Del Research Corporation, St. Louis 1986).
[3] Murakami: *Stress Intensity Factors Handbook* (Pergamon Press, 1987).
[4] W. Weibull: J. Appl. Mech. 18 (1951), p. 253.
[5] R. Danzer: J. Eur. Ceram. Soc. 10 (1992), p. 461.
[6] A. G. Evans: J. Am. Ceram. Soc 73 (1990), p. 187.
[7] A. G. Evans: J. Am. Ceram. Soc 65 (1982), p. 127.
[8] M. Rao, J. Sanchez-Herencia, G. Beltz, R. M. McMeeking, F. Lange: Science 286 (1999), p. 102.
[9] T. Lube, J. Pascual, F. Chalvet, G. De Portu: in press Journal of the European Ceramic Society (2006), p. .
[10] E. Orowan: Rep. Prog. Phys 12 (1949), p. 185.
[11] R. Danzer, L. Sigl: Veitsch Radex Rundschau 1994 (1994), p. 511.
[12] N. Claussen: J. Am. Ceram. Soc 61 (1978), p. 85.
[13] M. Rühle, N. Claussen, A. Heuer: J. Am. Ceram. Soc 69 (1986), p. 195.
[14] L. X. Han, R. Warren, S. Suresh: Acta mett mater 40 (1992), p. 259.
[15] F. Buresch: Material Science and Engineering 71 (1985), p. 187.
[16] R. W. Steinbrech, A. Reichl, W. Schaarwaechter: J. Am. Ceram. Soc 73 (1990), p. 2009.
[17] Y.-W. Mai, B. Lawn: J. Am. Ceram. Soc 70 (1987), p. 289.
[18] H. U. Marschall: *Rißbrücken in Aluminiumoxid*, Dissertation (Montanuniversität Leoben, Leoben 2001).
[19] R. Paar, J. L. Vallés, R. Danzer: Material Science and Engineering A 250 (1998), p. 209.
[20] K. T. Faber, A. G. Evans: Act. Metall. 31 (1986), p. 565.
[21] K. T. Faber, A. G. Evans: Act. Metall. 31 (1986), p. 577.
[22] M. Oechsner, C. Hillman, F. Lange: J. Am. Ceram. Soc 79 (1996), p. 1834.
[23] R. Danzer: Key Eng. Mat. 223 (2002), p. 1.
[24] E. Sharon, J. Fineberg: Adv. Eng. Mat. (1999), p. 119.
[25] R. Morrell: *Fractography of Brittle Materials. Measurement Good Practice Guide 15* (National Physical Laboratory, Teddigton 1999).
[26] M. Lugovy, V. Slyunyayev, N. Orlovskaya, G. Blugan, J. Kübler, M. Lewis: Act. Mater. 53 (2005), p. 289.

Key Engineering Materials Vol. 333 (2007) pp 87-96
online at http://www.scientific.net

Mechanical Properties of Ceramic Laminates

Tanja Lube

Institut für Struktur- und Funktionskeramik, Montanuniversität Leoben,
Peter-Tunner-Straße 5, A-8700 Leoben, Austria

tanja.lube@mu-leoben.at

Keywords: ceramic multilayers, residual stress, strength, R-curve, crack deflection

Abstract. The progress of research efforts on the mechanical properties of ceramic laminates is reviewed. Laminates with weak interface are described with respect to their failure mode and the criterions to achieve graceful failure. For laminates with strong interfaces basic principles concerning the residual stresses and their influence on crack propagation are introduced. The implications for strength, indentation strength, strength distributions, macroscopic R-curves and threshold strength are discussed.

Introduction

Ceramics have a variety of attractive properties (chemical and thermal stability, low density, high hardness and Young's modulus, specific physical behaviour...) for structural and functional applications. However, their inherent brittleness and lack of deformation limit their use under load-bearing conditions, yielding in most of the cases an unexpected and catastrophic failure. Their strength variability is associated with the intrinsic (processing) and/or extrinsic (machining) defects distribution. Hence, the increase of toughness or the creation of a non-brittle, defect tolerant failure mode is a principle goal of material development in the field of ceramics. Inspired by examples found in nature [1], and developed for other material classes (polymers, glasses [2]) ceramic structures have been developed, that have a layered macrostructure. The involved layers consist of different materials and/or have different microstructure. A characteristic feature of these laminates is the quality of the interface: they can have a low strength and/or toughness or be nearly ideally strong to promote perfect adherence. In the former case, a non-brittle failure mode is achieved by vast crack deflection, in the latter case, residual stresses may arise that are responsible for an enhanced toughness or a threshold strength [3].

Laminates with weak interfaces

The idea of introducing weak interfaces to toughen a brittle material has for a considerable time been a successful strategy for example in fibre reinforced polymer composites, where the technical realisation of this idea can easily be accomplished. However, this procedure does not appear straightforward for the case of ceramics. For instance, an addition of suitable fibres into a powder compact may result in problems during densification and sintering. Using the inspiration of a material found in nature [1] effort was directed towards the fabrication of laminated structures. For such multilayers, strong ceramic sheets are coated with a suitable material that will form a weak layer after sintering. Ideally, such systems do not develop residual stresses due to thermal expansion mismatch. Several systems have been attempted asucesfully [4-8]. Instead of using a different material, weak layers can also be of the same material as the rest of the components but with a higher porosity. The porosity is introduced by adding particles of a pyrolysable agent to the slurry [9-11]. After pyrolysis, the remaining pores should be so large that they do not sinter [12], and thus, since they are not constrained, the porous layer will shrink by the same amount as the surrounding dense matrix. Following the above ideas, any problems arising from chemical incompatibility of the involved materials or of differential shrinkage due to thermal expansion mismatch and the consequently resulting residual stresses can be avoided.

The beneficial effect of the weak interlayers can be exploited for cracks that grow normal to the interfaces of the laminate. During the fracture process the propagating crack kinks into the weak layer where it continues to grow to form a delamination crack between two adjacent layers of the thicker, strong ceramic. Fig. 1 schematically shows a load – deflection diagram of a bending test on a notched monolithic material and a notched laminate of the same material but containing weak interlayers. The first part of the laminate curve is identical with the curve of the monolith. At point * the first lamina breaks, but the load can be increased to drive the delamination crack forward. Due to the reduced compliance of the remaining cross section the slope is lower. Eventually, the growing through thickness crack is re-initiated to transverse the next lamina. Failure of this lamina occurs at the maximum peak of the curve (when the failure stress of the lamina is reached); the crack is again deflected into the interface. Failure continues in this manner to give the typical saw-tooth shaped load-defelction curve that is characteristic for the so-called graceful failure of laminates with weak interfaces. The apparent toughness of the laminate can be increased by a factor of ~4, raising the work required to break the sample by a factor of over 100, compared to the monolith [4]. After delamination, the crack in the adjacent strong ceramic layer does not initiate at the tip of the delamination crack. Instead, the crack is re-initiated at a flaw within the lamina, preferably close to the centre of the specimen, where the stress is high. This implies that the strength of the composite is determined by the strength of the ceramic layers [13].

In order to tailor the properties of such laminates, it is necessary to know the conditions for a crack to deflect (in the case of weak layers from a different material) or stay inside the layer (for porous layers), and to understand the role played by other mechanical properties like strength, Young's modulus, and strength dispersion of the laminae.

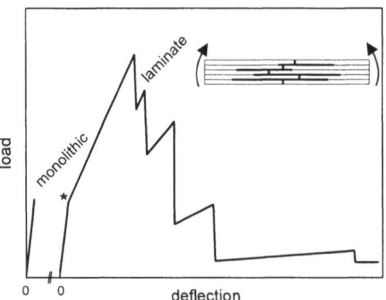

Fig. 1: (a) Typical load-deflection diagram of a notched bending specimen of a laminate with weak interfaces. Each load drop corresponds to the failure of an individual lamina. The behaviour of the same monolithic ceramic without weak interfaces is shown for comparison. A schematic side view of a broken specimen showing crack deflection into the weak layers and delamination is shown in the insert.

A widely accepted criterion for crack deflection at an interface between two elastically dissimilar materials was developed by He and Hutchinson [14]. They compared the energy release rates for a crack penetrating the interface from (the strong ceramic) material 2 into the (weak layer) material 1, G_p and a crack deflecting into the interface, G_d. A crack is likely to be deflected if $G_{c,interface}/G_{c,1} < G_d/G_p$. For materials with Young's moduli not too different, the critical ratio is $G_d/G_p \approx 0.25$. The overall graceful failure behaviour was quantitatively described by Phillipps and co-authors [15] and Folsom and co-authors [16]. Phillipps et al. [15] model the process in a way that laminae failure and growth of the deflected crack occur sequentially. Cracked layers are considered to carry no load. The energy for the growth of the delamination crack is provided by the strain energy released on failure of the preceding lamina. Failure of the laminae is considered to take place if the applied stress reaches the (dispersed) strength of the laminae material. The consequences for the design of strong laminates resulting from this description are: i) a high strength of the laminae, ii) a high toughness of the interface to enhance the energy consumption for the propagation of the delamination cracks, but low enough to guarantee crack deflection, iii) for the same reason as ii), a low Young's modulus of the weak layers. The model proposed by Folsom et al. [16] takes the interface toughness to be zero and focuses on the different characteristics of the behaviour of laminates with weak interfaces in bending and tension.

For laminates in which the weak interlayers are porous layers, the interesting question for laminate design is the level of porosity, which will promote the desired toughening effect. For these materials, the condition for successful toughening is that the deflected crack continues to grow inside the porous layer. It has been shown [17] that the condition for a crack to continue to grow in the interface towards a defect ahead of it is given by $R_i/R_0 < 0.57$, where R_i and R_0 are the fracture energies of the interface and the lamina respectively. For the situation of a crack in a porous layer, R_i is essentially the same as R_0 and it can be expected that the crack kinks out of the interface immediately. Experiments have although shown that above a certain level of porosity, the deflected crack continues to propagate in the porous layer. The above relation has to be modified [9] to say $R_{lig}/R_0 < 0.57$, where R_{lig} is the fracture energy of the ligament of material between the crack and the next pore. The interaction of the crack tip with the closest pore reduces the fracture energy of the material in the ligament and thus R_{lig} depends on the porosity, becoming smaller than R_0. The necessary level of

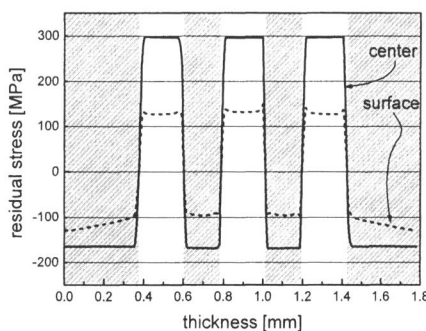

Fig. 2: FEM calculated stresses in the centre (through line) of a specimen and close to the edge (dashed line) of a laminate. The background of the diagram indicates the layers. The analytical solution gives values nearly similar to those in the centre of the specimen.

porosity can be estimated by geometric considerations concerning the spatial arrangement of pores. The resulting value of 37% porosity is in good agreement with experimental observations [9-11]. Experiments on laminates with layers with different porosities also give evidence, that an optimum level of porosity exists (~60%, [18]), beyond which the whole system will be weakened and the work of fracture will decrease.

Laminates with strong interfaces

Laminates with strong interfaces [19-30] are mainly produced to provoke residual stresses within the different layers. But there are also approaches, were the layers are made of essentially the same material [31, 32] or of materials that produce no significant differential shrinkage and/or have different microstructures [31, 33, 34]. In such laminates it is the different strength or the intrinsic R-curve behaviour of the layers rather than residual stresses what gives rise to enhanced mechanical properties.

Residual stresses in laminates. In most cases where dissimilar materials are joined together by a perfectly adherent (strong) interface that is able to transfer stresses and strains, residual stresses are present after the production process. These stresses arise due to a differential shrinkage during sintering [35], a thermal expansion mismatch, phase transformations [28, 29, 36] or a combination of those. They depend on a variety of properties: grain boundary phase viscosities, densification rates, thermal expansion coefficients, elastic constants, processing temperature, transformation strains, layer thickness,.... . Apart from being the desired result of the manufacturing process, the stresses may also cause damage to the sintered multilayer component in the form of edge cracks, tunnel or channel cracks [22, 24, 37], Fig. 2. To avoid this kind of damage and at the same time achieve high

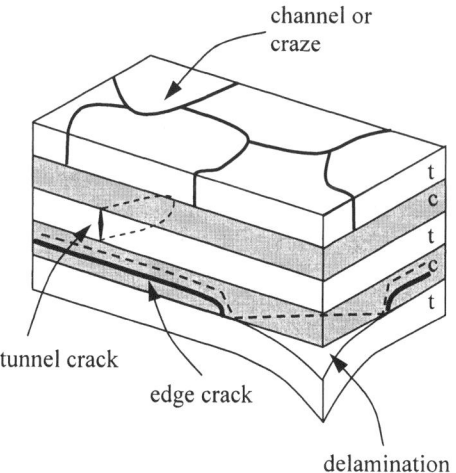

Fig. 3: Damage encountered in laminates with residual stresses. **t** denotes tensile layers, **c** compressive layers.

residual stresses through appropriate design is the art of laminate processing.

A crack that is propagating normal to the layers interacts with the residual stresses but also with the elastic modulus mismatch in the layer just ahead of the crack tip. Compressive stresses will lead to a deflection into the layer plane, while tensile stresses will cause the crack to deflect to a direction more perpendicular to the layer plane (compare also Fig. 7). An inclined crack coming from a more compliant material into a stiffer material will deflect away from the interface owed to the tension induced in the stiffer material by the stress intensity ahead of the crack tip [38]. Corresponding experimental evidence for this behavior was found [39]. Under certain conditions (high compressive stresses in rather thick layers), especially in the cases where edge cracking is observed [37], even crack bifurcation may take place within the compressive layers [36, 40, 41].

The magnitude of the stresses has a crucial influence on the mechanical behavior of the composites, but it cannot be easily determined. Common methods are X-ray diffraction [42], indentation methods [43] which both have a rather coarse spatial resolution or piezospectroscopy [44, 45] with a much smaller probe/beam area. These methods only test a limited material volume situated close to the surface, were stresses might differ essentially from the values inside the structure far away from edges [46, 47], Fig. 2, and even have an additional stress component [37]. The strip bending method [48] investigates the bulk material. An analytical estimation of the stresses far away from end regions is possible assuming a plain strain condition with zero stress throughout the thickness of the laminate [49]. For a symmetrical laminate the biaxial stresses in the layer plane are given by

$$\sigma_1 = \frac{E_1' \, \Delta\varepsilon}{1 + \dfrac{(n+1)t_1 E_1'}{(n-1)\, t_2 \, E_2'}} \quad \text{and} \quad \sigma_2 = -\sigma_1 \frac{(n+1)\, t_1}{(n-1)\, t_2} \tag{1}$$

where $E_i' = E_i / (1 - v_i)$, E is Young's modulus, v is Poisson's ratio, n the total (odd) number of layers and t_i the thicknesses of the individual layers of different materials 1 and 2, respectively. $\Delta\varepsilon$ is the differential strain either resulting from a phase transformation or from a thermal expansion mismatch: $\Delta\varepsilon = (\alpha_1 - \alpha_2)\, \Delta T_0$ with the temperature interval in which stresses develop, ΔT_0 and α_i the coefficients of thermal expansion (CTE) of the involved materials[1]. The material that undergoes a volume increase upon cooling from sintering temperature or that has the lower CTE will have compressive residual stresses. As it can be seen from Eq. (1), the magnitude of the stresses can be tailored by choosing the materials (with the limitation that they have to form strong interfaces) and by adjusting the layer thickness within the limits given by processing damage [35]. FEM calculation show that the in plane residual stresses reach a constant value at a distance of approximately 10% - 15% of the width inside from the surface [46]. The out of plane stress component declines substantially within one layer thickness inside from the surface [37].

Since one reason to design and produce ceramic laminates is to increase strength and reliability by influencing the behaviour of surface cracks (which will be present in almost every component due to machining) and also with respect to the orientation of the layers in a component, a test configuration as depicted in Fig. 4a with a crack growing perpendicular through the thickness of the laminae will be a realistic loading situation. Most of the following – if not stated otherwise – will refer to longitudinal loading.

---

[1] For the derivation of this equation it was assumed that all layers of the outer material 1 have the same thickness t_1 and that the same is true for the inner layer material 2. If this condition is not fulfilled (like for the multilayer in Fig. 2), the term $(n + 1)\, t_1$ should be replaced by $t_{1,\, tot}$, the total thickness of material 1 and $(n - 1)\, t_2$ by $t_{2,\, tot}$, the total thickness of material 2.

Strength, indentation strength and strength distributions. The effect of a residual compressive stress in a surface layer on the bending strength is obvious. It will lead to an increase in bending strength, as observed for

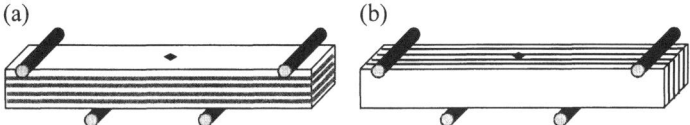

Fig. 4: Two common test configurations for laminates: (a) loading axis normal to the layer plane, longitudinal flexure, (b) loading axis parallel to the layer plane, transverse 4-point flexure.

example by Virkar et al. [50, 51]. By testing indented specimens, the influence of the residual stress on cracks with different size can be investigated. The strength σ_{IS} of laminated specimens indented with different loads P on the tensile surface will no longer follow the relation $\sigma_{IS} \propto P^{-1/3}$ [52], but have a different exponent that depends on the magnitude of the residual compressive stress, Fig. 5a. This behaviour has been confirmed by different authors [53, 54] and has also been reported for laminates that have a tensile residual stress in the surface layer. In some of these cases it can even become independent of the indentation load [55, 56], Fig. 5b.

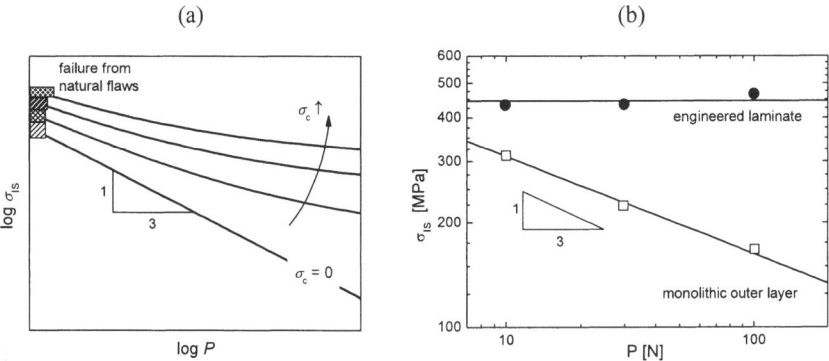

Fig. 5: (a) Theoretically predicted indentation strength behaviour for monolithic specimens and for specimens with different compressive residual stresses in the surface, (b) measured values for an engineered laminate from various Al_2O_3/mullite composites, compared to the monolithic outer layer material (data from [56]).

A compressive residual stress in the surface layer can be interpreted as a lower limit for the strength. Then the scatter of strength values can be described by a 3-parameter Weibull distribution [57] with the compressive stress as the lower limit for the strength σ_u, instead of a 2-parameter Weibull distribution for which $\sigma_u = 0$ [58]:

$$F = 1 - \exp[-\left(\frac{\sigma - \sigma_u}{\sigma_0}\right)^m] \quad .$$

(2)

F is the probability of failure at a given stress σ, m is the Weibull modulus, σ_0 the characteristic strength and σ_u the lower limit for the strength. It is important to note that the parameters σ_0 and m have different definitions for 2- and 3-parameter distributions! In general it is not possible to fit all three parameters of the distribution function to a set of data comprising only 30 or less values [59]. The 3-parameter distribution can only be fitted with a reasonable accuracy if σ_u is known. Even though a 2- and a 3-parameter distribution fitted to the same data-set will look the same, they differ when being extrapolated to small probabilities of failure as they a required for the design of

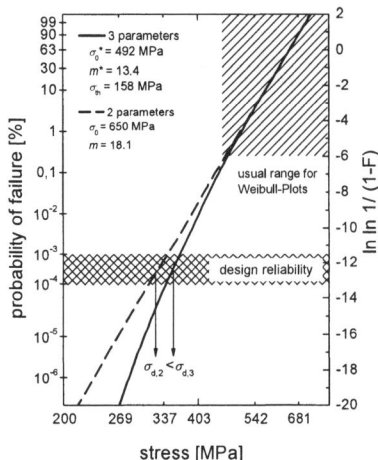

Fig. 6: 2-parameter and 3-parameter Weibull distribution functions fitted to the same data. The distribution functions are not discernable in the experimental range (dashed area) but differ significantly in the range of design probabilities (hatched area).

components (see Fig. 5). The use of a 2-parameter distribution for an extrapolation will result in a conservative (lower) design-stress σ_d.

Apparent R-Curve of laminates. As indicated by Tandon and Green [60] by analytical calculations, residual compression in a surface layer can lead to an apparent rising crack growth resistance curve (R-curve) throughout this layer. Such an R-curve is the effect of a reduction of the crack driving force and not an intrinsic property of the material. The R-curve resulting from the residual stresses in a multilayer can be evaluated quantitatively for certain specimen/crack configurations using the weight function method introduced by Bueckner [61].

This method can be seen as an approximation since the change of the elastic properties at the interface is not taken into account by the weight function. In general the analyses are performed for edge cracks in bend specimens, see Fig. 4a, in order to be compared with experimental results [20, 25, 26, 44, 62] Such calculations show that the apparent toughness increases with increasing crack length in the region where compressive residual stresses prevail and that it decreases for tensile residual stresses. Typical R-curves for laminates with compressive residual stresses in the surface layer are shown in Fig. 6.

Lakshminarayanan et al. [20] estimated that a residual compression of 500 MPa in the external layer of a 3-layer laminate could lead to an increase of the apparent fracture toughness by a factor of 7.5. Further investigations [63] showed that the maximum toughness for a given overall thickness is achieved for outer layers having ¼ of the overall thickness and for cracks extending to the first interface. Moon et al. [62] used the weight function approach to separate the intrinsic microstructural toughening effect from the macroscopic toughening due to the residual stresses, demonstrating that the main contribution to the toughness increase comes from the latter.

The studies of Pascual et al. [64, 65] investigated the influence of various parameters defining the architecture (layer thickness ratio, number of layers, total thickness) or the materials (thermal expansion mismatch and elastic mismatch) on laminates with compressive external stresses. Depending on the layer thickness ratio, an overall rising or decreasing tendency for the apparent R-curve was found. For a given material combination an optimal layer

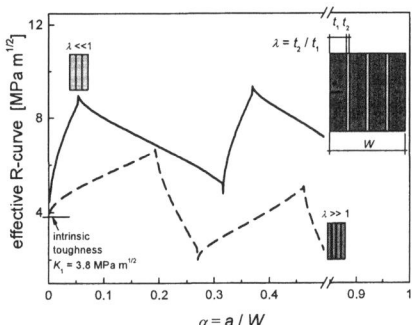

Fig. 7: Apparent R-curves for laminates with compressive residual stresses in the surface for an edge crack in bending calculated using the weight function method. The through line refers to a specimen with thin compressive layers t_1, the dashed line to a situation where the tensile layers t_2 are thin.

thickness ratio can be defined. The effect of the residual stresses is higher for a low number of layers and thick specimens. The optimal thickness ratio can be changed by variation of the elastic mismatch.

The analysis of Lugovy et al. [26, 66] showed that a threshold strength exists for certain architecture + crack length combinations. In laminates with a compressive residual stress in the surface all cracks with the tip inside the compressive layer fail at the same stress defined by the toughness at the first interface. Only if the compressive stress is very small, the condition for stable crack growth will not be achieved. In laminates with tensile residual stresses in the surface layer, all cracks longer than a certain length will fail at the same stress defined by the toughness at the second interface after having undergone unstable and subsequent stable crack growth. This behaviour was also observed experimentally by Bermejo et al. [29, 55].

The weight function method was also used to develop an integrated approach to the design of laminates with respect to a maximised apparent toughness [24]. Sglavo et al. [56] used the weight function to successfully develop a flaw tolerant engineered laminate consisting of layers of several (not only 2!) different alumina-zirconia or alumina-mullite composites having a residual stress profile with the maximum compressive stress at a certain depth beneath the surface.

As already stated, the weight function used for many of the above mentioned analyses [67] has been developed for elastic isotropy, a situation which is violated in the investigated laminates. Previous work [68] validates the application to inhomogeneous materials (an error of 10% is expected for a variation of the elastic modulus by 33%), but in order to achieve coincidence of experimental and calculated apparent toughness values, parameters (e.g. the temperature T_0 [26]) have to be adjusted. A different method, the material (or configurational) force method was applied to a laminated system [46]. For these computations, the crack driving force is calculated as the derivative of the generalized Gibb's potential with respect to the spatial coordinates. For an elastically homogeneous material this quantity is equal to the classical J-integral of fracture mechanics, for an elastic inhomogeneous material additional terms have to added [69]. The results were compared to those obtained with the weight function method [64]. It turned out that - in that particular case - the weight function method overestimated the calculated toughness curve by a factor of approx. $\sqrt{2}$, whereas experimental data fitted well to the results obtained with the material force method. Even though these differences seem to be not negligible it is believed that the tendencies for the apparent R-curves and the implications for laminate design will remain valid.

Laminates with a threshold strength. A strongly simplified method to investigate the behavior of cracks that are not situated in an outer layer of the laminate is transverse flexure in a set-up as shown Fig. 4b. This test geometry can be seen as an approximation (which can be readily performed experimentally) of the situation where uniform tension is applied parallel to the layers and defects in tensile layers are investigated.

Rao et al. [27] performed a fracture mechanical analysis for an internal crack situated in one of the tensile layers and derived an expression for the threshold stress σ_{thr}, that is necessary to drive the crack straight through the adjacent compressive layers:

$$\sigma_{thr} = \frac{K_c}{\sqrt{\pi \frac{t_2}{2}\left(1+\frac{2t_1}{t_2}\right)}} + \sigma_c \left[1-\left(1+\frac{t_1}{t_2}\right)\frac{2}{\pi}\sin^{-1}\left(\frac{1}{1+\frac{2t_1}{t_2}}\right)\right] \quad . \tag{3}$$

t_1 and t_2 are the thicknesses of the compressive and tensile layers respectively, σ_c is the compressive residual stress and K_c the intrinsic fracture toughness of compressive-layer material. The stress according to Eq. (3) defines a threshold, below which the specimen cannot fail. This prediction was verified by breaking indented specimens. The indentation strength did not depend on the indentation load [27, 28]. During straight propagation through the adjacent compressive layers, the crack encounters a rising R-Curve [70]. Numerical analysis showed that the conditions to obtain a high threshold strength are a high toughness (also evident from Eq. (3)), a high strain mismatch to give high compressive residual stresses and thin layers. The tensile layers should be made as stiff as possible, the compressive layers as compliant as possible [71]. Experimental studies investigating the influence of the magnitude of the compressive stress and the layer thicknesses revealed the limitations of these recommendations [40]. As i) the compressive stress increases, ii) the compressive layer thickness increases or iii) the tensile layer thickness decreases, an increased tendency for crack bifurcation is observed. Along with bifurcation, the measured threshold strength was higher than the predicted value. Derived from the critical thickness condition $t_{c, edge}$ for edge cracking [37], the minimum compressive layer thickness $t_{c,\,bif} = 2\,t_{c,\,edge}$ to promote bifurcation was found to be [40]

$$t_{c,bif} = \frac{K_c^2}{0.17\left(1 - \nu^2\right)\sigma_c^2} \quad , \tag{4}$$

where the symbols have the same meaning as above. The effect of Young's modulus was verified by producing compressive layers with different levels of porosity. While layers with porosity up to 40% behaved as predicted, a higher porosity completely inhibited any crack extension through the layers [72]. The stress given by Eq. (3) can be regarded as a possible lower limit for the threshold strength in this kind of materials. Apparently, deviations from the assumed failure mode (straight crack propagation through the compressive layer) lead to an even higher threshold strength [40, 72].

Crack deflection and crack bifurcation. Crack deflection due to compressive stresses and/or crack bifurcation are favoured in laminates where the compressive stresses are high in layers that have a certain minimum thickness [23, 36, 73]. In laminates tested in longitudinal flexure, an increase of the threshold stress is observed [40]. In laminates tested in transverse flexure, these mechanisms lead to an remarkable increase in the work of fracture: it can be up to six times higher than that of the corresponding outer layer monolithic material [55]. An important factor influencing the energy absorption during fracture seems to be the distance the deflected/or bifurcated crack travels parallel to the loading direction [74]. A step like fracture like the one shown in Fig. 7 would not lead to a dramatically increased work of fracture.

Fig. 8: Step like fracture path in a $Al_2O_3 +$ $Zr(3Y)O_2$ / $m\text{-}ZrO_2/Al_2O_3$ laminate. The crack deflects towards the layer plane as it enters the thin compressive layers and deflects back towards a direction more perpendicular to the layers as it enters the thick tensile layers (courtesy of R. Bermejo).

Summary

As compared to the situation ten years ago [3], recent research in mechanical properties of ceramic laminates seems to be more focused on laminates with strong interfaces and in-plane residual stresses. Several processing techniques are mastered to an extend that a variety of model material systems can be produced without structural difficulties. Research on these laminates is far beyond characterization of mechanical properties. Fracture mechanics and the weight function method are used to gain insight into the mechanisms that lead to the superior properties of the laminates, mostly if conditions for straight crack growth through the layers prevail. Laminates with a specific R-curve and crack growth behaviour can be designed from a-priori knowledge of the relevant properties of the involved materials. Other failure modes like crack deflection and bifurcation are less thoroughly understood, but add the interesting aspect of a remarkable increase in work of fracture to the failure behaviour. With respect to possible application as structural components, investigations of the influence of time (and/or cycles) and temperature on strength are indispensable. Research on these properties is still in its infancy.

With respect to laminates with weak interfaces, the most promising development is the introduction of porous layers to achieve crack deflection. This method can be applied to virtually all ceramics since no attention has to be paid to chemical compatibility. Porous layers with a wide spacing could be used to direct cracks into a direction were they are less harmful to the overall performance of the component.

Despite remarkable progress in understanding the mechanical behaviour of all kinds of ceramic laminates, the challenge for the future is still the exploitation of the concept in commercial applications.

Acknowledgements

The work was supported in part by the European Community's Human Potential Programme under the contract HPRT-CT-2002-00203, [SICMAC].
The author is pleased to acknowledge stimulating discussions with J. Pascual and R. Bermejo.

References

[1]   V.J. Laraia, A.H. Heuer: J. Am. Ceram. Soc. Vol. 72 (1989), p. 2177
[2]   R.H. Doremus, *Glass Science*, (Wiley-Interscience, New York 1994).
[3]   H.M. Chan: Annu. Rev. Mater. Sci. Vol. 27 (1997), p. 249
[4]   W.J. Clegg, et al.: Nature Vol. 347 (1990), p. 455
[5]   H. Liu, S.M. Hsu: J. Am. Ceram. Soc. Vol. 79 (1996), p. 2452
[6]   J.R. Mawdsley, D. Kovar, J.W. Halloran: J. Am. Ceram. Soc. Vol. 83 (2000), p. 802
[7]   E.J. Winn, I.-W. Chen: J. Am. Ceram. Soc. Vol. 83 (2000), p. 3222
[8]   J. She, T. Inoue, K. Ueno: J. Eur. Ceram. Soc. Vol. 20 (2000), p. 1771
[9]   K.S. Blanks, A. Kristoffersson, E. Carlström, W.J. Clegg: J. Eur. Ceram. Soc. Vol. 18 (1998), p. 1945
[10] J.B. Davis, A. Kristoffersson, E. Carlström, W.J. Clegg: J. Am. Ceram. Soc. Vol. 83 (2000), p. 2369
[11] J. Ma, H. Wang, L. Weng, G.E.B. Tan: J. Eur. Ceram. Soc. Vol. 24 (2004), p. 825
[12] R.M. German, *Sintering Theory and Practice*, (Wiley, New York 1996).
[13] W.J. Clegg: Acta Metall. Mater. Vol. 40 (1992), p. 3085
[14] M.-Y. He, J.W. Hutchinson: Int. J. Solids Struct. Vol. 25 (1989), p. 1053
[15] A.J. Phillipps, W.J. Clegg, T.W. Clyne: Acta Metall. Mater. Vol. 41 (1993), p. 805
[16] C.A. Folsom, F.W. Zok, F.F. Lange: J. Am. Ceram. Soc. Vol. 77 (1994), p. 689
[17] M.-Y. He, J.W. Hutchinson: Trans. ASME Vol. 56 (1989), p. 270
[18] J. Abanto-Bueno, J. Lambros: Eng. Fract. Mech. Vol. 69 (2002), p. 1695
[19] A. Tarlazzi, et al.: Wear Vol. 244 (2000), p. 29
[20] R. Lakshminarayanan, D.K. Shetty, R.A. Cutler: J. Am. Ceram. Soc. Vol. 79 (1996), p. 79
[21] T. Chartier, T. Rouxel: J. Eur. Ceram. Soc. Vol. 17 (1997), p. 299
[22] P.Z. Cai, D.J. Green, G.L. Messing: J. Am. Ceram. Soc. Vol. 80 (1997), p. 1929
[23] B. Hatton, P.S. Nicholson: J. Am. Ceram. Soc. Vol. 84 (2001), p. 571
[24] I.A. Gee, et al.: Adv. Appl. Ceram. Vol. 104 (2005), p. 103
[25] G. Blugan, et al.: Key Eng. Mat. Vol. 290 (2005), p. 175
[26] M. Lugovy, et al.: Acta Mat. Vol. 53 (2005), p. 289
[27] M.P. Rao, et al.: Science Vol. 286 (1999), p. 102
[28] M.G. Pontin, M.P. Rao, A.J. Sánchez-Herencia, F.F. Lange: J. Am. Ceram. Soc. Vol. 85 (2002), p. 3041
[29] R. Bermejo, et al.: Fatigue Fract. Engng. Mater. Struct. Vol. 29 (2006), p. 71
[30] N. Orlovskaya, et al.: J. Mat. Sci. Vol. 40 (2005), p. 5483
[31] C.J. Russo, M.P. Harmer, H.M. Chan, G.A. Miller: J. Am. Ceram. Soc. Vol. 75 (1992), p. 3396
[32] S. Bueno, R. Moreno, C. Baudin: J. Eur. Ceram. Soc. Vol. 25 (2005),
[33] K.-S. Cho, H.-J. Choi, J.-G. Lee, Y.-W. Kim: J. Mat. Sci. Vol. 36 (2001), p. 2189
[34] D.-S. Park, et al.: J. Eur. Ceram. Soc. Vol. 22 (2002), p. 535
[35] P.Z. Cai, D.J. Green, G.L. Messing: J. Am. Ceram. Soc. Vol. 80 (1997), p. 1940
[36] A.J. Sánchez-Herencia, C. Pascual, J. He, F.F. Lange: J. Am. Ceram. Soc. Vol. 82 (1999), p. 1512

[37]   S. Ho, C. Hillman, F.F. Lange, Z. Suo: J. Am. Ceram. Soc. Vol. 78 (1995), p. 2353

[38]   M.-Y. He, A.G. Evans, J.W. Hutchinson: Int. J. Solids Struct. Vol. 31 (1994), p. 3443

[39]   O. Prakash, P. Sarkar, P.S. Nicholson: J. Am. Ceram. Soc. Vol. 78 (1995), p. 1125

[40]   M.P. Rao, F.F. Lange: J. Am. Ceram. Soc. Vol. 85 (2002), p. 1222

[41]   M. Oechsner, C. Hillman, F.F. Lange: J. Eur. Ceram. Soc. Vol. 79 (1996), p. 1834

[42]   B.D. Cullity, S.R. Stock, *Elements of X-Ray Diffraction*, (Prentice Hall, New Jersey 2001).

[43]   K. Zeng, D.J. Rowcliffe: J. Am. Ceram. Soc. Vol. 77 (1994), p. 524

[44]   G. De Portu, et al.: Comp. Sci. Tech. Vol. 65 (2005), p. 1501

[45]   G. De Portu, S. Bueno, L. Micele, C. Baudin: J. Eur. Ceram. Soc. Vol. (in press 2006),

[46]   C.R. Chen, et al.: Acta Mat. Vol. (submitted 2006),

[47]   V. Sergo, D.M. Lipkin, G. de Portu, D.R. Clarke: J. Am. Ceram. Soc. Vol. 80 (1997), p. 1633

[48]   A.V. Virkar, J.F. Jue, J.J. Hansen, R.A. Cutler: J. Am. Ceram. Soc. Vol. 71 (1988), p. C148

[49]   H.J. Oel, V.D. Frechette: J. Am. Ceram. Soc. Vol. 50 (1967), p. 542

[50]   A.V. Virkar, J.L. Huang, R.A. Cutler: J. Am. Ceram. Soc. Vol. 70 (1987), p. 164

[51]   R.A. Cutler, J.D. Bright, A.V. Virkar, D.K. Shetty: J. Am. Ceram. Soc. Vol. 70 (1987), p. 714

[52]   P. Chantikul, G.R. Anstis, B.R. Lawn, D.B. Marshall: J. Am. Ceram. Soc. Vol. 64 (1981), p. 539

[53]   J.J. Hansen, R.A. Cutler, D.K. Shetty, A.V. Virkar: J. Am. Ceram. Soc. Vol. 71 (1988), p. C501

[54]   P.Z. Cai, D.J. Green, G.L. Messing: J. Eur. Ceram. Soc. Vol. 18 (1998), p. 2025

[55]   R. Bermejo, et al.: J. Eur. Ceram. Soc. Vol. (in press 2006),

[56]   V.M. Sglavo, M. Paternoster, M. Bertoldi: J. Am. Ceram. Soc. Vol. 88 (2005), p. 2826

[57]   W. Weibull, *A Statistical Theory of the Strength of Materials*, (Generalstabens Litografiska Anstalts Förlag, Stockholm 1939).

[58]   J. Pascual Herrero, F. Chalvet, T. Lube, G. de Portu: Mat. Sci. Forum Vol. 492-493 (2005), p. 581

[59]   S.L. Fok, B.C. Mitchell, J. Smart, B.J. Marsden: Eng. Fract. Mech. Vol. 68 (2001), p. 1171

[60]   R. Tandon, D.J. Green: J. Am. Ceram. Soc. Vol. 74 (1991), p. 1981

[61]   H.F. Bueckner: Z. Angew. Math. Mech. Vol. 50 (1970), p. 529

[62]   R.J. Moon, et al.: J. Am. Ceram. Soc. Vol. 85 (2002), p. 1505

[63]   A.J. Blattner, R. Lakshminarayanan, D.K. Shetty: Eng. Fract. Mech. Vol. 68 (2001), p. 1

[64]   J. Pascual, et al., Weight Function, J-Integral and Material Forces Approach to Apparent Toughness of Ceramic Multilayers, in: Proceedings of ECF16, Alexandropoulos, submitted 2006.

[65]   T. Lube, J. Pascual Herrero, F. Chalvet, G. de Portu: J. Eur. Ceram. Soc. Vol. (in press 2006),

[66]   J. Pascual, F. Chalvet, T. Lube, G. De Portu: Key Eng. Mat. Vol. 290 (2005), p. 214

[67]   T. Fett, D. Munz, *Stress Intensity Factors and Weight Functions*, (Computational Mechanics Publications, Southampton 1997).

[68]   T. Fett, D. Munz, Y.Y. Yang: Fatigue Fract. Engng. Mater. Struct. Vol. 23 (2000), p. 191

[69]   N.K. Simha, et al.: Int. J. Fract. Vol. 135 (2005), p. 73

[70]   M.P. Rao, J. Rödel, F.F. Lange: J. Am. Ceram. Soc. Vol. 84 (2001), p. 2722

[71]   K. Hbaieb, R.M. McMeeking: Mech. Mat. Vol. 34 (2002), p. 755

[72]   M.G. Pontin, F.F. Lange: J. Am. Ceram. Soc. Vol. 88 (2005), p. 376

[73]   M. Lugovy, et al.: Comp. Sci. Tech. Vol. 62 (2002), p. 819

[74]   D. Kovar, M.D. Thouless, J.W. Halloran: J. Am. Ceram. Soc. Vol. 81 (1998), p. 1004

Key Engineering Materials Vol. 333 (2007) pp 97-106
online at http://www.scientific.net
© 2007 Trans Tech Publications, Switzerland

Fracture of Ceramics with Near Surface
Gradient Residual Stresses

M. Anglada

Department of Materials Science and Metallurgical Engineering
Universitat Politècnica de Catalunya
Diagonal 647, 0828 Barcelona, Spain

marc.j.anglada@upc.edu

Keywords: Ceramic gradient materials, fracture, residual stress

Abstract. The fracture toughness and strength of ceramics can be improved with respect to monolithic ceramics by developing graded materials as laminates composed of periodic alternating layers of one material separated by layers of a second material. The second layer must contain residual compressive stresses which are induced during densification because of differential thermal contraction of the layers. The overall residual stresses increase the apparent fracture toughness of the laminate. However, most deleterious natural flaws and most of the damage induced in service by the environment, contact loading, wear, etc, are small cracks on the surface of the outer layer, so that the effect of the laminate residual stresses on these cracks should be rationalised to understand their behaviour. This work presents an analysis of the influence of the gradient residual stresses on the behaviour of surface cracks under bending and indentation in materials with outer layers either with tensile or compressive residual stresses.

Introduction

The contact of indenters on the surface of brittle materials is usually performed by means of either sharp or blunt indenters. Sharp indenters, like the Vickers indenter, are extensively used to study the fracture and deformation properties of polycrystalline ceramics. Sharp contacts provide deep insight into the fundamental nature of damage modes in brittle solids. It can be also used to induce controlled flaws from which different properties can be evaluated. Increase in crack length of these flaws in service by thermal shock, static or cyclic fatigue, etc, conducts to strength degradation from which the damage induced by these mechanisms can be assessed [1]. Indentations can be used to estimate the fracture toughness of intrinsically brittle ceramics with routine simplicity. In particular, Vickers indentation is now the most widely used test method for evaluating local qualitative fracture toughness in brittle materials. The toughness is calculated from the lengths of surface radial cracks at the indentation corners [2]:

$$K_c = \chi \frac{P}{c_0^{3/2}},$$ (1)

with

$$\chi = \xi \left(\frac{E}{H} \right)^{1/2},$$ (2)

and $\xi = 0.016$, E is the elastic modulus of the material and H is the indentation hardness. The constant ξ has an empirical value that represents the average indentation behaviour of different brittle materials studied by Anstis et al [2]. For one specific ceramic, there is considerable uncertainty in the exact value, which may lead to uncertainties of about 25% in the indentation fracture toughness even when environment effects are taken into account. In addition, if there are extrinsic toughening mechanisms which act on the crack wake, that is, R-curve behaviour, the toughness measured should be a value which lies between the intrinsic fracture toughness and the steady-state plateau toughness depending on the crack length and the mechanism of toughening.

On the other hand, because of the uncertainties in the measurements of fracture toughness by indentation methods, fracture mechanics techniques with straight cracks on four point bend specimens have been widely used to characterise the fracture toughness in ceramic materials. Recent developments in functionally graded ceramics and ceramic laminates have shown that producing multi-layer ceramic laminates with alternative layers under compressive and tensile residual stress can lead to significant improvements in toughness at a relatively low cost [3]. Most of these gradient layered materials are based on an optimal distribution of residual stresses in order to increase the fracture toughness, the highest value being reached after the crack has propagated relatively a long distance through the thickness.

The residual stress in the outer layers of functional gradient materials and laminates plays a dominant role on the usability of these materials. In other words, crack nucleation under contact loading and subsequent propagation inside these layers is of fundamental importance for most applications. The focus of this paper is on the study of the effect of gradient either tensile or compressive residual stresses near the free surface of ceramics.

Residual stresses in laminated and functional graded materials

It is well known that in multilayers and functionally gradient materials the mismatch of thermal expansion coefficients between different layers generates residual stresses during subsequent cooling [4,5]. In the presence of a crack induced by indentation, these residual stresses conduct to a laminate residual stress intensity factor, K_r, which can be added to the Vickers residual stress intensity factor induced by indentation, K_{ind}, so that the crack will grow until the equilibrium condition is reached, that is,

$$K_r + K_{ind} = K_c . \tag{3}$$

Thus, if a specimen is tested using Vickers indentation in a laminate under a given load, the indentation crack length that results is shorter or longer than in the same isolated monolithic material under the same load depending whether the residual stress in the laminate is compressive ($K_r < 0$) or tensile ($K_r > 0$). Then if Eq.(1) is used the measurement represents an apparent fracture toughness which may be higher (compressive stresses) or lower (tensile stresses) than the intrinsic toughness, K_c.

In addition, the exact value of the apparent fracture toughness will change along the crack front following the change in the residual stress depth profile, which depends itself of the position of the crack on the layer. Therefore, for a surface semicircular flaw (see Fig.1), the total stress intensity factor may have different values at the surface and at the deepest point, even if the surface effect is neglected [6].

As a consequence, during indentation the flaw will tend to grow more either at the surface (point B of Fig. 1(b)) or at the deepest point (A) depending on whether the stresses are compressive or tensile, respectively. If the indentation crack length is only measured at the surface and it is assumed to be semicircular, the apparent indentation fracture toughness may be either underestimated or overestimated according whether the stresses are compressive or tensile, respectively.

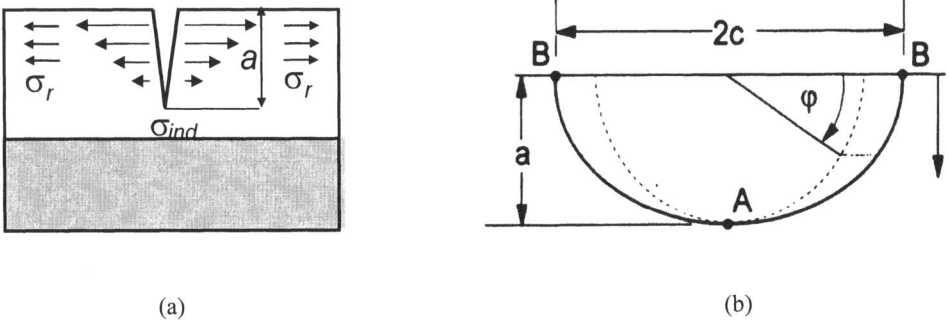

(a)   (b)

Fig. 1. a) Schematic representation of the residual stresses along the crack depth: b) Surface elliptical crack showing the definition of the different geometrical terms.

Estimation of the fracture toughness by indentation

However, in a first step and in order to simplify the treatment, it will be assumed that the residual stress distribution is uniform over the depth of the surface cracks, so that a rough estimation of the stress intensity factor for a semicircular crack ($a=c$) can be given as [7]:

$$K_r = F_{res}(a/c, \varphi)\sigma_0 \sqrt{\pi c} , \tag{4}$$

where σ_0 is the constant residual stress in the outer layer and F_{res} are functions that depend on a/c and on the position along the crack front. By using Eqs. (1) and (4) in Eq. (3),

$$F_{res}\sigma_0\sqrt{\pi c} + \chi \frac{P}{c^{3/2}} = \chi \frac{P}{c_0^{3/2}}, \tag{5}$$

where c and c_0 are the surface indentation crack lengths with and without residual stresses respectively, and from which an expression for σ_0 can be obtained as:

$$\sigma_0 = \frac{1}{F_{res}\sqrt{\pi c}} K_c \left(1 - \left(\frac{c_0}{c}\right)^{3/2}\right). \tag{6}$$

This is a useful and simple equation in order to estimate the level of residual stress and to compare the results of measurements on different points of the same material.

The normalized total and residual stress intensity factors are plotted in Fig 2 in terms of c/c_0 for different values of $K_r^0 = \sigma_{res}\sqrt{\pi c_0}$. Note that the effect of a tensile residual stress will be to increase the length of indentation cracks considerably. If the level of tensile residual stress is high enough, the crack may reach the condition for unstable growth during the indentation process, which takes place when K_r^0 reaches $0.47K_c$ and c/c_0 approaches 2.5. Higher values of K_r^0 will induce crack extension at least until the neighbouring compressive layer, where it may stop if the residual compression is high.

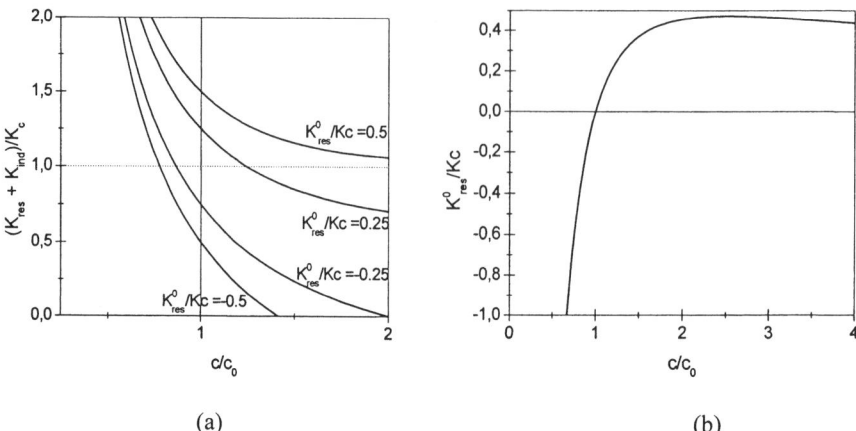

(a)   (b)

Fig. 2 a) Normalised total (a) and residual (b) stress intensity factors in terms of normalised indentation crack length.

The effect of a compressive residual stress will be to reduce crack growth in the indentation process by an amount different to the extension in case of tensile residual stresses. Thus, for a compressive stress with the same absolute value as the tensile stress that induces popping in at the tensile outer layers, the indentation crack length will reach a length of only about 0.78 c_0. If the indention load is high and the compressive layer is thin, the indentation crack could reach the tensile neighbouring layer and induce unstable fracture.

Residual stress field with a surface gradient

The above simple analysis does not take into account those situations in which residual stresses in laminates or in functional gradient materials are not exactly constant close to the surface of the specimens because of a concentration gradient, or edge effects in laminates, asymmetric layers, etc. Thus the residual stress in the depth direction (perpendicular to the plates) will be assumed as having a gradient near the surface: the residual stress at the surface, σ_0, increases in the depth direction until reaching a constant value, $\sigma_0 + \Delta\sigma_0$ at a depth s. For simplicity it will be assumed that the residual stress can be represented by a linear relationship, that is:

$$\sigma_r(x) = \sigma_0 + \Delta\sigma_0(x/s) \qquad x \le s, \tag{7}$$

$$\sigma_r(x) = \sigma_0 + \Delta\sigma_0 \qquad x \ge s. \tag{8}$$

It is possible then to calculate the apparent fracture toughness of a straight edge crack which could be considered as the limit of a semicircular surface crack when $c \gg a$ (Fig. 1). This hypothesis is better realised in the case of the existence of a gradient of residual stress close to the surface so that the apparent fracture toughness at the surface, point B, is much smaller than at point A. In order to obtain analytical solutions we shall consider the material homogenous and isotropic and that all natural or indentation cracks can be assimilated to edge cracks.

Let us assume that the stress intensity factor of the residual stress in the laminate increases with depth according to Eq. (7). The stress intensity factor can be calculated by the weight function method and assuming that the crack is in an infinite body,

$$K_r = 2 \int_0^a \sigma_r(x) \sqrt{\frac{a}{\pi(a^2 - x^2)}} dx . \tag{9}$$

If the crack length is smaller than s then by using Eq. (7) and integrating this equation we obtain:

$$\frac{K_r^I(z)}{\sigma_0 \sqrt{\pi s}} = z\left(1 + \beta \frac{2}{\pi} z^2\right) \qquad z \leq 1, \tag{10}$$

where $K_r^I(z)$ is the residual stress intensity factor for a crack of length $a<s$, $\beta \equiv \Delta\sigma_0/\sigma_0$ and $z \equiv (a/s)^{1/2}$. If a is equal or larger than s, then by using Eqs. (7) and (8) in (9) and integrating we have:

$$\frac{K_r^{II}(a)}{\sigma_0 \sqrt{\pi s}} = 1 + \frac{2}{\pi}\beta + z(1+\beta)\left(1 - \frac{2}{\pi}\sin^{-1}(\frac{1}{z^2})\right) \qquad z \geq 1, \tag{11}$$

$K_r^{II}(z)$ is the residual stress intensity factor for $a \geq s$.

Apparent fracture toughness

The apparent fracture toughness close to the surface is given by the intrinsic fracture toughness of the outer layer, K_c, plus or minus the above expressions. They are added or subtracted according to the sign of the residual stress along the crack path, that is, whether they are compressive or tensile. Fig. 3 shows the plots of the normalised apparent fracture toughness in terms of the square root of the crack length normalised by $s^{1/2}$ for $\sigma_0(\pi s)^{1/2} = K_c$. In these figures the indentation and the applied bending stress intensity factors, both normalised by $\sigma_0(\pi s)^{1/2}$, are also plotted. For the indentation residual stress intensity factor, K_{ind}, it is assumed that it can be written as in Eq.(1), so that the normalised indentation residual stress intensity factor can be expressed as,

$$\frac{K_{ind}}{\sigma_0 \sqrt{\pi s}} = K_C (\frac{z_0}{z})^3 , \tag{12}$$

where $z_0=(a_0/s)^{1/2}$ and a_0 is the crack length in the monolithic material under the same indentation load as in the laminate. In the case of an applied bending stress, the stress intensity factor, K_b, is given by,

$$\frac{K_b}{\sigma_0 \sqrt{\pi s}} = \frac{\sigma}{\sigma_0} \sqrt{z} . \tag{13}$$

If the near surface region under compression a clear apparent R-curve behaviour appears (Fig.3). Therefore, when applying a bending stress there will be stable crack growth.

The intercepts of the indentation residual stress curves with the apparent fracture toughness curves give the final indentation crack length. It is clear that when applying an indentation the resulting cracks will be shorter than in the monolithic materials as shown previously for a constant residual compressive stress.

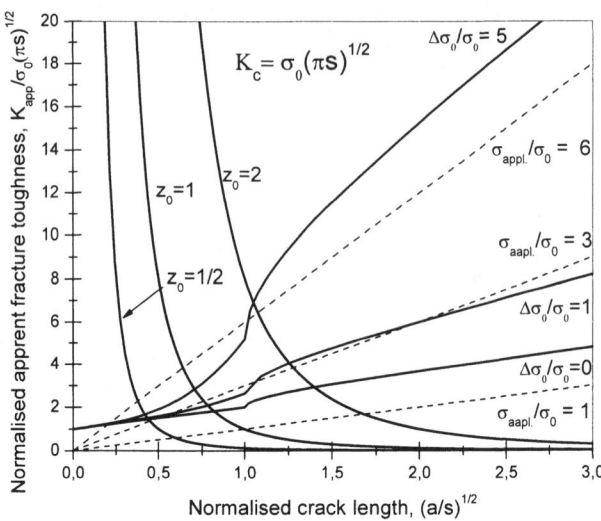

Fig.3. Apparent fracture toughness under compressive residual stress for different gradients of residual stress. Normalised K_{ind} and K_b have also been plotted

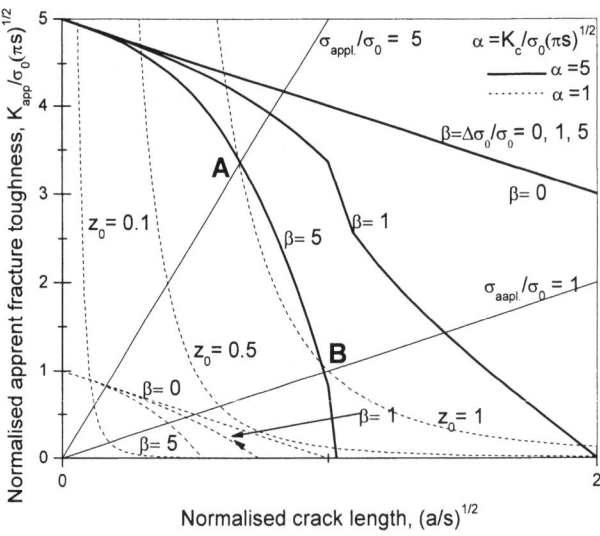

Fig. 4. Apparent fracture toughness for tensile residual stresses.

In the case of an outer tensile layer (Fig.4), the fracture toughness decreases with crack length and only cracks produced under small loads will be able to stop and not to run inside the specimen, as it would be the case for $z_0 = 0.1$ (see Fig. 4). There is an indentation load that will easily propagate the crack until meets neighbouring compressive regions where it may stop. It can be seen in Fig 4. that this load is close to the value that induces cracks with z_0 around 0.5 in the monolithic material.

In bending, the stress intensity factor acting on edge surface cracks increases according to the straight lines shown in Figs. 3 and 4. The intercept with the apparent fracture toughness curves will give the fracture condition. Unstable crack extension and complete fracture of this layer will occur when reaching the unstable fracture condition, that is, for a small crack extension the increase in K_b is larger than the increase in *Kapp*. From this figure it is also clear that unstable fracture is controlled by the apparent fracture toughness for $a>s$. Then the condition for unstable fracture for $a>s$ can be written as:

$$K_{apparent} = K_c + K^{II}(z) = K_b,\tag{14}$$

$$\frac{dK_{apparent}}{da} = \frac{dK_b}{da}.\tag{15}$$

By solving Eqs.(14) and (15) and using Eqs. (13) and (11) the critical crack length in bending can be found in an analytical form:

$$z_{crit} = \frac{8}{\pi^2}\left(\frac{1+\beta}{1+(2/\pi)\beta+\alpha}\right)^2 \pm \sqrt{\frac{64}{\pi^4}\left(\frac{1+\beta}{1+(2/\pi)\beta+\alpha}\right)^4 +1}.\tag{16}$$

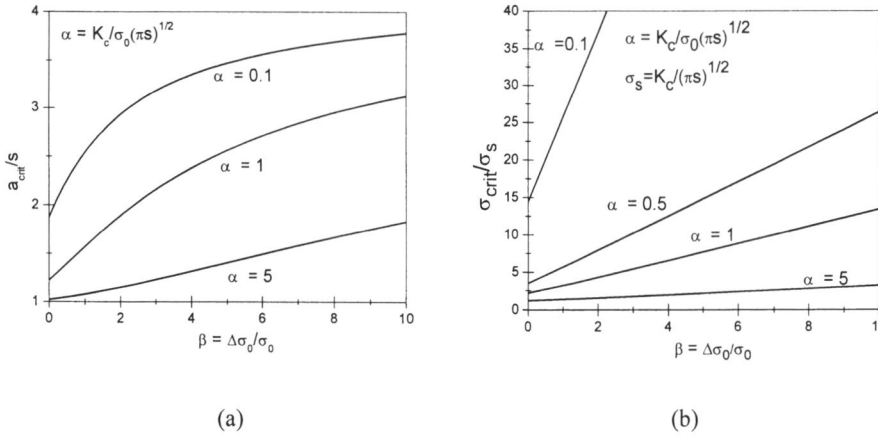

(a)   (b)

Fig. 5. Critical crack length (a) and strength (b) of the outer compressive layer in terms of gradient residual stress field and normalised by s and by the strength of the monolithic material with a straight edge crack of length equal to s, respectively.

The length of the critical crack (see Fig. 5(a)) increases with σ_0 (smaller values of α) and, for a given α, increases with the surface gradient. Also, as is apparent in the curves of Fig. 3, the critical crack length is always larger than s. Then, any crack of length shorter than the critical value will grow in a stable manner until the critical value is reached. The critical stress for this compressive outer layer (Fig. 5(b)) shows an increase with respect to the monolithic material. It can be seen that all cracks shorter than s will increase and extend until reaching the critical condition beyond s=1. As may be expected, very large residual compression stresses along thick layers, that is, α<<1, induce the largest critical stress.

Tension in the outer layer

In the case of tensional outer layers the K_r residual stress intensity factor must be subtracted to the intrinsic fracture toughness to find out the apparent fracture toughness, which, in this case, will decrease with crack length (Fig. 4). If the tensional residual stress is high enough so that the apparent fracture toughness reaches zero for a given crack length, cracking of this layer will take place during fabrication before applying any external load. For this reason the apparent fracture toughness in Fig. 4 has only been plotted for α equal to 1 and 5, since for α = 0.1 it is already negative. Residual stresses can drive the crack directly to the compressive neighbouring layer. For example, if s is equal to the indentation crack length in the monolithic material, z_0=1, and it can be seen that under the same indentation load the crack will run in an unstable manner at least until next compressive layer is reached where it may stop if the compressive residual stress is high enough. This is because in this case K_{ind} is always higher than K_{app} as it happens when α is equal to 1. On the other hand, when the tensional residual stress is low, for example for α = 5, the crack will stop when K_{ind} becomes equal to K_{app}. This is reached at points A and B in Fig. 4 for α = 5, β=0 and z_0 = 1. The indentation crack in principle may stop at A but not at B, since, at difference with A, B is an unstable position as K_{ind} grows faster than $K_{app.}$.

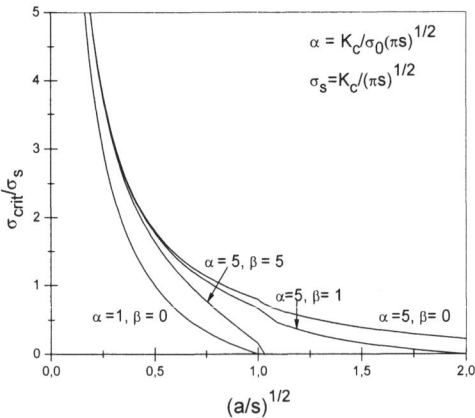

Fig. 6. Strength of the outer layer with residual tensile stresses in terms of crack length

Under bending only unstable crack growth will occur. The normalised stress to propagate the crack to the neighbouring compressive layer is shown in Fig. 6 in terms of the normalised crack length. As the normalising stress is equal to the strength of the monolithic material (without residual stresses) for a semicircular crack of depth equal to s, it becomes clear that surface cracks can be easily extended to the next neighbouring compressive layer with a relatively low stress.

Conclusions

Conditions for extension of straight edge cracks and Vickers indentation cracks in the outer layer of ceramics with increasing residual stresses in the depth direction have been analysed by means of a simple approach in which the material is considered homogenous. The analysis shows clearly that the strength and apparent fracture toughness can be considerably increased or decreased in relation to the monolithic material.

Acknowledgements

The author is grateful for the support of the European Community's Human Potential Programme under contract HPRN-CT-2002-00203, [SICMAC] and by the Spanish Ministry of Science and Culture, through grant MAT-2005-01168.

References

[1] B. R. Lawn: *Fracture of Brittle Solids;* Ch. 8. (Cambridge University Press, Cambridge, U.K., 1993).

[2] G. R. Anstis, P. Chantikul, D. B. Marshall and B. R. Lawn, J. Am. Ceram. Soc., 64 [9] (1981), p. 533

[3] M. Lugovy, V. Slyunyayev, N. Orlovskaya, G. Blugan, J. Kuebler and M. Lewis: Acta Materialia, 53 (2005), p. 289.

[4] W.J. Clegg, K. Kendall, N.M. Alford, T.W. Button, and J.D. Birchall: Nature, 347 (1990), p.455.

[5] D.B. Marshall, J.J. Ratto and F.F. Lange: J. Am. Ceram. Soc. 74 (1991), p.2979

[6] T. Fett: Eng. Fract Mech. 66 (2000), p.349

[7] Newman and I.S. Raju: Eng. Frac. Mechanics 15 (1982), p.185

Key Engineering Materials Vol. 333 (2007) pp 107-116
online at http://www.scientific.net
© 2007 Trans Tech Publications, Switzerland

Instrumented indentation of layered ceramic materials

E. Jiménez-Piqué[a] Y. Gaillard[b] and M. Anglada[c]

Material Science and Metallurgical engineering department, Universitat Politècnica de Catalunya

Avda. Diagonal 647 (ETSEIB), 08028 – Barcelona, SPAIN

[a]emilio.jimenez@upc.edu, [b] yves.gaillard@upc.edu, [c]marc.j.anglada@upc.edu

Keywords: Nanoindentation, Instrumented Indentation, ceramic, coatings, multilayers

Abstract. This paper gives a short review of the instrumented indentation technique when applied to ceramic layered materials. The main causes of error are commented, together with the effect of the substrate, the microstructure and the possible fracture events.

Introduction

Depth-sensing instrumented indentation is nowadays a novel yet well-established technique for the mechanical characterization of materials at a wide range of scales, from few nanometres (where the technique is also known as nanoindentation) to micron and millimetre range.

In an instrumented-indentation test, similarly to the way a hardness test is performed, an indenter of known geometry is pressed into a material and the penetration into the material, or penetration depth, (h) is constantly recorded against the applied load (P) for both the loading and the unloading cycle. Figure 1 shows a typical P-h curve performed with a sharp indenter, where it is appreciated that the unloading (generally associated to a pure elastic deformation) is different from the loading curve (associated to an elasto-plastic deformation) as it returns to a residual depth (h_r) different from zero, indicating a permanent deformation which results in a hardness imprint.

By recurring to different analysis methodologies of the P-h curves it is possible to extract material parameters, principally instrumented indentation testing hardness (H) and instrumented indentation Young's modulus (E), without the visualization of the residual imprint. This is one of the main advantages of this technique, as imprint visualization is difficult at very low scales. In addition, instrumented techniques probe a small volume of material, allowing for the determination of local properties and small volumes of materials (such as single grains, phases or thin films). Moreover, this technique gives information of the mechanical response of the surface which is a crucial information in order to understand the behaviour of material in applications where concentrated surface loadings are expected.

The scope of this paper is to give a short review of the most usual instrumented indentation techniques, with special focus on the specific problems arised when testing ceramic materials and layered ceramic materials. Because ceramic materials do not present strong time-dependant responses at room temperatures (viscosity, creep,...), these effects are not considered in this

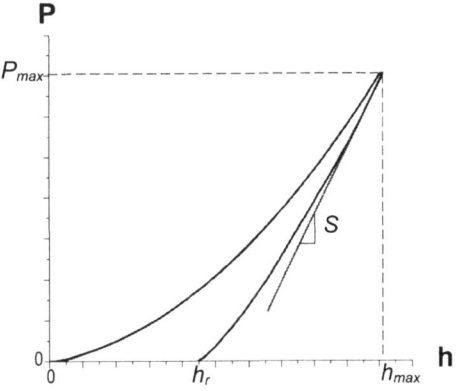

Figure 1.- Typical load (P) vs. penetration depth (h) curve of a load-unload instrumented indentation test, where it can also be appreciated the maximum load (P_{max}), maximum penetration depth (h_{max}), residual depth (h_r) and unload contact stiffness (S).

work. For a more complete understanding of the instrumented indentation techniques, the reader is addressed to some of the excellent reviews available in the literature [1, 2, 3, 4, 5].

Contact Mechanics

Types of probes: As in traditional hardness testing, the response of the material will strongly depend on the shape of the indenter, which can be divided in two main types: blunt indenters and sharp indenters. Blunt indenters comprise spherical indenters, flat punches, sphero-conical indenters, among others, and they are characterized by the fact that during the first stages of the contact with the material, they produce an elastical response on it. That is, in an instrumented indentation test the unloading curve is equal to the loading curve and there the material is not permanently deformed. Upon further loading, the response of the material becomes elasto-plastic producing a residual imprint. Sharp probes, on the other hand, produce an elasto-plastically imprint from the beginning of the loading cycle, due to the fact that, in theory, the indenter tip has a null area, producing a stress singularity in the material. Usually sharp probes present a constant face angle, so the response of the material is self-similar for all penetration depths. Sharp tips are usually conical or, specially, pyramidal, being Vickers (4-side), Knoop (4-side asymmetrical), Cube-corner (3-side), Berkovich (3-side) the most usual ones (figure 2). Three-side pyramidal indenters are the most usual ones because of technical fabrication reasons: three sides always converge into a single point, whereas four sides normally produce an apex. Among them, the Berkovich indenter is the most employed because it presents the same projected area that the Vickers indenter (they have he same equivalent cone area function).

However, it has to be taken into account that actually a sharp tip has a finite radius (due to the physical impossibility of producing a null area) and present a response similar to blunt tips at very low loads. In addition, the tip is worn out as it is used, specially if hard materials are tested or the tip is used in sliding contact tests, accentuating the sphericity of the tip. Typically, the curvature radius of the tip of a sharp indenter is from some few tenths of nanometers for a new tip to around a micron for an extensively used tip. Tip rounding may have a negligible effect for traditional hardness testing and instrumented indentation at large loads, but it is an important factor at nanoindentation levels. The indenters are usually made of diamond or other hard materials, such as sapphire or ruby.

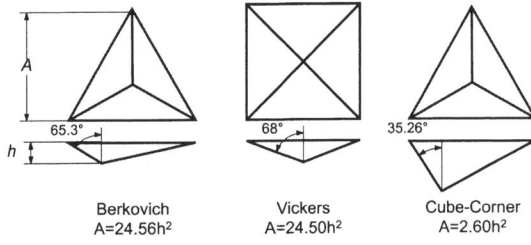

Figure 2.- Angles and relationship between area and indentation depth of the most employed sharp indenters in nanoindentation: Berkovich, Vickers and cube-corner.

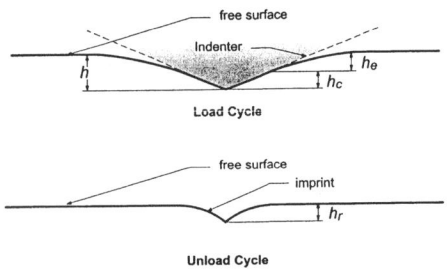

Figure 3.- Scheme of the contact between indenter and material in the loading and the unloading cycle. Contact depth (h_c), elastic depth (h_e) and residual depth (h_r) are marked

Loading curve

The loading response of the material will show, therefore, whether if the indentation probe is blunt or sharp, producing an elastic initial response or an elasto-plastic response. Once the loading is sufficiently high, both indenters will produce similar elasto-plastic response in the material. It has to be taken into account that all real sharp indenters have a tip curvature which translates into a blunt indentation at the initial contact depths.

The analysis of the *P-h* curves depends if it is a loading curve (which can be elastic or elastoplastic) or a unloading curve, where the material is usually deformed.

a) Elastic spherical contact: The elastic contact between two bodies was first described by Hertz in 1882, and the equations describing this response are usually known as Hertzian equations. Hertz shown that, when a sphere of radius R is pressed against a flat solid, the relationship between the contact radius (a) and the load (P) for small penetration depths equals to [6]:

$$a^3 = \frac{3}{4}\frac{PR}{E_{eff}}$$

(1)

where E_{eff} is the effective modulus between the sample and the indenter, which it is defined as:

$$\frac{1}{E_{eff}} = \frac{1-v^2}{E} + \frac{1-v_i^2}{E_i}$$

(2)

with the subscript i indicating the indenter material (E_i=1140GPa and v_i=0.07 for diamond). For the analysis of an instrumented indentation curve, it is more convenient to express equation 1 as a function of *P-h*:

$$P = \left(\frac{4}{3}E_{eff}\sqrt{R}\right)h^{3/2}$$

(3)

That is, by fitting an elastic *P-h* curve by a curve of the type $P \propto h^{1.5}$, and by knowing the radius of curvature of the indenter, it is possible to extract the elastic modulus of the indented material.

The mean pressure between the indenter and the material (p_0) is:

$$p_0 = \frac{P}{\pi a^2} = \frac{4E_{eff}}{3\pi}\left(\frac{a}{R}\right)$$

(4)

In the case of spherical indentation, by plotting p_0 against a/R it is possible to display the stress-strain contact relationship, as shown in figure 4 for a ceramic coating on a metal substrate. Moreover, by loading in the elastic regime, it is possible to characterize other mechanical parameters of the material, such as stress corrosion cracking [7] or contact fatigue resistance [8].

The maximum tensile stress (σ_{tm}) is produced along the contact perimeter of the contact, and equals to [6]

$$\sigma_{tm} = \frac{1}{2}(1-2v)p_0$$

(5)

The maximum shear stress (τ_m) is produced beneath the indentation axis at a depth close to $0.5a$, and equals to:

$$\tau_m = 0.46p_0$$

(6)

In most materials, and according to the Tresca criterion, when the value of τ_m reaches a certain limit ($\tau_m = \sigma_y/2$ where σ_y is the yield strength of the material), the material will start to flow and create a permanent deformation. This criterion has been also invoked to explain the nucleation of dislocation during pop-in events [9], and the material will make the transition from an elastic response to an elasto-plastic response [10]. In this case, p_0 will converge to the hardness of the material. However, it has to be taken into account that, if no yield is produced in the material, the unloading curve will be equal to the loading curve and the same equations will hold (figure 5).

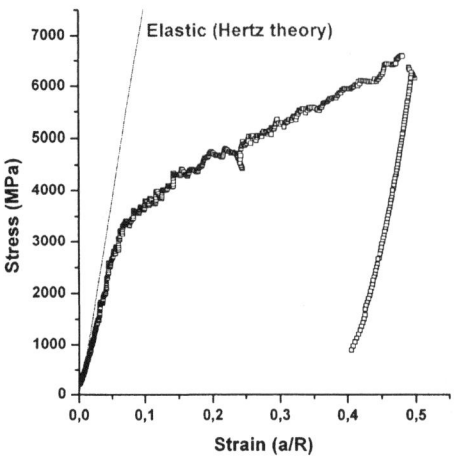

Figure 4: Typical indentation strain-stress curve obtained on air plasma sprayed (APS) thermal barrier coating (with a spalt layered microstructure). A quasi-elastic stage of deformation can be distinguished during the first steps of indentation.

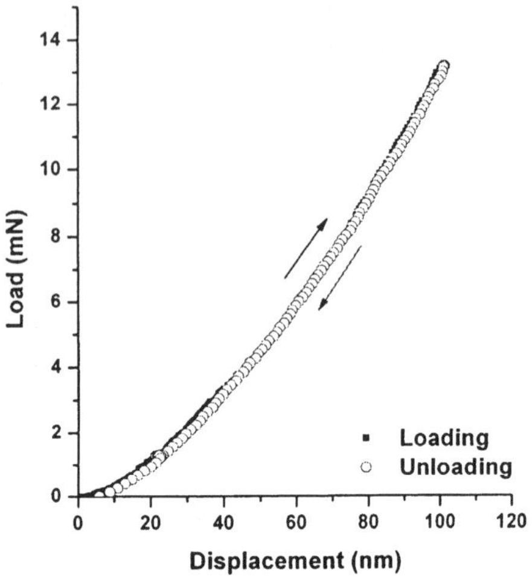

Figure 5: Spherical indentation (R=25μm) in sapphire in the elastic range, where it can be observed that the unloading curve is the same as the loading curve, as no permanent deformation in produced in the material.

b) Sharp plastic contact: If the contact is fully plastic, the relationship between load and depth of penetration becomes parabolic $P \propto h^2$, taking the form for a sharp transition from elastic to plastic regime [1]:

$$P = \frac{E_{eff}}{\left[\frac{1}{\sqrt{\pi} \tan \alpha} \sqrt{\frac{E_{eff}}{H}} + \left(\frac{2(\pi - 2)}{\pi} \right) \sqrt{\frac{\pi}{4}} \sqrt{\frac{H}{E_{eff}}} \right]^2} h^2 \qquad (7)$$

Figure 6 presents a _P-h_ curve made with a Berkovich indenter with a rounded tip of approximately 750 nm in a polycrystalline zirconia sample, where it can be appreciated that at the beginning of the loading the contact remains elastic ($h^{1.5}$) up to a given load where dislocation glide is activated and then the contact becomes elasto-plastic (h^2).

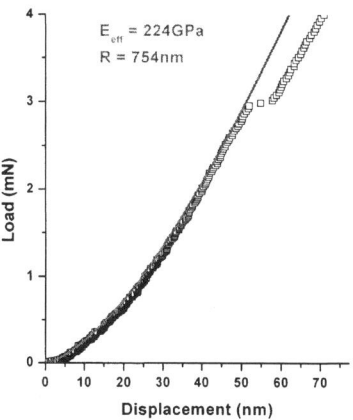
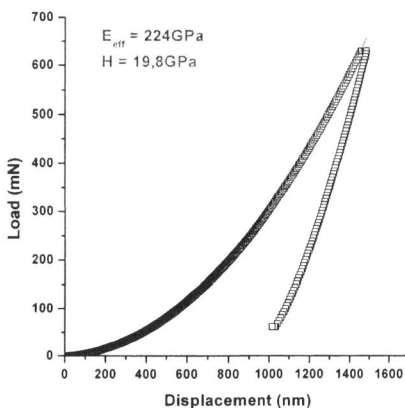

Figure 6: Indentation curve performed on polycrystalline zirconia. Equation 3 and 7 are used to fit the loading curve respectively during the elastic part before the activation of plasticity (where the Berkovich indenter can be assimilated to a sphere with a tip radius of 754nm)-left- and during the elasto-plastic stage (where the indenter can be assumed to be fully pyramidal) –right-.

Unloading curve: analysis with the Oliver & Pharr method (1992): Most of the methodologies for evaluation of mechanical properties relay in the analysis of the unloading portion of P-h curves. In fact, this portion of the curve is generally associated to purely elastic deformation (if a reloading curve is performed just after the unloading the two curves appear to be superimposed in most of the materials), and allow to apply the classic elastic mechanics [11] to determine the contact depth and consequently the projected contact area, which is then used to determine the hardness and modulus of the material. It has to be reminded than the contact depth (h_c) is different from the h_{max} and h_r, and can not be directly extracted from the raw data of the P-h curve. The most usual methodology for extraction of the projected contact area is the one of Oliver & Pharr [12], because it yields relatively accurate results with a simple analysis. In their analysis, they determined that the contact depth equals to:

$$h_c = h_{max} - \varepsilon \frac{P}{S} \tag{8}$$

where ε is a parameter approximately equal to 0.75 for a spherical or a Berkovich indenter and 0.72 for a conical indenter, and S is the unloading contact stiffness at maximum depth (see figure 1):

$$S = \frac{dP}{dh}\bigg|_{h=h_{max}} \tag{9}$$

Once the contact depth is calculated, the projected contact area (A) can be determined from geometrical considerations (see figure 2). In the case of a perfect Berkovich $A=24.56h_c^2$. However, in reality the projected contact area differs from this value because the faces angles may be somehow different from the nominal ones and the tip has been worn out. This implies that the projected contact area will depend on the depth of penetration, and should be calibrated. This is done by indenting a material with known elastic properties at different penetration depths, and adjusting the projected contact area to a polynomial depending with h, expressed as:

$$A(h) = \sum_{i=0}^{8} C_{i-1} h^{2^{1-i}} = C_0 h^2 + C_1 h + C_2 h^{1/2} + ... + C_8 h^{1/128} \tag{10}$$

where C_0 correspond to the shape of a perfect indenter and the others C_i are adjustable numerical parameters allowing to take into account the tip defect. Once the correct area function is known, the hardness is calculated by dividing the load by the projected contact area at maximum load:

$$H = \frac{P_{max}}{A(h_c)} \tag{11}$$

This hardness, defined as instrumented indentation hardness, differs from the one of a Vickers indentation because equation (11) uses the projected area of contact and it is calculated at maximum load. The definition of Vickers Hardness employs the true contact area and is measured upon unloading. The other difference is that an error is made observing the residual imprint because of the effect of elastic relaxation occurring during or after the unloading. This phenomenon is particularly important in ceramics materials. Elastic modulus is then calculated by:

$$E_{eff} = \frac{1}{\beta} \frac{\sqrt{\pi}}{2} \frac{S}{\sqrt{A(h_c)}} \tag{12}$$

being β equal to 1 for spherical indentation and approximately equal to 1.034 for a Berkovich indentation. This parameter takes into account the lack of symmetry of the Berkovich indenter, as equations 8-12 are developed for an axisymmetric indenter.

Common errors produced in instrumented indentation of ceramic materials

There are many sources of error and inaccuracies related with instrumented indentation test due mainly to the very low length scales in which the tests are performed. A complete review of most of these possible sources of errors can be found in the literature [2,13], though the most usual sources of uncertainty for ceramic materials are here briefly commented. The reader is addressed to the literature for more insight into the different sources of error.

a) Instrument compliance: Because of the finite stiffness of the indentation instrument the displacement measured by the machine (h') will be the displacement into the material plus a contribution from the deformation of the machine (h_{mach}), which is equal to its compliance (C_f) multiplied by the applied load.

For low loads and soft materials, the compliance of the instrument is normally small, however, for hard materials, and specially at large loads the frame compliance contribution has to be properly accounted for in order to calculate the real value of h. In order to perform a frame stiffness calibration, several indentations are done in a hard material (for example, sapphire) different from the one used for tip shape calibration, and plotting dh'/dP against the contact depth, and extracting the value of C_f as the axis interception at $h_c=0$:

$$h' = h + h_{mach} = h + C_f P \Rightarrow \frac{dh'}{dP} = \frac{1}{S} + C_f = \left(\sqrt{\frac{\pi}{A(h_c)}} \frac{1}{2E_{eff}} \right) + C_f \tag{13}$$

b) Actual projected contact area: The analysis by Oliver & Pharr described above was based on a purely elastic material, and it always assumes that the contact area is below the surface sample. However, in materials with a large amount of plasticity (that is, with a large E_{eff}/σ_y ratio) they can pile-up around the indentation and, consequently, have a contact area larger than the one estimated by the method [14]. In most of ceramic materials, pile-up will not be an issue, but it may be worth measuring correctly the indentation imprint (by, e.g., AFM). In addition, because of the high hardness of the ceramic, the strain suffered by the indenter will slightly change the contact area [15].

   c) Improvement of the method: The parameters ε and β described in equations (8) and (12) were assumed to be constant and only dependant on the type of indenter used. However, these parameters have some dependence on the exact shape of the unloading curve (parameter ε) [16,17, 18] and several material parameters, specially Poisson's ration (parameter β) [19]. This last parameter may affect to a large extent the correct evaluation of the mechanical properties of hard materials, as it captures not only the lack of axisymmetry of the indenter but also the radial displacements of the material [3].

   d) Material related issues: Most of the instrumented indentation analysis methods model the solid as a perfectly flat half space, so it is extremely important to indent polished plane-parallel surfaces in order to acquire proper *P-h* data. In addition, it may be possible that the behaviour of the material varies with different length scales (size-effect) because the deformation mechanisms are different. It has to be taken in mind that, although a large number of materials present size-effects, specially with increasing values of hardness for small penetration depths, this is not an universal phenomenon and other common methods have to be discarded (specially incorrect tip calibration) before being sure that the material presents a real size-effect.

Measurement of ceramic coatings and laminates

   The main difference of performing an instrumented indentation on a ceramic coating or laminate will be that at a given penetration depth the response will not only be given by the first indented layer, but also the substrate or the subsequent layers will start to influence the indentation response. In addition, coatings are generally deposited by techniques that do not produce an homogeneous microstructure and normally have a preferential grown direction. Therefore, it may happen that the deformation sensed by the instrumented indentation test is due to, not only to the elastic and plastic deformation of the material, but also to other deformation and fracture mechanisms that are activated during the test.

   a) Measurement of hardness and elastic modulus on laminates: As it is said before, the response of a layered system will be given by a combination of the response of the coating and the substrate (or first and second layer in a laminate). For relative hard systems, it is generally accepted that an instrumented indentation with a maximum depth of less than 10% of the coating thickness will yield the hardness of the coating without any influence of the substrate [13]. However, it is sometimes not easy to indent to a penetration depth less that this value because the coating is too thin or because of experimental limitations. In case that a measurement of coating hardness without the influence of the substrate is not possible, there are several models for fitting the composite hardness response of the coating-substrate as a function of the penetration depth [1, 13]. These models allow for extraction of the hardness of the coating by knowing the hardness of the substrate and the composite at different depths by extrapolating to zero depth. It has to be taken into account that the model of choice depends on the relative mechanical properties of the coating and the substrate, particularly their relative ductility and stiffness. One of the models most used for hard coatings on hard substrates is the one of Tuck et al. [20], which states that the apparent hardness (H_a) can be expressed as a function of the hardness of the coating (H_c) and substrate (H_s) as a function of penetration depth, h, coating thickness, t, and two numerical parameters κ and α.

$$H_a = H_s + \frac{H_c - H_s}{1 + \kappa\left(\dfrac{h}{t}\right)^{\alpha}} \qquad (14)$$

   Figure 7 presents the hardness values obtained by a instrumented indentation experiment of a TiN coating on two different substrates of WC-Co with different hardness ($H_s' = 24$ GPa and H_s''

= 13GPa). It can be appreciated that for small penetration depths (less than 10%), the hardness value is constant for both systems, and equals to the value of the coating (H_c = 30GPa). As the penetration depth is increased, the apparent hardness becomes more influenced by the substrate and decreases faster in the system with the less hard material. A fit of the data with equation (15) is also presented, were it can be appreciated that the model fits relatively well the experimental data.

The measurement of Young's modulus in coatings and laminates presents the same problems as the measurement of hardness, but in this case the substrate presents an influence in the response at much lower penetration depths due to the fact that the elastic fields of an indentation

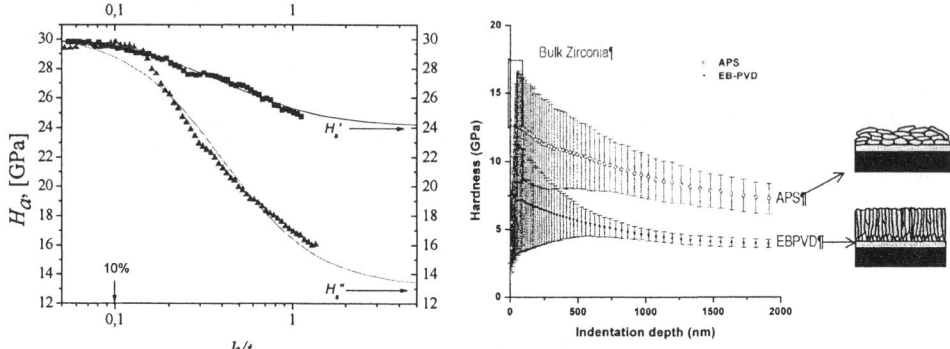

Figure 7.- Hardness values vs. penetration depth of a TiN coating on two different substrates with different hardness values

Figure 8.- Hardness vs. penetration depth of two different thermal barrier coatings with different microstructure and equal chemical compositon (8%Y–PSZ)

have a larger volume than the ones that produce plastic deformation. Therefore, for characterizing the response of the coating alone, smaller penetration depths are required, which will mean that, if this penetration depths are in the order of some tenths of nanomenters, a careful calibration of the indentation tip has to be made. There are several models in the literature for extraction of E after the analysis of the evolution of E for different penetration depths [1, 13].

Alternatively, the use of spherical indenters instead of sharp indenters allow for characterizing the elastic modulus at much lower penetration depths. Therefore, the use of this geometry of indenters is a powerful tool, especially if the sample is probed with different spheres of different radii, in order to correctly calibrate the influence of the substrate.

b) Microstructural effects While most of the methodologies of analysis of instrumented indentation test assume that the materials are homogeneous and continuous, this is not the case for real material, as the continuum is broken at a given length (size-effect). The microstructure length of the material has to be taken into account when analysing the *P-h* curves of any material. However, in the case of coatings, these effects are much more present, because the microstructure strongly depends on the deposition process and it is normally heterogeneous, forming layers, columns, splats, among others.

Therefore, it is necessary to carefully observe which deformation mechanisms are activated during the indentation of a highly heterogeneous coating in order to differentiate the mechanisms of deformation. For example, figure 8 presents the hardness values of two 200 μm thick zirconia doped with 8% yttria thermal barrier coatings on a nickel superalloy substrate produced by two different deposition methods: Atmospheric Plasma Spray (APS), which produces a splat layered microstructure, and Electron-Beam Physical Vapour Deposition (EB-PVD), which produces a columnar microstructure. It can be appreciated that, although both coatings are made of the same material, the response is different between each other, and different to that of bulk zirconia. In addition, it can be seen that the hardness diminishes almost from the beginning. This fact implies that the coating deforms by several deformation mechanisms, which can be identified as column sink-in in the case of EB-PVD or splat fracture in the case of APS, and not only from the plasticity of the zirconia. [21].

c) Residual Stress: Instrumented indentation is, a priori, a good technique for local evaluation of the residual stresses, and, for that purpose, several methods have been developed, specially for metals with relatively low yield stress and that do not work harden appreciably when indented with sharp indenters. Most of these methods are based on the change in contact areas with and without residual stress, such as in the works of Suresh and Giannakopoulos [22], Tsui et al [23] or the change in the elastic penetration depth, as the work of Xu and Li [24]. Swadener et al. [25] used spherical indenters instead of sharp indenters, and showed that blunt indentation is more sensitive to residual stress than sharp indentation. However, these tests work well when the residual stress measured is in the order of magnitude of the hardness of the material. In ceramic materials, this is seldom the case because of the high hardness of these materials. In this case, the differences obtained in the *P-h* curve from a stressed to an unstressed material are to small to quantitatively extract a value of residual stress, though some qualitative information may be inferred in some materials [26].

Fracture

Ceramics are generally brittle, and prone to generation of cracks when indented. Toughness estimation by microfracture Vickers indentation is a well-known and broadly employed technique in ceramic materials. This technique consists on the application of a Vickers (or Berkovich) indenter at a given load to the material sufficiently high to nucleate cracks at the corners of the imprint, and latter measure the crack lengths produced at the corners of the imprint (*c*), in order to evaluate the fracture toughness K_{Ic}. In the case of a penny-shape crack, one of the most used expression, developed by Anstis et al, is [27]. However, there is a load threshold below which no cracks of sufficient length will be generated. In case that the evaluation of toughness of very small volumes of material are required (e.g. grains, phases, coatings), obtuse indenters like the Berkovich do not produce cracks large enough for a correct estimation of K_{Ic}. In this case, it is common to use a cube-corner indenter, a three face indenter which a much more acute face angle and able to generate cracks at much smaller loads [2]. In the case of spherical indenters, the type of fracture will depend also on the radius of the indenter. For small spheres, the material will deform and then produce radial cracking similar to the one produced by a pyramidal indenter. For larger spheres, the material will fracture in the elastic regime by a ring crack close to the contact perimeter that will develop into a cone crack into the material. Further loading will produce radial cracking [6].

Recently several methods for fracture toughness estimation without visualization of the crack length have been proposed, for example the works of Field et al. [28], where they relate the crack length produced to the pop-in behaviour during loading or the work of Dahami et al. [29] where they related the crack length with the total penetration depth for fused silica. However, it has to be taken into account that it is still not clear that these methods will have universal validity for all materials, because several cracks systems may be activated apart from radial cracking [30, 31].

When indenting a ceramic laminate for determining the fracture toughness, the produced cracks should be away from the other layer, as otherwise the crack could be strongly influenced by the elastic mismatch with the other layer material [32]. In addition, other fracture events, such as delamination or radial cracking at the interface, may be activated in case that the indentation field comprises both materials [33]. As in the case of radial cracking, the delamination event will only be detected in an instrumented indentation test for certain materials.

Conclusions

Instrumented indentation is a powerful technique to extract material properties at very low scales, once correctly calibrated. In testing ceramic materials, special care has to be taken in the data acquisition and analysis due to the high hardness and stiffness of these materials. When indenting ceramic laminates, the influence of the underling material has to be taken into account, as well as the possible effects of the microstructure, which is generally very heterogeneous in coating materials. Finally, in the case that fracture events are present during instrumented indentation, it is always recommended to visualize the imprint in order to correlate the different fractures with the *P-h* response.

Acknowledgements

The authors would like to thank E. Tarres, prof. L. Llanes and prof. J. Pavón for helpful discussions. Work supported in part by the European Community's Human Potential Programme under contract HPRN-CT-2002-00203, [SICMAC].

Bibliography

[1] Fischer-Cripps, A.C. "Nanoindentation." (Spinger New York, 2002)
[2] Pharr, G.M. Mat. Sci. Eng. A. 253 (1998): 151
[3] Oliver, W.C., and Pharr, G.M. J. Mater. Res. 19.1 (2004): 3
[4] Fischer-Cripps, A.C. Vacuum 58 (2000): 569
[5] Ullner, C., Beckmann, J., and Morrell, R. J. Eur. Ceram. Soc. 22 (2002): 1183
[6] Lawn, B.R. J. Am. Ceram. Soc. 81.8 (1998): 1977
[7] Pavón, J., Jiménez-Piqué, E., Anglada, M., López-Esteban, S., Saiz, E., and Tomsia, A.P. J. Eur. Ceram. Soc. 26.7 (2006): 1159
[8] Jiménez-Piqué, E., Ceseracciu,L., Chalvet, F., Anglada, M. and de Portu,G. J.Eur.Ceram. Soc. 25.15 (2005): 3393
[9] D. Lorenz, A. Zeckzer, U. Hilpert, P. Grau, H. Johansen, H. S. Leipner, Phys Rev B., 67 (2003): 172101.
[10] Gaillard Y., Tromas C., Woirgard J., Phil. Mag. Let., 9 (2003): 553
[11] Sneddon, I. N., Int. J. Engng Sci.,3 (1965): 47
[12] Oliver W.C., Pharr G.M., J. Mater. Res. 7 (1992): 1564
[13] Fischer-Cripps, A.C. Surf. Coat. Tech. 200 (2006): 4153
[14] Bolshakov, A., and Pharr, G.M. J. Mater. Res. 13.4 (1998): 1049
[15] Gong, J., Miao, H., Peng, Z. Mat. Lett. 58 (2004): 1349
[16] Pharr, G.M., Bolshakov, A. J. Mater. Res. 17.10 (2002): 2660
[17] Troyon, M., Huang, L. J. Mater. Res. 20.3 (2205): 610
[18] Woirgard J., Dargenton J-C., J. Mater. Res., 12 (1997): 2455.
[19] Hay, J.C., Bolshakov, A., and Pharr, G.M. J. Mater. Res. 14.6 (1999): 2296
[20] Tuck, J. R., Korsunsky, A. M., Bhat, D. G., Bull, S. J., Surf. Coat. Technol. 139 (2001) 63
[21] Gaillard, Y., Jiménez-Piqué, E. and Anglada, M. Phil. Mag. (2006), in press
[22] Suresh, S. and Giannakopoulos, E. Acta Mater. 46.16 (1998): 5755
[23] Tsui, T.Y. ; Oliver, W.C. and Pharr, G.M. J. Mater. Res. 11.3 (1996): 752
[24] Xu, Z.-H., and Li, X. Acta Mater. 53 (2005): 1913
[25] Swadener, J.G., Taljat, B., and Pharr, G.M. J. Mater. Res. 16.7 (2001): 2091
[26] Jiménez-Piqué, E., Ceseracciu, L., Gaillard, Y., Brach, M., de Portu, G. and Anglada, M. Phil. Mag (2006) in press
[27] Anstis, G.R., Chantikul, P. and Lawn, B.R.. J. Am. Ceram. Soc. 64 (1981):533
[28] Field,S., Swain, M.V., and Dukino, R.D. J. Mater. Res. 18 (2003): 1412
[29] Dahmani, F, Lambropoulos, J.C, Schmid, A.W., Burns, S.J., and Pratt, C. J. Mat. Sci. 33.19 (1998): 4677
[30] Cook, R.F. , Pharr, G.M. J. Am. Ceram. Soc. 73. 4, (1992): 787
[31] Morris, D.J., Myers, S.B., Cook, R.F. J. Mater. Res. 19.1 (2004): 165
[32] Pavón, J. "Fractura y fatiga por contacto de un recubrimiento de vidrio sobre Ti6Al4V para aplicaciones biomédicas" Ph. D. Thesis. Universitat Politécnica de Catalunya, Barcelona, Spain, 2006
[33] Pavón, J., Jiménez-Piqué, E., Anglada, M., Saiz, E., Tomsia, A. P. Acta Mater. (2006), in press

Key Engineering Materials Vol. 333 (2007) pp 117-126
online at http://www.scientific.net
© *2007 Trans Tech Publications, Switzerland*

Development of Failure Tolerant Multi-layer Silicon Nitride Ceramics: Review from Macro to Micro Layered Structures

J. Kuebler[1, a], G. Blugan[1], M. Lugovy[2], V. Slyunyayev[2], N. Orlovskaya[3] and R. Dobedoe[4]

[1]Empa, Materials Science and Technology, Laboratory for High Performance Ceramics, Ueberlandstrasse 129, 8600 Duebendorf, Switzerland

[2]Institute for Problems of Materials Science, 3 Krzhizhanovskii Street, 03142 Kiev, Ukraine

[3] Michigan Technological University, Department of Materials Science and Engineering,

1400 Townsend Drive Houghton, MI 49931, USA

[4]University of Warwick, Department of Physics, Centre for Advanced Materials, Coventry CV4 7AL, United Kingdom

[a]jakob.kuebler@empa.ch

Keywords: silicon nitride, composite, laminate, residual stress, fracture toughness

Abstract. Recent developments have shown that producing multi-layer ceramic laminates with alternative layers under compressive and tensile stress can lead to significant improvements in toughness at a low cost. However, in many cases the improvements in fracture toughness is associated with the presence of surface "edge cracks" in the compressive layers or the use of porous interfaces between the layers. At the same time such effects can limit the performance of ceramics when used in harsh environments. This review covers the development of silicon nitride based laminate structures and characterisation of these multi-layer structures. The work presents the results of macro-layered laminates with layers greater than 150 μm thickness. The apparent fracture toughness of different designs is measured and the conditions for failure tolerant effects, including crack deflection, bifurcation and edge cracking, are shown and discussed. The structural and processing limitations of the macro-layered laminates are also presented. The development of a weight function analysis as an effective design tool for developing micro-layered laminates with layers of approximately 50 μm thickness is discussed along with the apparent fracture toughness results from these micro-laminates. The failure tolerant behaviour as well as the ease of producing micro-layered laminates with a toughness of 2-3 times higher than that of silicon nitride is shown.

Introduction

Silicon nitride, (Si_3N_4) is typical of many engineering ceramics in that it possesses a combination of favourable material properties including high strength, Young's modulus and hardness at room and elevated temperatures as well as high wear and chemical resistance. However, like the majority of technical ceramics, it is relatively brittle especially when compared to metals. The imposition of a controlled residual stress distribution is well known to significantly improve the mechanical properties (namely strength and toughness) of glasses, glass-ceramics and ceramics. The use of residual stress to improve the mechanical properties of a brittle material normally involves the formation (by physical or chemical means) of compressive stresses in a finite layer on the surface, compensated by balancing tensile stresses in the interior of the material. The strength is increased because the compressive stress must be overcome before sufficient tensile stress can be concentrated in a flaw to initiate crack propagation. This approach is particularly successful with glass due to the sensitivity to surface flaws and the lack of internal defects and can result in both improved strength and toughness.

In the present work a method of designing structural laminated ceramics where the residual stress distribution and properties are optimized by fracture mechanics based weight function analysis is presented. Fracture mechanic considerations are used to make other parasitic failure modes energetically unfavorable. The laminate geometry allows development of ceramic composite structures since analytical solutions of residual stress distributions and test procedures (e.g. SEVNB, DCB) are available enabling designs to be tested and optimised. The approach used does not require a weak interface or interphase (as pioneered by Clegg and others [1]. Residual stress enhanced ceramic laminates can be designed to optimize strength, to obtain a minimum threshold strength, display significantly increased apparent fracture toughness, have increased resistance to sub-critical crack growth, or to display a degree of "graceful failure" [2-4]

Theory

The mismatch of thermal expansion coefficients between different layers inevitably generates residual stresses during subsequent cooling of layered ceramics with strong interfaces [5]. The expressions for residual stresses in layered structures are given in [6]. A schematic presentation of a general case of two-component multi-layered sample is shown in Fig. 1, where t_i is the thickness of the i-th layer, w is the total thickness of the specimen, and N is the total number of layers. In general, the condition for crack growth onset is

$$K_a+K_r=K_c \qquad (1)$$

(Fig. 2), where K_a is the applied stress intensity factor that can be measured, K_r is the stress intensity factor due to a residual stress, and K_c is the intrinsic fracture toughness of a material in the layer. If a condition of a crack growth onset is fulfilled then $K_a=K_c-K_r$ is the apparent fracture toughness. If the residual stress is compressive, then $K_r < 0$ and K_a increases. Note that small surface cracks have K_a close to K_c due to the stress gradient at the surface of the outer layer approaching zero.

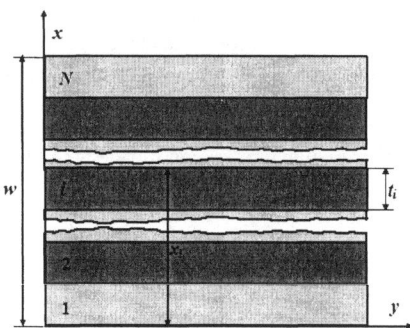

Surface under tension

Fig. 1. A scheme of a two-component multilayer specimen: numbers of layers and layer boundary coordinates.

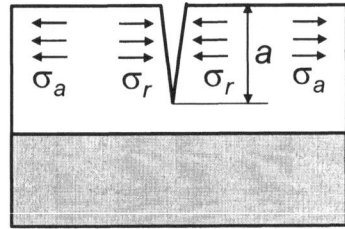

Fig. 2. Factors affecting the apparent fracture toughness in a laminate: σ_a is the applied stress; σ_r is the residual compressive stress; a is the crack length.

A weight function analysis, (WFA) has been used to estimate the apparent fracture toughness in laminates with residual stresses [7]. A schematic presentation of the analyzed crack location in the layered specimen is presented in Fig. 3, where a is the crack length and n is the number of layers crossed by the crack. The geometry of the multi-layered material analyzed here is such that the problem can be reduced to one dimension and that analytically tractable solutions can be used [6].

The apparent fracture toughness, ($K_{Ic\ app.}$) of a layered composite can be calculated analytically by [6,8]:

$$K_{Ic\ app.} = \frac{6Y(\alpha)a^{1/2}\left(I_{L1}^2 - I_{L0}I_{L2}\right)\left(K_{Ic}^{(i)} - Kr\right)}{w^2\left\{E_{n+1}'\int_{xn}^{a} h\left(\frac{x}{a},a\right)\left[I_{L0}x - I_{L1}\right]dx + \sum_{i=1}^{n} E_i' \int_{x_{i-1}}^{x_i} h\left(\frac{x}{a},a\right)\left[I_{L0}x - I_{L1}\right]dx\right\}}$$ (2)

where $K_{Ic}^{(i)}$ is the intrinsic fracture toughness of the i-th layer material, K_r is the stress intensity due to the residual stresses, $Y(\alpha) = \dfrac{1.99 - \alpha(1-\alpha)(2.15 - 3.93\alpha + 2.7\alpha^2)}{(1+2\alpha)(1-\alpha)^{3/2}}$ where $\alpha = a/w$ and $h(x/a,\alpha)$ is the weight function for an edge-cracked sample [9], x_i is the coordinate of an upper boundary of the i-th layer (Fig. 1), $E_i' = E_i/(1-v_i)$, and E_i and v_i are the elastic modulus and Poisson ratio of the i-th layer, respectively. The expression for I_{Lj} ($j=0,1,2$) was obtained from Ref. [6] as follows:

$$I_{Lj} = \frac{1}{j+1}\sum_{i=1}^{N} E_i'(x_i^{j+1} - x_{i-1}^{j+1}),$$ (3)

where $\widetilde{\varepsilon}_i$ is the strain in the i-th layer, which is not associated with any stress. The stress intensity factor due to residual stresses is given in [6,8].

An unstable crack growth occurs (Fig. 4) if the slope of the straight line corresponding to the stress intensity factor at constant applied stress is greater than or equal to the slope of the tangent line to the fracture resistance curve at the same point, in coordinates, apparent fracture toughness K_{app} minus crack length parameter $\tilde{a} = Y(\alpha)a^{1/2}$ [6,8]. Also the applied stress intensity factor becomes higher than the fracture resistance of the material. Modelling demonstrates the toughness increases in the layers with a compressive stress as the crack length increases, and it decreases in the layers with a tensile stress as the crack continues to grow [10].

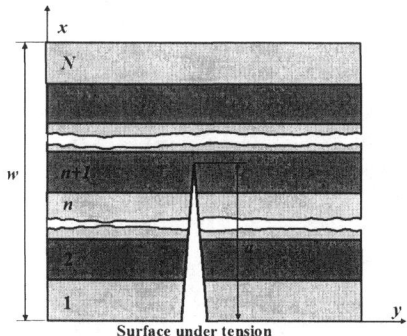

Fig. 3. An analyzed crack location in a layered sample.

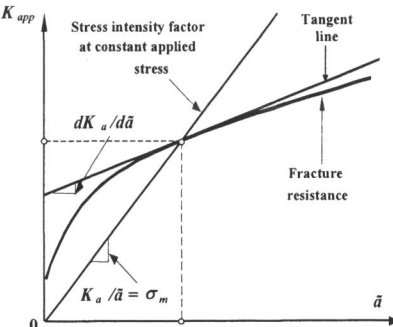

Fig. 4. General criterion of stable/unstable crack growth in a brittle material.

Experimental

In the development of Si_3N_4 based laminates three techniques were used to prepare the required layers of different compositions and thickness. For individual sintered layers of thickness of ~150 μm when sintered, calendering of a ceramic dough was used [11]. For thinner layers, tape casting was used to produce individual sintered layers of ~50 μm thickness (and multiples thereof), and for thicker layers greater than 500 μm conventional die pressing was used. A composition of Si_3N_4 with YAG sintering additives was used as the starting material. Additional compositions with 10, 20, 30, 35, 50 and 70 wt.% TiN powder added to the Si_3N, were also prepared, in addition laminates with layers of 100% TiN were also prepared. (All given compositions of the layers are in wt.%.) From the tapes, discs of 50 mm diameter were punched and stacked as required for the different designs. The stacked layers were hot pressed and four point bend bars with lengths of ~48 mm were produced with final machining being done as specified in EN 843-1 [12].

For all laminate designs that were successfully manufactured the $K_{Ic\ app.}$ was measured as a function of the inserted notch length. The $K_{Ic\ app.}$ was determined using the SEVNB method [13]. Initially the $K_{Ic\ app.}$ was evaluated using a soft universal testing machine (Zwick 005), later tests were performed using a highly stiff machine which had been carefully constructed to allow exact control and measurement of crack propagation in ceramic materials [14].

Results and Discussion

Macro-laminates: Initial development of laminates was based on designs with Si_3N_4 to Si_3N_4+ X% TiN layer stacking of 1:4, with thin Si_3N_4 layers (of ~150 μm) under compressive stresses. Alternative designs with external layers under compression and external layers under tension were evaluated [11]. A typical design of such a laminate system used to determine $K_{Ic\ app.}$ as a function of notch length is shown in Fig. 5. From initial results it was determined that a TiN content greater than 40% in the tensile layers often resulted in multiple channel cracks in the tensile layers. Fig. 6 shows a $Si_3N_4/\ Si_3N_4 + 50\%$ TiN laminate system with extensive channel cracks visible in the tensile $Si_3N_4 + 50\%$ TiN layers. These cracks have been termed as tunnel or tensile cracks and Ho and Suo have shown for given residual tensile stress there is a critical tensile layer thickness below which no tensile/tunnel cracks can occur [15]. In the Si_3N_4 laminate systems it was shown that increasing the TiN content in the tensile systems and hence the CTE mismatch resulted in increased tensile cracking.

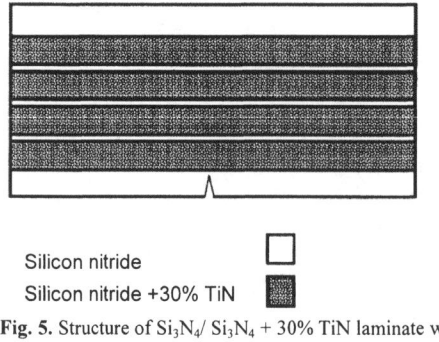

Silicon nitride   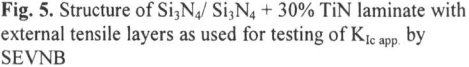
Silicon nitride +30% TiN

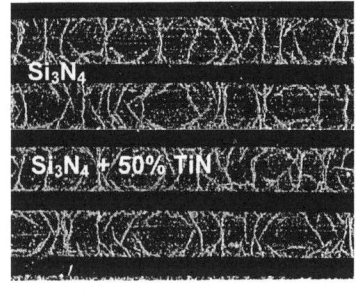

Fig. 5. Structure of $Si_3N_4/\ Si_3N_4 + 30\%$ TiN laminate with external tensile layers as used for testing of $K_{Ic\ app.}$ by SEVNB

Fig. 6. An example of sever transverse tensile cracking in a $Si_3N_4/\ Si_3N_4 + 50\%$ TiN laminate.

For the design illustrated in Fig. 5, the $K_{Ic\ app.}$ tests were performed on a soft testing machine using a number of specimens with sharp notches of various lengths inserted in the outer compressive layer.

Laminates of this design were produced with tensile layers of the composition of $Si_3N_4 + 10\%$ TiN, $Si_3N_4 + 20\%$ TiN and $Si_3N_4 + 30\%$ TiN. Initial results of the laminate system showed significant increases in the $K_{Ic\ app.}$ with increasing TiN content and with increasing notch length as the notch approaches the first tensile layer where maximum compressive stresses are placed on the notch tip. The results for the $Si_3N_4/\ Si_3N_4 + 30\%$ TiN laminate system are presented in Fig. 7 and a maximum $K_{Ic\ app.}$ of >17 MPa m$^{1/2}$ was measured.

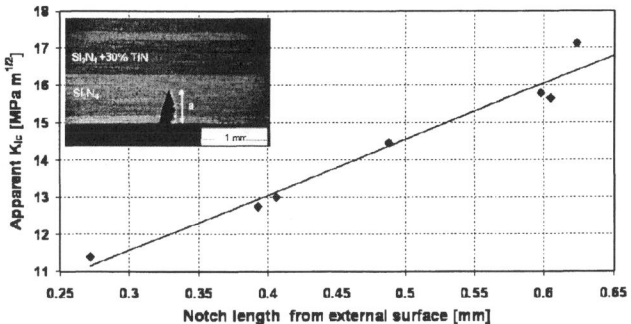

Fig. 7. Dependence of $K_{Ic\ app.}$ as function of notch length inserted in the outer compressive layer in the $Si_3N_4/\ Si_3N_4 + 30\%$ TiN laminate

When an alternative design of laminate system using the same layer ratios but having external tensile layers was used, a different failure behaviour was observed. The $K_{Ic\ app.}$ that was observed was lower, this was expected when the notches were inserted into layers under tensile stress, however now during K_{Ic} testing "pop-in" steps were observed in the load vs. time curves and these were verified by acoustic emissions data that showed "pop-in" steps and acoustic bursts to occur when the crack tip interacted with the compressive layers. The results of the $K_{Ic\ app.}$ as a function of crack length are shown in Fig. 8 where the $K_{Ic\ app.}$ was calculated using both the pop-in load and the failure load. The $K_{Ic\ app.}$ calculated using the "pop-in" load are the correct values, the crack "pop-in" phenomena is described in more detail elsewhere [8].

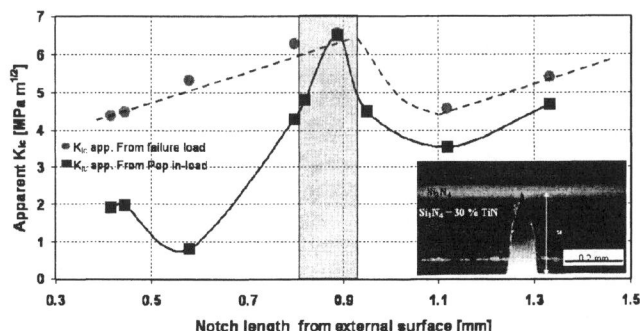

Fig. 8. Dependence of $K_{Ic\ app.}$ as function of notch length inserted in the $Si_3N_4/\ Si_3N_4 + 30\%$ TiN laminate with external tensile layers.

Careful analysis of the failure loads showed that all specimens that had a notch in the first two layers had a failure stress of 116 ± 2 MPa and the two specimens notched in the third layer had a failure stress of 71 ± 1 MPa. These results indicate a threshold strength occurs in the laminates which is controlled by the amount of compressive stress in the next adjacent compressive layer to the notch tip. Another feature observed in this design of ceramic laminate was the presence of edge-cracks in the centre axis of the compressive layers. These edge cracks would interact with a propagating notch leading to crack deflection and crack bifurcation.

However the design of these laminates was on the borderline of stable stress state and often delamination of the ceramic test specimen along an edge crack was observed during or after machining (Fig. 9). As a result it was decided to design balanced micro-laminates using the weight function analysis as described previously and controlling the layer thicknesses to avoid transverse cracking and edge cracking. In addition the goal was to have "graceful" and controlled failure.

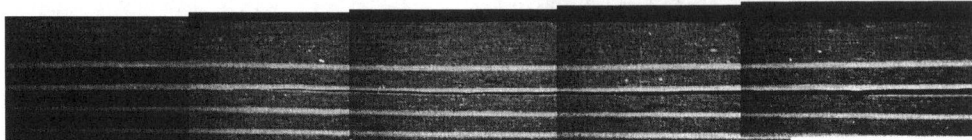

Fig. 9. Delamination in a Si_3N_4/ Si_3N_4 + 30% TiN laminate with external tensile layers due to edge cracks in the Si_3N_4 layers.

It was therefore extremely important to have accurate materials information on the Si_3N_4 and Si_3N_4+TiN composites, the Young's modulus, K_{Ic} and CTE were determined for a number of suitable compositions. The results are presented in Table 1 and described in more detail elsewhere [16]. In addition high temperature K_{Ic} tests up to 1400°C were carried out to accurately determine the joining temperature of the Si_3N_4 / Si_3N_4+TiN laminates, this was determined to be ~1100°C [17].

Table 1. Materials properties of Si_3N_4 and Si_3N_4 + TiN composites.

Material	E Modulus [GPa]	CTE [x $10^6 K^{-1}$]	K_{Ic} (s.d.) [MPa m$^{1/2}$]
Si_3N_4	303	2.1	4.26 (0.09)
Si_3N_4 + 10% wt. TiN	311	2.7	4.47 (0.03)
Si_3N_4 + 20% wt. TiN	317	3.1	4.61 (0.11)
Si_3N_4 + 30% wt. TiN	330	3.7	4.71 (0.05)

Micro-laminates: Two designs of micro-laminates with a rising R-curve behaviour were successfully designed and manufactured. Thin compressive Si_3N_4 layers of 50 μm thickness were used to control the residual stresses and hence the apparent fracture toughness in the micro-laminates. No "edge cracking", crack deflection or crack bifurcation was observed in the two designs of micro-laminates studied. Previously, Si_3N_4-TiN based laminates with compressive layers of 150 μm thickness had exhibited edge cracks in the compressive layers leading to bifurcation during mechanical testing. Although some of the conditions for bifurcation to occur have been defined, to date satisfactory modelling of the relationship between threshold strength and bifurcation occurring has not been completed.

The first design consists of external tensile layers (Si_3N_4 + 30% TiN) of ~300 μm thickness followed by four buried thin compressive Si_3N_4 layers of ~50 μm thickness each separated by tensile Si_3N_4 + 30% TiN layers of ~100 μm thickness. This is followed by a tensile Si_3N_4 + 30%TiN core of ~3 mm thickness before the micro-laminate stack is repeated for a balanced symmetry. The first design has been shown schematically in Fig. 10. This design produces a rising apparent fracture toughness as the crack tip extends from the surface layer into the micro-laminate. For the design a tensile external surface was used to reduce the risk of possible spontaneous cracking. The first design of micro-laminate was found to have a significant and defined "threshold strength" or load which was controlled by the thermal compressive stress in the final (4th) compressive layer (where the highest stress exists). For specimens of dimensions of 4.5 mm x 4.0 mm the average (final failure) stress was 197 MPa with a standard deviation of only 3.8 MPa with an inserted sharp notch of nearly 1 mm length. The maximum possible apparent fracture toughness obtainable by this design was ~10.5 MPa m$^{1/2}$. The K_{Ic} values at the "threshold/failure" load were calculated with a notch length extending to the end of the last compressive Si_3N_4 layer. Five such K_{Ic} values are shown on Fig 11. Hence such a design of micro-laminates with a rising R-curve behaviour produces a threshold safeguard for real ceramic applications. The apparent fracture toughness was also calculated using the load at the crack "pop-ins" and the original crack length. The values of apparent fracture toughness against notch length are plotted in Fig. 11 and are overlaid against the theoretical calculations. In general the apparent fracture toughness fits the WFA model very well. This was illustrated even better when $K_{Ic\ app}$ testing was performed on a stiff testing machine. With the notch inserted in the outer tensile layer it was possible to slowly increase the load and measure the crack propagation to calculate $K_{Ic\ app}$ as a function of crack length, on a single specimen, as the crack tip extended through the micro-laminate structure [18].

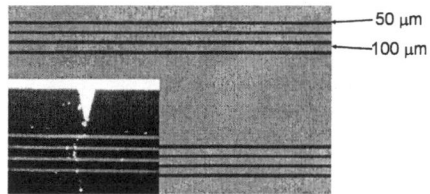

Fig. 10. Micro-laminate design 1 with external tensile layers: insert is SEVNB specimen (compressive Si_3N_4 layers=50 μm thick).

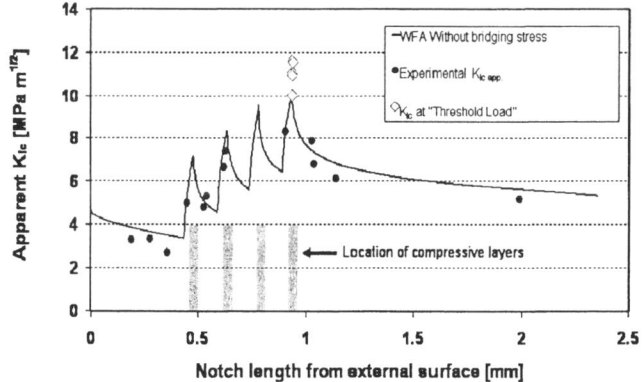

Fig. 11. Apparent K_{Ic} as a function of crack length for the first micro-laminate design (with external tensile layers) superimposed onto the WFA model

The fracture toughness of the micro-laminates could be increased further by the use of a compressive external layer as shown in design two. The second design of micro-laminate is shown in Fig. 12. Apparent fracture toughness and predicted R-curve behaviour of the design are shown in Fig. 13. In the second design the external tensile layers are replaced by external compressive layers of Si_3N_4 of ~150 μm thickness, this is followed by a similar stack with 100 μm thick tensile layers

(Si_3N_4 + 30% TiN) and the ~50 μm thick compressive layers and then the tensile core of ~3 mm. This leads to a maximum possible apparent fracture toughness of 18 MPa m$^{1/2}$.

In this design the "threshold strength" effect with a defined final failure load at the final compressive layer does not appear. The $K_{Ic\ app.}$ results in Fig. 13 are from two individual test specimens, the first with a notch in the outer compressive layer and the second with a notch in the first compressive "micro" layer. It was not possible to obtain a graceful failure in this design of composite, yet. To obtain graceful failure in the micro-laminates we need to ensure that the apparent fracture toughness, at successive compressive layers, increases faster than the square root of the crack length.

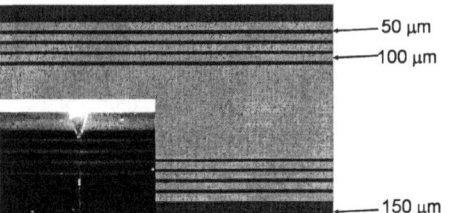

Fig. 12. Micro-laminate design 2 with external layers under compression: insert is SEVNB specimen (compressive layers Si_3N_4=dark layers).

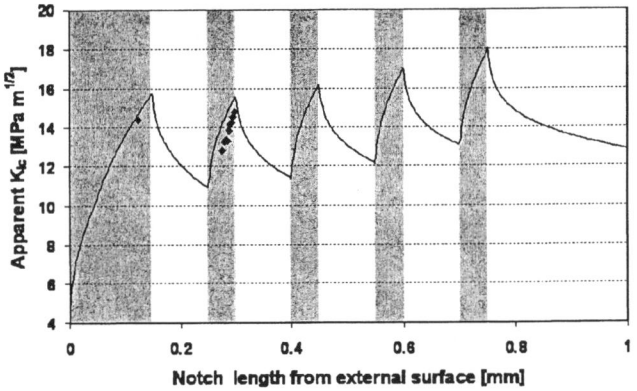

Fig. 13. Apparent K_{Ic} as a function of crack length of the second micro-laminate design with external compressive layers superimposed onto the WFA model.

Summary and Outlook

The micro-laminate designs with increasing R-curve behaviour have been shown to be an interesting toughening mechanism for ceramic materials and structures. The rising R-curve behaviour shows that a high apparent fracture toughness can be obtained even with significant crack or flaw sizes. It has been shown that micro-laminates can be produced using standard processing techniques of tape casting and die-pressing, hence it would be relatively simple to make simple structural ceramic components on an industrial scale from two compositions.

There exists the possibility for further design improvements of these types of micro-laminates, including graded structures with more than two components. Such a design from five different compositions is shown in Fig. 14. The WFA modelling predicts a high apparent K_{Ic} combined with graceful failure behaviour. The predicted threshold strength as a function of inserted notch length is shown in Fig 15, as can be seen high threshold strength is retained even with large cracks inserted into the micro-laminate system.

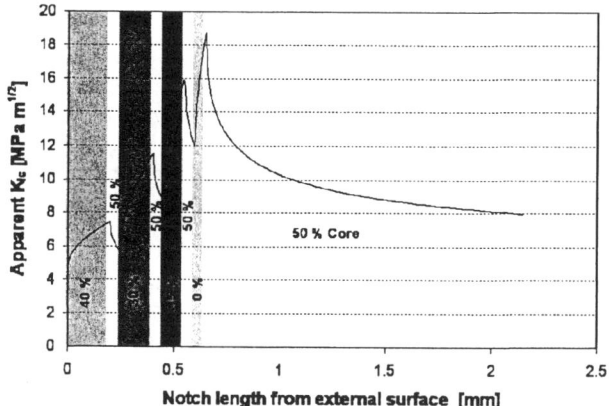

Fig. 14. Suggested design and $K_{Ic\ app.}$ behaviour of micro-laminate design with layers of five different compositions.

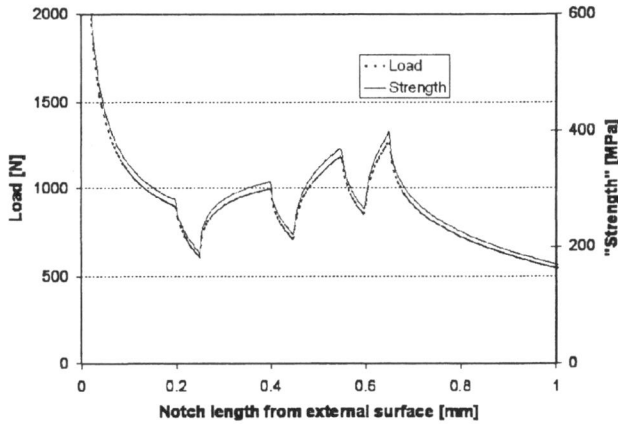

Fig. 15. Predicted threshold strength of micro-laminate design shown in Fig. 14.

References

[1]   W. J. Clegg, K. Kendall, N. M. Alford, T. W. Button, and J. D. Birchall: Nature, 347 (1990), p. 455.

[2]   I. A. Gee, R. S. Dobedoe, R. Vann, M. H. Lewis, G. Blugan, and J. Kuebler: Advances In Applied Ceramics, 104 (2005), p. 103.

[3]   G. Blugan, R. Dobedoe, M. Lugovy, S. Koebel, and J. Kuebler: Composites Part B: Engineering, In Press, Corrected Proof.

[4]   R. S. Dobedoe, I. Gee, M. Lewis, R. Vann, G. Blugan, and J. Kuebler. A method of controlling fracture in brittle materials; U. P., Ed. UK, 2004.

[5]   T. Chartier, D. Merle, and J. L. Besson: Journal of the European Ceramic Society, 15 (1995), p. 101.

[6]   M. Lugovy, V. Slyunyayev, V. Subbotin, N. Orlovskaya, and G. Gogotsi: Composites Science and Technology, 64 (2004), p. 1947.

[7]   R. J. Moon, M. Hoffman, J. Hilden, K. Bowman, K. Trumble, and J. Rodel: Journal of the American Ceramic Society, 85 (2002), p. 1505.

[8]   M. Lugovy, V. Slyunyayev, N. Orlovskaya, G. Blugan, J. Kuebler, and M. Lewis: Acta Materialia, 53 (2005), p. 289.

[9]   T. Fett, and D. Munz: Journal Of Materials Science Letters, 9 (1990), p. 1403.

[10]   N. Orlovskaya, M. Lugovy, J. Kuebler, S. Yarmolenko, and J. Sankar: Chapter 7. Design of tough ceramic laminates by residual stresses control, in: Ceramic Matrix Composites: Microstructure/Property Relationship, (2006), p. 178.

[11]   G. Blugan, N. Orlovskaya, M. Lewis, and J. Kuebler: Key Engineering Materials, High-Performance Ceramics III, 280-283 (2005), p. 1863.

[12]   EN 843-1: Advanced technical ceramics. Monolithic ceramics. Mechanical properties at room temperature. Determination of flexural strength, (1995).

[13]   J. Kübler: Fracture Resistance Testing of Monolithic and Composite Brittle Material, ASTM STP 1409, (2002), p. 93.

[14]   H. Jelitto, F. Felten, C. Hausler, H. Kessler, H. Balke, and G. A. Schneider: Journal of the European Ceramic Society, 25 (2005), p. 2817.

[15]   S. Ho, and Z. Suo: Journal of Applied Mechanics-Transactions of the Asme, 60 (1993), p. 890.

[16]   G. Blugan, M. Hadad, J. Janczak-Rusch, J. Kuebler, and T. Graule: Journal of the American Ceramic Society, 88 (2005), p. 926.

[17]   G. Blugan, R. Dobedoe, I. Gee, N. Orlovskaya, and J. Kuebler: Key Engineering Materials, 290 (2005), p. 175.

[18]   J. Kuebler, G. Blugan, H. Jelitto, G. A. Schneider, and R. S. Dobedoe: Key Engineering Materials, CICC-4 in press.

Key Engineering Materials Vol. 333 (2007) pp 127-136
online at http://www.scientific.net
© *2007 Trans Tech Publications, Switzerland*

Quantitative Assessment of Crack-Tip Stress Field in Semiconductor GaN Using Electrostimulated Piezo-Spectroscopy

Giuseppe Pezzotti

Ceramic Physics Laboratory & Research Institute for Nanoscience, Kyoto Institute of Technology, Matsugasaki, Sakyo-ku, Kyoto 606-8585, Japan

pezzotti@chem.kit.ac.jp

Keywords: Piezo-spectroscopy, cathodoluminescence, high spatial resolution, stress

Abstract. The piezo-spectroscopic (PS) effect, which may be defined as the shift in wavelength of a spectroscopic transition in a solid in response to an applied strain or stress, may occur both in crystalline and in amorphous structures, regardless of the particular spectroscopic transition involved (*e.g.*, luminescence or Raman spectrum), and independent of the specific mechanism of luminescence emission (*i.e.*, including spectra generated from substitutional impurities, optically active point defects, etc.). The PS effect can be monitored on electro-stimulated spectra when the scale on which the needed characterization lie is of a nanometer length. This effect, being a physical property of the studied material, should be calibrated case by case. The high scanning speed (and computer control) of the electron beam, which can be easily obtained with scan coils, is unsurpassed. Since the most recently developed optoelectronic devices have active areas of sub-micron dimensions and many of them less than 100 nm, no obvious choice is left but urgently developing an electro-stimulated probe for nano-scale residual stress assessments. In this paper, we show the feasibility of nano-scale stress assessments in the neighborhood of the tip of a crack propagating in GaN, selected as a paradigm semiconductor material.

Introduction

Semiconductor devices may internally develop significant residual stresses when cooled from the temperature at which they are processed as a result of different cooling rates between core and surface parts, and of thermal expansion mismatches when joints between different materials are involved [1-3]. Extensive examples in electronic technologies can be found in diodes and dielectric films. Among the possible failure modes in devices, delamination at the interface with the substrate or between different layers will likely occur when in the presence of a stress state of residual compression. On the other hand, parts of the device subjected to a residual tensile stress field may fail by the formation of microcracks. Understanding of residual stress fields in ceramic materials may be of significant interest to technology because manufacturers are interested in how to improve the reliability of the device. From this perspective, the development of a quantitative and, possibly, non-destructive method of stress analysis applicable to semiconductor devices is highly desirable. Raman and fluorescence PS has been used to assess residual stress fields in ceramics [4-8] in the past two decades and it is now regarded as a useful tool for material development and quality assurance. However, in many electronic devices, the technique lacks of spatial resolution and it is influenced by a number of intrinsic and extrinsic factors and the results of the spectroscopic assessment are usually regarded as only qualitative.

A straightforward way to solve both *in-depth* and *in-plane* spatial resolution issues in stress assessments is to adopt an electron probe, rather than a laser probe, as the excitation source for luminescence. This practice can be made in conventional scanning electron microscopy (SEM) devices coupled with a spectrometer and it is called cathodoluminescence (CL) piezo-spectroscopy. CL has traditionally been a very important method for studying the local chemistry and the physical structure of materials into the SEM. CL spectra have been useful to provide a variety of information on defects and crystallographic discontinuities (*e.g.*, grain-boundaries, precipitates and inclusions)

[9]; also variations in impurity concentration [10], states in the energy gap and excess carrier diffusion lengths have been evaluated by monitoring the voltage dependence of the intensity of selected CL bands [11]. However, the CL spectrum of semiconductor compounds also contains, in addition to features related to the chemistry and to the crystallographic structure of the material, information about the stress state through the PS effect. This information can be made available by first building on a conventional SEM images an additional (hyperspectral) image of wavelength shift, and then by translating this image into a stress map by means of suitable PS calibrations, which convert wavelength shift into a stress magnitude. Therefore, in addition to already established CL techniques (*i.e.*, based on luminescence intensity assessments), dealing with crystal defects and chemistry, we propose to employ CL PS in local stress assessments and to push this practice towards a nanometer scale, namely to a spatial resolution which is not achievable by conventional laser spectroscopies [12].

This paper will first highlight the use of a calibration method for the quantification of the *in-plane* probe response function of selected CL bands. A deconvolution procedure is proposed for extracting "true" residual stress distributions and the procedure applied to quantitatively assess the residual stress field stored at the tip of a crack propagating in GaN. To visualize the superior spatial resolution of the CL probe, a two-dimensional stress map is shown and the effect of increasing electron beam energies on the spatial resolution of the stress measurement examined and rationalized.

Experimental procedures

A GaN epilayer was used for PS calibration. The epilayer was 4.0 μm in thickness, grown on (0001) sapphire substrates with a thickness of 0.5 mm by low-pressure metalorganic chemical vapor deposition (LP-MOCVD). Trimethylgallium (TMGa) was used as the Ga source and ammonia (NH_3) as the N source. CL experiments were performed on a CL detector unit (MP-32FE, Horiba) attached to a FEG-SEM (JSM-6500F, JEOL) at room temperature. The emitted light was collected by a grating monochromator (Triax-320, Jobin-Yvon, Horiba) equipped with a liquid-nitrogen cooled charge-coupled-device detector with a spectral resolution of 0.03 nm. A ball-on-ring bending jig [13] was used to add a controlled (biaxial) stress field on the GaN/sapphire sample. The stress distribution, developed along an arbitrary diameter selected on the tensile surface of the plate sample, was used for band-shift calibration, as schematically shown in Fig. 1. The biaxial bending jig was put into the FEG-SEM chamber to measure *in situ* CL spectra on the stressed film. Irradiation by electrons was performed at 5.0 kV (1.8 nA), thus generating a rather shallow probe (250 nm electron scattering length in depth as estimated according to the Kanaya-Okayama equation [14]) as compared to the sample thickness. According to the shallow morphology of the electron probe, we have neglected in this study the *in-depth* (linear) stress distribution developed in the sample loaded in ball-on-ring configuration. Further details about a fully three-dimensional stress analysis of thin plate samples loaded in the present jig by photo-stimulated piezo-spectroscopy are given elsewhere. [13]

Although a highly focused beam was used in the SEM as the source for electron stimulation, diffusion of electrons occurs in the material. Therefore, the electron probe can be greatly enlarged with increasing electron accelerating voltage Fig. 2(a). An evaluation of the *in-plane* probe geometry or probe response function can be pursued by selecting a straight and atomically sharp interface between two different materials and by scanning with the electron beam across the interface. This procedure, schematically shown in Fig. 2(b), can be eventually repeated at different electron accelerating voltages, V, in order to retrieve from CL intensity variation the dependence of the probe morphology on electron beam impingement conditions. The intensity variation of the selected CL band can be collected as a function of distance, x_0, between the interface and the location of the center of the electron beam. The differential contribution of each portion of volume of the luminescent probe (*i.e.*, the intensity variation as a function of distance between the interface and the irradiated center) can be then obtained from plotting the first derivative of the CL intensity function as a function of the abscissa, x_0.

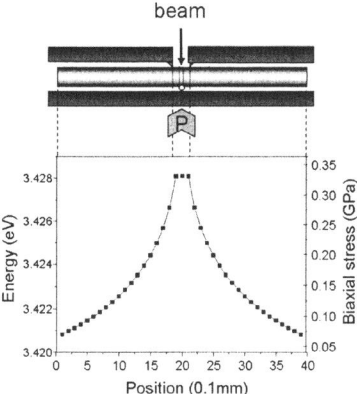

Fig. 1: Schematic draft of ball-on-ring biaxial stress calibration and related stress distribution.

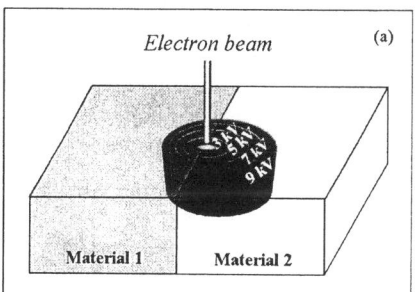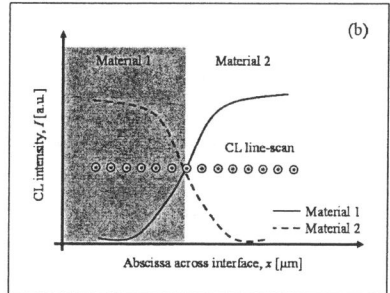

Fig. 2: (a) Method of determination of electron probe response function across a sharp interface; (b) CL intensity change across interface between different materials.

All the fitting procedures followed in this paper were automatically performed by means of commercially available software (SPECTRA CALC, Galactic Industries, Salem, NH). The equations used in this work were solved with the aid of a commercially available computational software package (Wolfram S. MATHEMATICA5 - Wolfram Research, Inc., Champaign, IL, 2000).

Theoretical and computational backgrounds

Quantitative analysis of *in-plane* electron probe/sample interaction. Assuming that each point in space (x,y,z) gives rise to a (local) intensity contribution to the electron stimulated spectrum of a given form, the observed CL spectrum will depend on the energy distribution of the incident electrons along their diffusional path within the material. When the incident beam with accelerating voltage, V, is focused at an arbitrary point (x_0,y_0,z_0), the energy distribution can be represented by the probe response function, $G(x,y,z,x_0,y_0,z_0,V)$ and the observable CL spectrum can be expressed according to the following triple integral:

$$I_{obs}(\lambda, x_0, y_0, z_0) = \int_{-\infty}^{\infty}\int_{-\infty}^{\infty}\int_0^{\infty} I(A, \lambda, \lambda_p, x, y, z) G(x, y, z, x_0, y_0, z_0, V) dx dy dz , \qquad (1)$$

where, $I_{obs}(\lambda, x_0, y_0, z_0)$ is the band shape and $I(A, \lambda, \lambda_p, x, y, z)$ is the local spectrum (A_p and λ_p are the band width and CL spectral wavelength at maximum, varying with position) emitted from any given point within the region irradiated by electrons. Many CL stress assessments can be limited to an *in-plane* configuration, given the very shallow nature of the electron probe; therefore, it can be assumed that the abscissa variables, z and z_0, along the *in-depth* direction can be ignored. In the case of a CL line scanning across an interface, as the probe is translated, for example, along the x-axis (cf. Fig. 2), the collected CL intensity, $I_{obs}(\lambda, x_0, y_0)$, can be expressed by a simplification of Eq. 1, as follows:

$$I_{obs}(\lambda, x_0, y_0) \propto \int_{-\infty}^{\infty}\int_{-\infty}^{\infty} G(x, y, x_0, y_0, V) dx dy . \qquad (2)$$

The intensity variation of the selected CL band can be collected as a function of distance, x_0, between the interface and the location of the center of the electron beam. Note that, when scanning across the interface from different directions, there might be some asymmetry in the experimentally retrieved CL intensity curves, due to property (*i.e.*, electron-diffusion-length) difference between the two adjacent materials. However, the probe-related parameters can be generally derived from the best fitting of the normalized experimental intensities at locations within the investigated material. The differential contribution of each portion of volume of the luminescent probe (*i.e.*, the intensity variation as a function of distance between the interface and the irradiated center) can be then obtained from plotting the first derivative of the CL intensity, I_{obs}, as a function of the abscissa, x_0.

Experimental plots of *in-plane* probe response functions collected in a homogeneous crystal plane at different accelerating voltages can be fitted to a degree of precision to Gaussian curves, according to the following equation:

$$G(x, y, x_0, y_0, V) = \exp{-\frac{(x-x_0)^2}{W_1 V^n}} \times \exp{-\frac{(y-y_0)^2}{W_2 V^n}} , \qquad (3)$$

where the parameters W_1, W_2, and n are intrinsic parameters of the investigated material. For simplicity's sake, we shall assume a complete material isotropy with respect to electron propagation ($W_1 = W_2$) and consider only an unidimensional probe, whose probe response function is $G(x, x_0, V)$. According to the experimental knowledge of the function, $G(x, x_0, V)$, the size of the luminescent probe (or CL activation length), can be also calculated as a function of the selected V value. At different accelerating voltages, the electrons are expected to show different diffusion characteristics and, thus, also different probe morphologies are expected. The luminescent probe size, R_{CL}, can be defined according to the following equation:

$$\frac{\int_{x_0-R_{CL}}^{x_0+R_{CL}} G(x, x_0, V) dx}{\int_{-\infty}^{+\infty} G(x, x_0, V) dx} = 0.9 . \qquad (4)$$

Equation 4 can be numerically solved for obtaining the lateral probe size, R_{CL}. Note that in Eq. 4 the choice of the numerical value 0.9 is arbitrary. However, it is reasonable to assume that, independent of the nature of the spectrum observed, any intensity contribution <10%, arising from remote portions of the probe, will give a negligible contribution to the overall band spectral shift.

Quantitative analysis of biaxial stress fields. The stress state in a homogeneous crystal can be related to the wavelength shift, $\Delta\lambda$, of a given CL band with respect to its unstressed state, according to the following tensorial equation (also referred to as the Grabner's formalism): [15]

$$\Delta\lambda = \Pi_{ij}\sigma_{ij} = \Pi^*_{ii}\sigma^*_{jj} \qquad (i,j=1,2,3), \qquad (5)$$

where the repetitive index notation applies to the stress, σ_{ij}, and piezo-spectroscopic, Π_{ij}, second-rank tensors. The asterisk in Eq. 5 denotes the choice for both stress and PS tensors of a set of Cartesian axes coincident with the principal stress directions and crystallographic axes of the material. According to Eq. 5, the PS shift is an invariant with respect to the selected system of coordinates and an appropriate choice of the Cartesian axes may directly relate the observed shift to the *principal* components of the stress tensor (as reflected by the same repetitive index used in the right hand of Eq. 5). In crystals with planar symmetry (or in polycrystals with randomly oriented grains), the piezo-spectroscopic coefficient can be considered to be constant to a degree of precision along any crystallographic direction in the plane of symmetry. Accordingly, for a general bi-axial stress state, $\sigma_{ij} \equiv \sigma^*_{ii}$, applied in the basal plane of the sample, Eq. 5 simply reduces to a scalar proportionality equation:

$$\Delta\lambda = \begin{vmatrix} \Pi_{11} & \Pi_{12} \\ \Pi_{21} & \Pi_{22} \end{vmatrix} \begin{vmatrix} \sigma_{11} & \sigma_{12} \\ \sigma_{21} & \sigma_{22} \end{vmatrix} \equiv \begin{vmatrix} \Pi^*_{11} & 0 \\ 0 & \Pi^*_{22} \end{vmatrix} \begin{vmatrix} \sigma^*_{11} & 0 \\ 0 & \sigma^*_{22} \end{vmatrix} = \Pi < \sigma^*_{ii} >, \qquad (6)$$

where $\Pi = (\Pi_{11} + \Pi_{22})/2$ and $< \sigma^*_{ii} > = \sigma^*_{11} + \sigma^*_{22}$ is the trace of the principal stress tensor. In other words, a measurement of spectral shift gives direct access to the trace of the planar stress tensor. It follows that the traces of an unknown residual stress tensor locally stored in a plane of crystallographic symmetry can be measured according to Eq. 6, provided that the PS coefficient, Π, can be *a priori* evaluated by introducing into the material a known bi-axial stress state.

Results and Discussion

Piezo-spectroscopic *in-plane* analysis of CL bands. Figure 3 shows the typical room-temperature near-band-edge CL band of the GaN film. As seen, the band is asymmetric and experiences a pronounced shoulder at lower energy [16]. Despite such an asymmetrical morphology, the overall band clearly shifts towards lower band energies from the applied-stress free position (3.430 eV) when a biaxial tensile stress field is applied. A plot was collected along a diameter on the tensile face of the bending plate (as shown in Fig. 1) with and without an applied stress (load $P=55$ N). In order to determine the PS coefficient, CL spectra were collected at 40 locations across the selected diameter (4.0 mm) of the stressed GaN sample (corresponding to the hole/window left open above the sample in the loading jig for electron scanning; cf. Fig. 1). The collected bands were fitted to two Voigt sub-bands (labelled FX and D^oX; [16] in Fig.3(a)) and the energy spectral position of the main band recorded as a function of the radial distance from the center of the sample, r. The relationship between (experimental) spectral shift and (theoretical) biaxial stress plot is shown in Fig. 3(b). The stress distribution on the surface of the thin plate sample loaded in the ball-on-ring jig has been described elsewhere.[13] The slope of the plot, $\Delta\lambda(< \sigma^*_{ii} >)$, which represents the PS coefficient of the near-band-gap luminescence of basal plane GaN, led to an estimate of $2\Pi = -25.8 \pm 0.2$ meV/GPa, according to a linear least square fitting procedure.

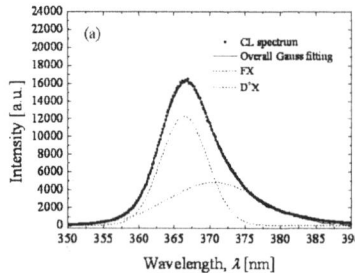 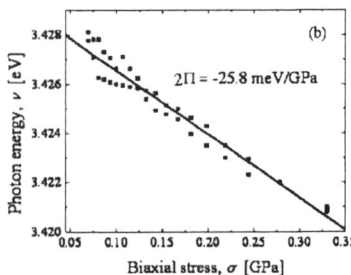

Fig. 3: (a) Room temperature GaN CL spectrum with related deconvolution into a main band and a shoulder (*FX* and *D⁰X*, respectively); (b) Plot of peak shift vs. biaxial stress, as collected for *FX* band.

Table 1. Salient Probe response parameters for GaN

AV [kV]	W_1	n	Probe size [μm]
3	3×10^{-5}	3.5	0.041
5	3×10^{-5}	3.5	0.101
7	3×10^{-5}	3.5	0.177
9	3×10^{-5}	3.5	0.272

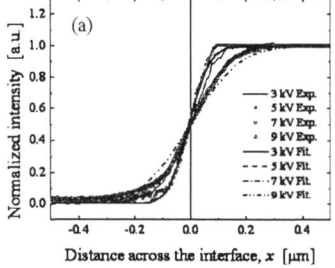 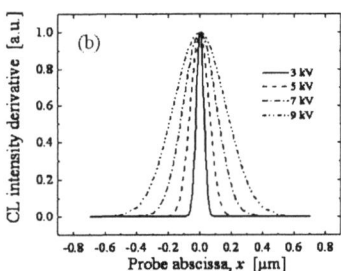

Fig. 4: (a) Intensity scan through a sapphire-GaN interface collected at different accelerating voltages. (b) Probe response functions obtained as the derivative of the experimental fitting curves in (a).

Several line scans were carried out at different accelerating voltages along the same line across a sharp interface between GaN and sapphire, monitoring the intensity of the CL bands in the GaN region. These plots are shown in Fig. 4(a), while their derivatives, representing the probe response functions, are given in Fig. 4(b). Upon decreasing the accelerating voltage, and thus the CL probe size, the CL intensity profile became steeper in proximity of the interface. Plots in Fig. 4(b) were rationalized according to Eq. 3, while the CL probe size, R_{CL}, was obtained according to Eq. 4. The salient probe response parameters for GaN are listed in Table 1.

In-plane stress analysis of crack-tip stress field in GaN. The local residual stress induced ahead of an indentation crack can be described by fracture mechanics equations as a function of polar coordinates, (x, θ), with origin at the crack tip:

$$\begin{pmatrix} \sigma_{11} \\ \sigma_{22} \\ \tau_{12} \end{pmatrix} = \frac{K_I}{\sqrt{2\pi x}} \begin{cases} \cos\theta \left[1 - \sin\frac{\theta}{2}\sin\frac{3\theta}{2}\right] \\ \cos\theta \left[1 + \sin\frac{\theta}{2}\sin\frac{3\theta}{2}\right] \\ \sin\frac{\theta}{2}\cos\frac{\theta}{2}\cos\frac{3\theta}{2} \end{cases}, \tag{7}$$

where K_I is the crack-tip stress intensity factor. Along a line straight ahead of the crack path $(\theta = 0)$, Eq. 7 reduces to:

$$\begin{pmatrix} \sigma_{11} \\ \sigma_{22} \end{pmatrix} \equiv \begin{pmatrix} \sigma_{11}^* \\ \sigma_{22}^* \end{pmatrix} = \frac{K_I}{\sqrt{2\pi x}}. \tag{8}$$

The determination of the spectral shift distribution, $\Delta\lambda\,(x)$, ahead of the crack tip with respect to an arbitrary position far away from the crack may allow the determination of the crack-tip stress field (or stress intensity factor) provided that the PS coefficient, Π, of the material is known. Eventually, the K_I value operative at the crack tip can be also obtained by a preliminary measurement of crack opening displacement (COD), $U(r)$, from high-resolution SEM images.[17] For a half-penny shaped crack configuration, K_I and $U(r)$ can be related through the following equation:

$$U(r) = \frac{4K_I}{\pi E'} Y\left(\frac{r}{a}\right), \tag{9}$$

where the weight function $Y\left(\dfrac{r}{a}\right)$ for an indentation crack is given by:[17]

$$Y\left(\frac{r}{a}\right) = \sqrt{b}\left[A\left(\frac{r}{a}\right)^{1/2} + B\left(\frac{r}{a}\right)^{3/2} + C\left(\frac{r}{a}\right)^{5/2}\right], \tag{10}$$

$$A = \left(\frac{\pi a}{2b}\right)^{1/2}, \tag{11}$$

$$B \cong 0.011 + 1.8197\ln\left(\frac{a}{b}\right), \tag{12}$$

$$C \cong -0.6513 + 2.121\ln\left(\frac{a}{b}\right), \tag{13}$$

where r is an abscissa taken along the crack path (with a direction opposite to crack propagation, i.e., $r = -x$), a is the distance between the crack tip and the center of the indentation print, b is half the diagonal of the indentation print, and E' is the plane stress Young's modulus of the material. Eq. 8 and Eq. 10-12 can be rearranged to relate the local stress ahead of the crack tip to the observed wavelength shift, as follows:

$$\begin{pmatrix} \sigma_{11}^* \\ \sigma_{22}^* \end{pmatrix} \equiv \frac{<\sigma_{ii}^*>}{2} = \frac{\pi E' U(r)}{4\sqrt{2\pi x}\cdot\sqrt{b}}\cdot\left\{A\left(\frac{r}{a}\right)^{\frac{1}{2}} + B\left(\frac{r}{a}\right)^{\frac{3}{2}} + C\left(\frac{r}{a}\right)^{\frac{3}{2}}\right\}^{-1} = \frac{\Delta\lambda}{2\Pi}. \tag{14}$$

Fig. 5 shows an explicative draft of the indentation stress field and related parameters.

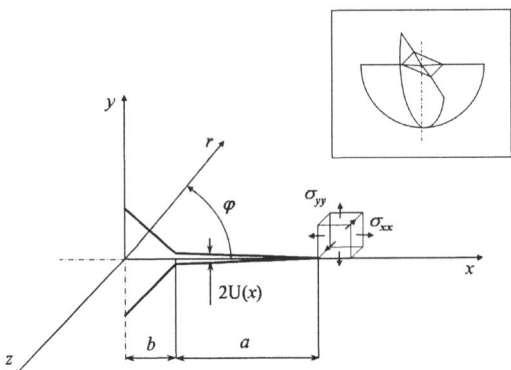

Fig. 5: Schematic of a Vickers indentation crack.

Note that, to hold validity, Eq. 14 needs to be applied to data collected with a highly focused electron probe, because of the highly steep stress gradient involved with the zone in the neighborhood of the crack tip.

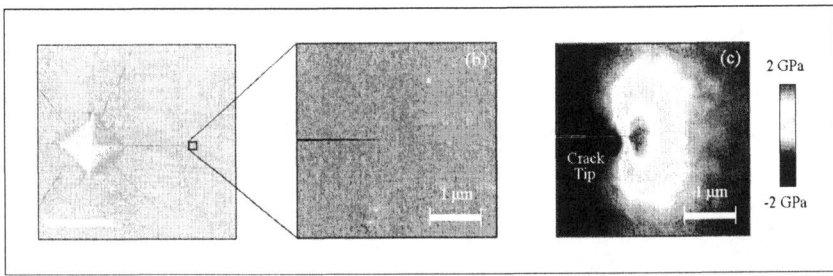

Fig. 6: (a) SEM micrograph of an indentation print impinged on the (0001) crystallographic plane of GaN. (b) and (c) represent an enlarged SEM micrograph of the tip neighborhood of an indentation crack and the related nm-scale stress map, respectively.

In-plane **stress analysis at the crack tip of GaN crystal.** Fig. 6(a) shows a SEM micrograph of an indentation print impinged on the (0001) crystallographic plane of GaN. The insets (b) and (c) represent an enlarged SEM micrograph of the tip neighborhood of an indentation crack and the related nm-scale stress (trace of principal stress tensor) map, respectively. The map, which was obtained from spectral band shift according to the Π value given a previous section, was collected with an *in-plane* resolution of about 20 nm and represents a vivid example of the significant improvement in spatial resolution achievable by using electro-stimulated luminescence. Linear plots collected at four different electron accelerating voltages across the crack-tip location along the line $r = 0$ are shown in Fig. 7(a). Both plots show the highest peak shift nearby the crack tip; however, the higher V the lower the observed maximum spectral shift magnitude.

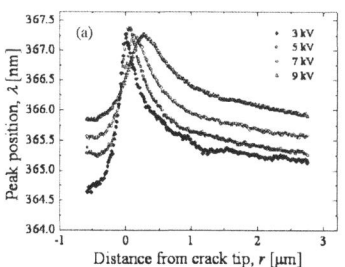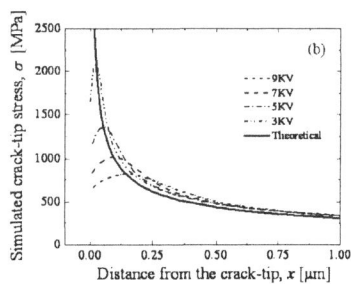

Fig. 7: (a) Peak position of the *FX* peak as collected along the crack propagation axis starting from a given location behind the crack-tip at different accelerating voltages. (b) Effect of the probe on the true stress trend at different accelerating voltages.

The inconsistency between the piezo-spectroscopic plots obtained at different voltages arises from the high gradient of crack-tip stress, $< \sigma_{ii}^* >$, compared with probe size. The wavelength shift is actually observed as a consequence of a convolution in the finite volume of the electron probe, thus giving rise to a convoluted stress value, $< \overline{\sigma_{ii}^*} >$. This latter stress value can be calculated at any location of the electron probe, x_0, by solving the following integral:

$$< \overline{\sigma_{ii}^*} > (x_0) = \frac{\int_{-\infty}^{+\infty} G(x; x_0, V) \cdot < \sigma_{ii}^* > (x) \, dx}{\int_{-\infty}^{+\infty} G(x, x_0, V) \, dx}. \tag{15}$$

Provided that the probe response function $G(x, x_0, V)$ is known, Eq. 15 can be numerically solved to assess the true stress distribution ahead of the crack tip. Note that the character of the $< \sigma_{ii}^* > (x)$ is known from Eq. 8. The results of this calculation are shown in Fig. 7(b). As can be seen, data points collected at different voltages lead to the same crack-tip stress plot, which leads to a crack-tip stress intensity, K_I=0.82 MPa x m$^{1/2}$. Therefore, we have shown that a procedure for deconvoluting the stress magnitude, developed into the electron probe, enables one to retrieve the actual crack-tip K_I value. This value can be compared with the K_I value directly obtained from COD measurements.

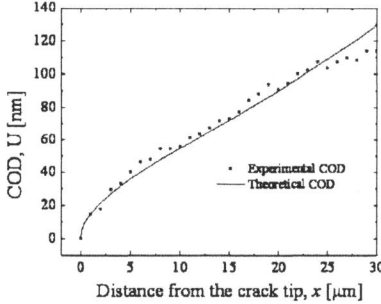

Fig. 8: COD profile retrieved from SEM observation and its best fit according to Eqs. 7-14.

Fig. 8 shows the COD profile retrieved from SEM observation and its best fit according to Eq. 7-14. The fitting curve satisfies a value K_I=0.78 MPa x m$^{1/2}$, thus demonstrating that the stress distribution obtained from CL measurement reflects to a degree of precision (within an error of 5 %) the actual crack-tip stress field. It should be noted that, in the present assessment, only a simplified calculation was presented, which was based on a unidirectional spatial deconvolution procedure (*i.e.*, only considering the polar abscissa, *x*). Further improvements in reducing the involved error can be achieved by including in Eq. 15 the effect of the polar angle, θ.

Summary

Electro-stimulated piezo-spectroscopic stress analyses using spatially resolved probe technique has been shown for GaN semiconductor material. *In-plane* probe response functions were deduced from scanning methods across a straight and sharp interface. Taking into account the pertinent probe response function, highly graded stress fields could be resolved as, for example, the planar stress field developed ahead of a crack tip in a semiconductor crystal. Despite some deviation, due to the simplified computational approach selected in this paper for brevity's sake, the deconvolution procedures turned out to be accurate to a degree of precision and confirmed the validity of the probe response function approach for quantitative stress assessment. In conclusion, we have clearly showed here not only the possibility of measuring stresses on the nanometer scale, but also that future efforts should be directed to extend the piezo-spectroscopic technique to practical applications.

References

[1] A. N. Itakura, T. Narushima, M. Kitajima, K. Teraishi, A. Yamada and A. Miyamoto: Appl. Surf. Sci.Vol. 159 (2000), p. 62

[2] A. Leto, A. Loreto, T. Hosokawa, Y. Yabuuchi, T. Valente, and G. Pezzotti: *New Frontiers of Process Science and Engineering in Advanced Materials (PSEA '04) Part 2*, edited by High Temperature Society of Japan publisher, Kyoto, (2004)

[3] S.Fuke, K. Sano, K.Kuwahara and Y.Takano: J.Appl.Phys. Vol. 77 (1995), p.420

[4] L. Colombi Ciacchi, G. Gregori, V. Lughi, et al.: Recent Res. Develop. Appl. Spectrosc. Vol. 2 (1999), p. 243

[5] G. Pezzotti: Composites Sci. Tech. Vol. 59 (1999), p. 821

[6] E. Gheeraert, A. Deneuville and A. M. Bonnot: Diamond and Related Materials, Vol. 1 (1992), p. 525

[7] H. Mohrbacher, K. Van Acker, B. Blanpain, P. Van Houtte and J-P. Celis: J. Mater. Res. Vol. 11 (1996), p. 1776

[8] Q. Ma, and D. R. Clarke: J. Am. Ceram. Soc. Vol. 77 (1994), p. 298

[9] G. Koschek and E. Kubalek: J. Am. Ceram. Soc. Vol 68 (1985), p. 582

[10] B. G. Yacobi and D. B. Holt: J. Appl. Phys. Vol. 59 (1986), p. R1-R24

[11] A. Gustafsson, M.-E. Pistol, L. Montelius and L. Samuelson: J. Appl. Phys. Vol. 84 (1986), p.1715

[12] G. Pezzotti: Microscopy & Analysis, Vol. 33 (2003), p. 5

[13] D. K. Shetty, A. R. Rosenfield, P. Mcguire, G. K. Bansal and W. H. Duckworth: Am. Ceram. Soc. Bull. Vol. 59 (1980), p. 1193

[14] K. Kanaya and S. Okayama: J. Phys. D. Appl. Phys. Vol. 5 (1972), p. 43

[15] L.Grabner: J.Appl.Phys. Vol.49 (1978), p.582

[16] S. Bloom: J. Phys. Chem. Solids Vol. 32 (1971), p. 2027

[17] T. Fett: Eng.. Fract. Mech. Vol. 52 (1995), p.773

Key Engineering Materials Vol. 333 (2007) pp 137-146
online at http://www.scientific.net
© 2007 Trans Tech Publications, Switzerland

Developments in Processing of Ceramic Top Coats of EB-PVD Thermal Barrier Coatings

Bilge Saruhan[1, a], Uwe Schulz[1, b] and Marion Bartsch[1, c]

[1]German Aerospace Center, Institute of Materials Research, Linder Hoehe, 51147 Cologne, Germany

[a]bilge.saruhan@dlr.de [b]uwe.schulz@dlr.de, [c]marion.bartsch@dlr.de

Keywords: Thermal Barrier Coatings, Electron-Beam Physical Deposition, Thermal Conductivity.

Abstract. Partially Yttria Stabilized Zirconia (PYSZ) based Thermal Barrier Coatings (TBC) manufactured by EB-PVD process are a crucial part of a system, which protects the turbine blades situated at the high pressure sector of aero engines and stationary gas turbines under severe service conditions. These materials show a high strain tolerance relying on their unique coating morphology, which is represented by weakly bonded columns. The porosity present in ceramic top coats affects the thermal conductivity by reducing the cross sectional area through which the heat flows. The increase in thermal conductivity after heat-treatment relates to the alteration of the shape of the pores rather than the reduction of their surface-area at the cross section. The studies carried out by the authors demonstrate that the variation of the parameters during the EB-PVD processing of PYSZ based top-coats alters the columnar morphology of the coatings. Consequently, these morphological changes affect primarily the thermal conductivity and eventually the Young' Modulus which are the key physical properties of this material group.

New ceramic compositions covering zirconia coatings stabilized with alternative oxides, pyrochlores and hexaluminates are addressed. Failures occurring in ceramic top coats mark the lifetime of TBC system and therefore, it is necessary that their performance should go beyond that of the-state-of-the-art materials. This context summarizes the research and developments devoted to future generation ceramic top coats of EB-PVD TBCs.

Introduction

Hot gas temperatures in modern gas turbines exceed the melting temperature of Ni-based alloys of turbine blades. In order to withstand the extreme thermal loading and to improve the engine efficiency, turbine blades and other hot-structure components of stationary and aircraft gas turbines are protected with TBCs based on PYSZ. The required properties for TBCs are low thermal conductivity, high thermal expansion coefficient, thermal shock resistance, as well as phase and morphology stability under long-term and high temperature service conditions. With application of new generation lower thermal conductivity coatings, the engine performance will be increased by improvement of the combustion efficiency (higher turbine entry temperature), the specific fuel consumption, internal cooling, the metallic component temperature will be reduced, and the component lifetime will be extended.

The state-of-the-art thermal barrier coatings consist of a PYSZ top-coat, a metallic (Pt/Al or NiCoCrAlY) Bond Coat (BC) and between these two, a Thermally Grown Oxide (TGO) based on Al_2O_3. Top coats of TBCs are typically manufactured by Electron-Beam Physical Deposition (EB-PVD) and Air Plasma Spraying (APS). Both EB-PVD and APS top coats display highly porous microstructures, which lowers the intrinsic thermal conductivity. Thus, the differences in the thermal conductivity values between EB-PVD and APS coatings are caused by the differences in shape, orientation and distribution of their porosity [1,2].

Current state of the art EB-PVD TBCs (PYSZ) are processed with a microstructure that is strain tolerant i.e. a columnar structure for EB-PVD TBCs. Thermo mechanical tests of these TBCs revealed significant property changes due to their limited phase and microstructure stability and low sintering resistance [3,4]. Increases of modulus and thermal conductivity by more than 200% can be manifested. Moreover, these changes occur upon exposure at moderate/high temperatures (< 1200°C) for short times (100-200 hrs). Such lengths of time are too short for the requirements of the next generation engines. The turbine entry temperatures of future gas turbine engines are to be increased for further gains in efficiency. Attempts to increase the phase stability of existing TBC systems involve partial or full replacement of yttria (Y_2O_3) by other stabilizing oxides [3]. Some efforts are pursued to replace/modify the present standard top coat by new ceramics [5-9]. This requires, beside the new chemistries, the identification and optimization of new process parameters. Alternatively, the morphology and density of the columnar microstructure of PYSZ top coats can be varied by changing the deposition parameters such as chamber pressure, substrate temperature and rotation speed [10-12], leading to decreased thermal conductivity and better stability.

Failures of TBC systems are often initiated by cracks which propagate close to the bond coat/ceramic interface due to the high thermal stresses originating from the thermally grown oxide. Consequently employment of a tough material in this highly stressed area is beneficial. Requirements for new top coats are: high toughness, ultra low thermal conductivity, and high temperature phase and microstructure stability, which associate with high resistance to sintering induced volume changes and no phase transformation on service.

The particular coating morphology of PYSZ top coat of EB-PVD TBCs is represented by individually grown weakly bonded columns and by inter-columnar spacing between these columns. Growth of these columns occurs in a preferred crystallographic orientation producing inter-columnar gaps in between them. Furthermore, the open porosity is enhanced by the presence of voids between feather-like sub-columns which are inclined toward their growing axis surrounding the primary column surfaces. Finally, an additional intra-columnar closed-porosity is produced within the primary columns by rotation of the substrates during the vapor deposition process.

During high temperature loading of zirconia-based TBCs in service, the microstructural features change heavily and fast. Inter-columnar *sintering* increases the Young's modulus of the ceramic by formation of contact points between the columns and thus leading to stress increase at the ceramic top coat during service. The additional elastic energy stored in the ceramic provides further driving force for crack initiation and propagation and promotes spalling of the coating, thereby reducing the favorable strain tolerance of EB-PVD coatings.

During the last decade research activities aimed at improved ceramic top coats either by modifying the morphology of PYSZ top coats or by employing new stabilizers and chemistries. This context demonstrates the achievement of improvements by EB-PVD processing of ceramic top coats yielding reduced thermal conductivity.

Methods of Processing and Characterization

Electron-Beam Physical Vapour Deposition (EB-PVD): Manufacturing of TBCs has been carried out by following the typical steps as indicated in Fig. 1. For the EB-PVD processing of the coatings, a semi-industrial size dual-source 150 kW "von Ardenne" EB-PVD coater has been used. The evaporation source, i.e. ingots are supplied by a commercial producer and manufactured by using powder route (Transtec, USA) in dimensions of 62.5 mm diameter and 150 mm length. The ingot was bottom fed in a water cooled copper crucible for evaporation. The electron beam power on the source was equilibrated to a certain power adequate for each composition during deposition at constant focus and beam pattern conditions and by keeping the evaporation rate constant. The substrates were rotated during coating at 12 min^{-1}.

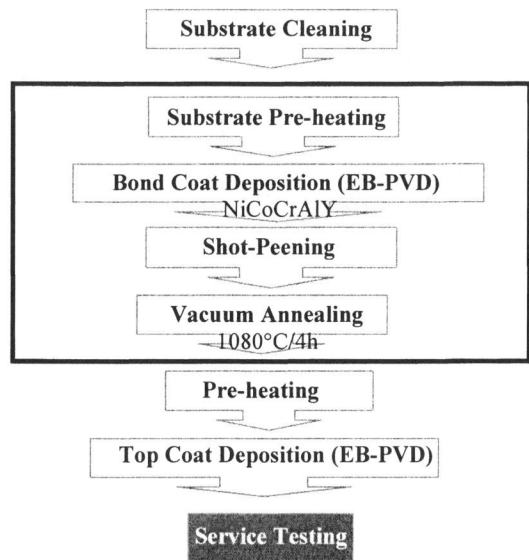

Figure 1: Typical manufacturing steps for EB-PVD PYSZ TBCs

Thermal conductivity: Thermal conductivity K [W·m⁻¹·K⁻¹] was calculated by using the experimental results obtained from the thermal diffusivity α [m²·s⁻¹], density ρ [kg·m⁻³] and heat capacity C_P [J·kg⁻¹·K⁻¹] measurements, according to the equation:

$$K = \alpha \cdot \rho \cdot C_P \tag{1}$$

The thermal diffusivity (α) of the Pt-coated free-standing specimens was measured employing a Netsch-LFA 427 instrument. The corresponding thermal conductivity was calculated through the formula $\lambda = \alpha \cdot \rho \cdot C_p$, where λ is the thermal conductivity, ρ represents in this case the bulk density of the free-standing coatings measured by the Archimedes Method and C_p is the specific heat measured by Differential Scanning Calorimeter (DSC).

The obtained model-fitted values were correlated via observation of Scanning Electron Microscope micrographs (FE-SEM, LEITZ LEO 982) of the polished sections.

Results and Discussion:

Improved morphologies of EB-PVD PYSZ TBCs: During the EB-PVD coating process, primary columns start to grow on the substrate surface following a nucleation stage. The columnar growth occurs in a preferred direction, typically perpendicular to the plane of the substrate. Inter-columnar gaps and feather-arm features are created due to the shadowing effect of the neighbouring column tips which impede the vapour flux to reach the bottom of the valley between the columns [13].

The microstructure of EB-PVD TBCs is easily changed by varying the rotational speed [8-13]. Generation of club-like columnar structure by additionally modifying the process deposition temperature results in TBCs with very low Young's modulus and tendency for sintering [3]. These effects can be summarised as follows:

(a) The sunshine-sunset shadowing effect of the neighbouring columns tips results in different morphological sequences

(b) The substrate temperature regulates the diffusion of atoms at the growing surface and

(c) The rotation speed influences the solid material growth and yields the "banana" shape pores due to shadowing after each completed rotation movement.

Figure 2: Columnar structure of the three analysed EB-PVD TBCs: (a) Intermediate, (b) Coarse, and (c) Feathery.

Table 1: EB-PVD manufacturing conditions and designation for the three coating microstructures

Morphology	Chamber Pressure (mbar)	Substrate Temperature (°C)	Rotation Speed (rpm)
Feathery	$8x10^{-3}$	850	30
Intermediate	$8x10^{-3}$	950	12
Coarse	$8x10^{-3}$	1000	3

Three coatings are analysed displaying typical columnar EB-PVD PYSZ microstructure and enclosing evident differences in the column diameter as well as significant variations in the gaps between feather-arm and intra-columnar features configuration, depending on the process parameters as given in Table 1. The microstructure "coarse" showed larger column diameters and the microstructure "feathery" more defined feather-arm features (Fig. 2). The corresponding total pore volume was measured with the Archimedes Method for each microstructure, which is 27.5% for the "feathery", 27.1% for the "intermediate", and 22.6% for the "coarse" respectively.

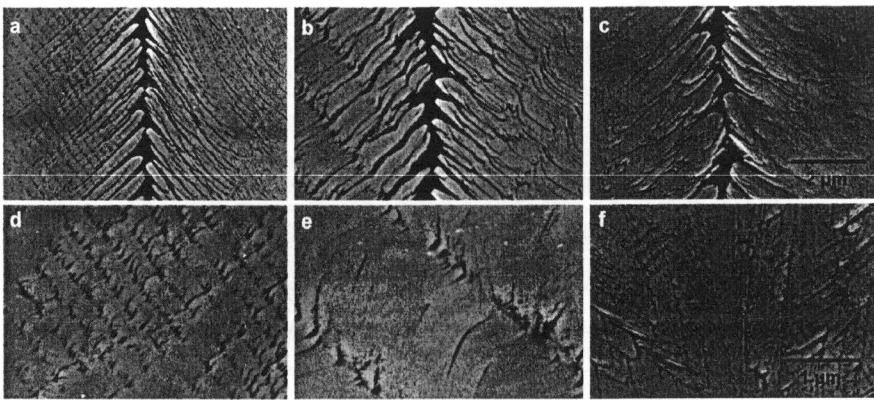

Figure 3: Configuration of the gaps between feather arms: (a) Intermediate, (b) Coarse, and (c) Feathery; and intra-columnar pores: (d) Intermediate, (e) Coarse, and (f) Feathery of the three analyzed EB-PVD TBCs in as coated conditions.

The measured thermal conductivity values of the analysed EB-PVD PYSZ TBCs with different microstructures are given in Fig. 4. According to the thermal conductivities of the different pore-group domains, it is likely that the inter-columnar gaps will contribute scarcely in reducing the thermal conductivity due to their orientation parallel to the heat flux (k_{coa}/k_{col}). Furthermore, the low values of the thermal conductivity at the gaps between feather arms domain (k_{fa}) are primarily determined by the stereometric characteristics of the deepest sector of the feather arms. These gaps interrupt the phonons flux by inserting oblique discontinuities (slices) within the columns, reaching varied deepness in each microstructure depending on the substrate temperature, rotation speed, and the shadowing sunshine-sunset effect of the neighbouring column tips. Finally, the cylindrical section of the intra-columnar pores (k_{ia}) represents the principal factor governing the thermal conductivity of the columns (k_{col}) due to its higher volume, aspect ratio, and distribution in the cross-section area (Fig. 5).

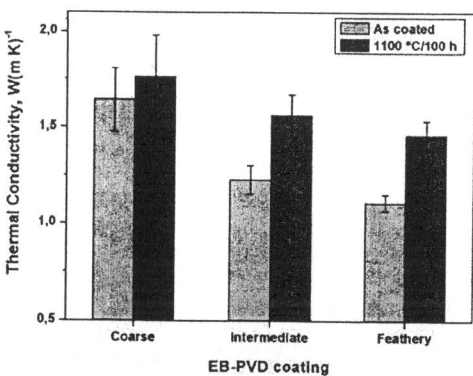

Figure 4: Measured thermal conductivity values of three analysed top-coats in as-coated and aged (1100°C/100h) conditions.

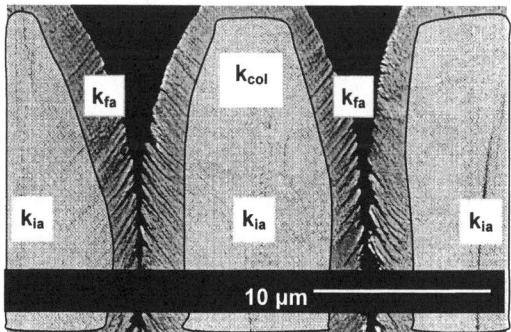

Figure 5: Different pore group domains in EB-PVD top coats influencing the thermal conductivity: k_{col}: column effect (inter-columnar), k_{ia}: intra-columnar pore effect, k_{fa}: effect of feather-arm features.

EB-PVD TBCs deposited at high chamber pressure and low substrate temperature posses a low density with a microstructure of low density TBCs characterized by larger gaps between the columns and an increasing column diameter with thickness [10, 13]. The effect of temperature on density is quite obvious: the lower the substrate temperature the lower is the ability for diffusion of the condensed particles into stable lattice positions, leading to a disordered and imperfect microstructure with low density. The correlation between density and thermal conductivity is nearly linear, offering a reduction potential of about 15% by microstructural tailoring. Larger reduction in thermal conductivity and improved cyclic lifetime have been found for TBCs deposited at high chamber pressure of 1.5×10^{-3} mbar by usage of mixtures of oxygen and an inert gas [14]. Generating inclined columns is by nature an inherent possibility for PVD line-of-sight grown films [15]. Any deviation from a perpendicular vapor incidence on a substrate leads to inclined columns. The deviation may be caused by axis tilting [16], off-axis part positioning, or substrate tilting [17, 18]. So-called "Zig-zag" or "Herringbone" structures (Fig. 6) provide reductions in thermal conductivity up to 40% [19]. A similar approach to provide a microstructural improvement by layering at a much finer scale (i.e. layers of about 1µm thickness) is to periodically switch on and off a strong BIAS voltage within a plasma atmosphere. This procedure changes the density of the alternating layers and provides up to 37-45% reduction in thermal conductivity at room temperature [20]. Similarly, layered structures with likewise reduced conductivity are obtained with multilayered structure developed by two different chemistries. In both cases however care has to be taken to stabilize the layered structure. Usually at higher temperatures an increase of thermal conductivity is observed which is attributed to the disappearance of the phase boundaries due to dissolution of the multilayer structure.

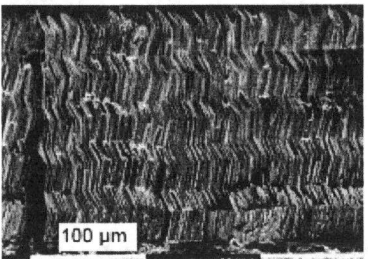

Figure 6: Zig-zag structured EB-PVD TBC

Alternative TBC compositions: A compositional change in zirconia-based TBCs can be realized mostly by either increasing the total amount of stabilizer or by stabilization with rare-earth metal oxides. These lead to higher disordered crystalline lattice by either introduction of oxygen vacancies or by substitution of Y and/or Zr ions with ions of different ionic radius and/or mass. Ceria stabilized zirconia (CeSZ) is considered as a potential EB-PVD candidate material, providing a good corrosion resistance and superior phase stability at high temperature. Furthermore, the thermal conductivity is found to be lower than for PYSZ, and benefits for lifetime and thermocyclic resistance are reported [21 - 25]. As the vapor pressures of zirconia and ceria differ considerably, the evaporation from one source containing both zirconia and ceria turns out to be critical. Dual-source evaporation was identified to offer a possibility to overcome the problem. Other rare earth oxide stabilizers such as dysprosia (DySZ) and ytterbia (YbSZ) behave similar to yttria: 4 mol% additions create a metastable tetragonal phase while 12mol% additions create a stable cubic lattice. As shown in Fig. 7, a reduction of up to 40% in thermal conductivity was achieved with an optimized version of 12 mol% DySZ [26, 27]. The cyclic lifetime of these new EB-PVD TBCs was comparable to the Hafnia containing TBCs but easier to manufacture by EB-PVD. The crystal lattice of zirconia and hafnia is isomorphous; a complete solubility exists. Although the vapor pressure of zirconia is slightly higher than that of hafnia, no major problems have been found in terms of composition up to now. Larger additions of hafnia, e.g. 40wt% zirconia+40wt% hafnia + 20wt% yttria, reduce thermal conductivity further, but the largest effect of 30% reduction at high temperature was reported for zirconia free 27wt% yttria-stabilized hafnia [28]. The latter showed a much denser and fine columnar microstructure and was less susceptible to sintering.

Similar favorable lower shrinkage rates have been found for EB-PVD 7.5wt% yttria-hafnia that was not rotated during deposition [29]. Own experiments with 32wt% yttria-hafnia (FYSHf) has shown similar results. Evaporation from one source was possible without notable problems with nearly the same composition in both ingot and TBC.

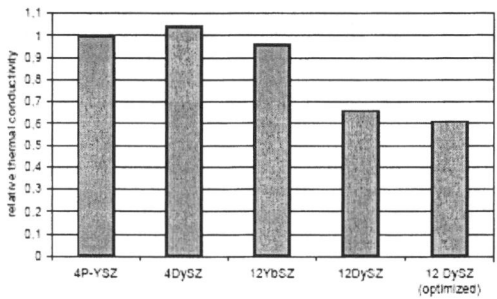

Figure 7: Measured thermal conductivity values of zirconia based top-coats stabilised with alternative oxides in as-coated condition.

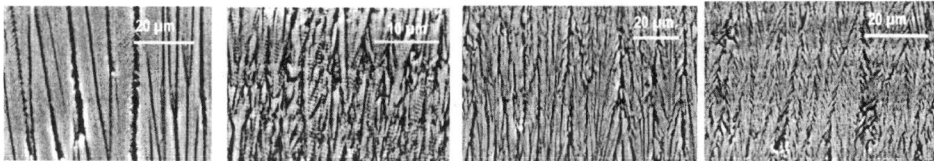

Figure 8: SEM-images of the as-coated EB-PVD-coated 8YSZ (a) undoped $La_2Zr_2O_7$ (b), 3 wt. % Y_2O_3-doped (c) and 10 wt. % Y_2O_3-doped $La_2Zr_2O_7$

Oxides other than zirconia having low thermal conductivity are good alternatives. Rare-earth zirconates, with other words pyrochlores (e.g. $La_2Zr_2O_7$) have reasonable potential for TBC application. Reduced thermal conductivity as well as improved sinter resistance has been found for EB-PVD pyrochlore TBCs, especially for $Gd_2Zr_2O_7$ [30] and $Sm_2Zr_2O_7$ [31].

Although not easy to manufacture and some fluctuation in composition encountered, a recent study has shown that doping $La_2Zr_2O_7$ with 3-10% yttria reduces the compositional scatter during evaporation [32]. Morphologically $La_2Zr_2O_7$ based coatings are similar to cubic structured EB-PVD TBCs (Fig. 8). Although the microstructure changed significantly after 100h of annealing at 1300°C, thermal conductivity remained low at about $1.4 Wm^{-1}K^{-1}$ [33]. Due to the likely reaction of the pyrochlores with the alumina of the TGO, it may be necessary that these coatings are deposited on relatively thin (50 - 100 µm) PYSZ bottom layers, which act as reaction barrier. A combination of the morphology approach by microlayering with new compositions yields multi-layers of different chemistry. Alternating layers of, e.g., zirconia and alumina [34, 35], PYSZ and yttria stabilized hafnia [36], PYSZ and yttria-ceria or PYSZ and gadolinia stabilized zirconia [37], PYSZ and dysprosia stabilized zirconia [24] have been investigated. For these systems at least two factors contribute to a reduced thermal conductivity: firstly, thin films have been reported to exhibit lower thermal conductivities than the respective bulk materials, and secondly, additional boundaries are incorporated perpendicularly aligned to the heat flux. Especially the long term high temperature stability of the thin multilayers has to be tested.

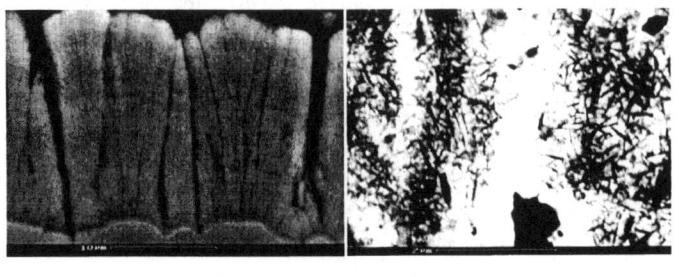

(a)   (b)

Figure 9: Morphology of as-coated MODH (a) and BS-image after heat-treatment at 1100°C/1h (b).

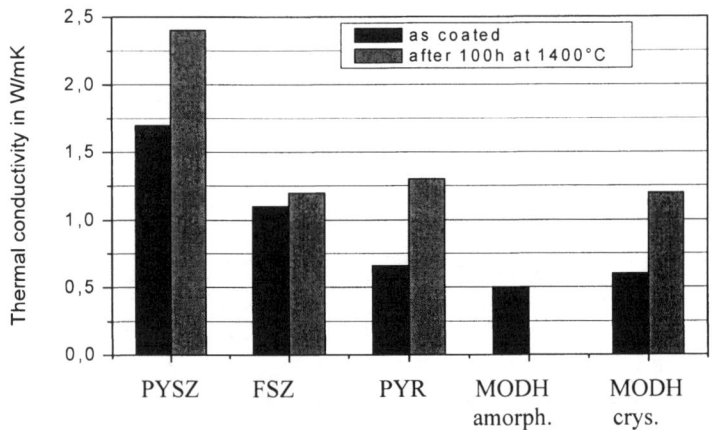

Figure 10: Measured thermal conductivity values of top-coats with new chemistries in comparison with PYSZ top coat in as-coated and aged (1400°C/100h) conditions. FSZ: Full stabilised zirconia, PYR: Pyrochlore, MODH: Metal oxide doped hexaluminate.

Non-zirconia chemistries involve metal oxide doped hexaluminates which also have very low intrinsic thermal conductivities. Problems related to amorphous phase present in the as-coated coatings, which contributes to low conductivity, however, results in a higher increase in thermal conductivity on heat-treatment (Fig. 9). It is a rather challenging task to achieve fully crystalline aluminate based EB-PVD top coats. Once, this burden is overcome, hexaluminates outstands with their low thermal conductivity, phase and microstructural stabilities (Fig. 10).

Summary

Microstructure of EB-PVD coatings can be varied in a wide range by controlling the processing parameters, leading to changes in thermal conductivity since the porosity distribution in top coat microstructure has a strong influence on heat flux diffusivity. High temperature ageing changes thermal and mechanical properties of TBCs. Two-source EB-PVD coating enables processing of specific microstructures and coatings with new chemical compositions. New top coat chemistries show a great potential for reduced thermal conductivity in long term service.

References

[1] D. D. Hass, A. J. Slifka, and H. N. G. Wadley, Acta Mater., 49, 973-83 (2001).

[2] J. R. Nicholls, K. J. Lawson, A. Johnstone, and D. S. Rickerby, Surface and Coatings Technology, 151-152, 383-91 (2002).

[3] K. Fritscher, F. Szücs, U. Schulz, B. Saruhan, W. A. Kaysser, Ceramic Engineering and Science Proceedings, 23 (4) Part B, 341-352 (2002).

[4] U. Schulz, K. Fritscher, C. Leyens, M. Peters, Ceramic Engineering and Science Proceedings, 12(4) Part B, 347-356 (2001).

[5] E. R. Andrievskaya, L. M. Lopato, J. of Mater. Sci., Vol. 30 (1995), S. 2591-2596.

[6] R. Subramanian, U.S. Pat. No. 6 258 467 B1, Jul. 2001

[7] M. J. Maloney: U.S. Pat. No. 6 177 200 B1, 2001.

[8] R. Vaßen, X. Cao, F. Tietz, D. Basu, D. Stöver, Journal of the American Ceramic Society, Vol. 83[8] (2000), S. 2023-2028.

[9] R. Gadow and M. Lischka, Surface & Coatings Technology, 151-152 (2002) 392-399.

[10] U. Schulz, J. Münzer, and U. Kaden, Ceramic Engineering and Science Proceedings, 23 (4), 353-360 (2002).

[11] U. Schulz, K. Fritscher, C. Leyens, M. Peters, and W. A. Kaysser, Materialwissenschaft und Werkstofftechnik, 28, 370-76 (1997).

[12] K. Wada, N. Yamaguchi, and H. Matsubara, Science and Coatings Technology, 191 (2005).

[13] S.G. Terry, Evolution of microstructure during the growth of thermal barrier coatings by Electron-Beam Physical Vapor Deposition, Materials Department, *Dissertation*, University of California, Santa Barbara, 2001, p. 197.

[14] D.V. Rigney, et al., U.S. Pat. No. 6 447 854 (2002).

[15] J.M. Nieuwenhuizen, H.B. Haanstra, Philips Techn.Rev. 27, 87 (1966).

[16] K. Fritscher, W. Bunk: *"Density Graded TBCs Processed by EB-PVD,"* in: FGM 90, ed. M. Yamanouchi, J. Sendai, FGM Forum, 91-96 (1990).

[17] K.J. Lawson, J.R. Nicholls, D.S. Rickerby: *Thermal conductivity and ceramic microstructure*, in "High temperature surface engineering", ed. D.R.J. Nicholls, D. Allen, The Institute of Materials, London, (1997).

[18] D.L. Youchison, et al., Surface and Coatings Technology 177-178 (2004) 158-164.

[19] W. Beele, G. Marijnissen, E. Vergeldt, Nice, Forum of Technology, 1-7, (2002).

[20] J.R. Nicholls, K. Lawson, A. Johnstone, D. Rickerby, "Low Thermal Conductivity EB-PVD TBCs" Materials Science Forum, 369-372 (2001) 595-606.

[21] U. Schulz, K. Fritscher, M. Peters, 82 (1996) 259-269.

[22] U. Schulz, K. Fritscher, M. Peters, J. Eng. Gas Turbines and Power 119 (1997) 917-921.

[23] B.A. Nagaraj, D.J. Wortmann, ASME J. Eng. Gas Turbine Power, 112 (1990) 536.

[24] U. Schulz, K. Fritscher, C. Leyens, Surface and Coatings Technology, 133-134 40-48 (2000).

[25] C. Leyens, U. Schulz, K. Fritscher, Materials at High Temperatures, 20 (2003) 475-480.

[26] U. Schulz, B. Saint-Ramond, O. Lavigne et al., Ceramic Engineering and Science Proceedings, 25(4), (2004) 375-380.

[27] S. Alperine, V. Arnault, O. Lavigne, R. Mevrel, 2001, U.S. Pat. No. 6333118, EP Pat. No. 1085109.

[28] J. Singh, et al., Journal of Materials Science, 39 (2004) 1975-1985.

[29] K. Matsumoto, Y. Itoh, T. Kameda, Science and Technology of Advanced Materials, 4 (2003) 153.

[30] M. Maloney, 2001, U.S. Pat.No. 6177200, U.S. Pat. No. 617560.

[31] R. Subramanian, 2002, U.S. Pats. No. 6258467 and 387539.

[32] B. Saruhan, P. Francois, K. Fritscher, U. Schulz, Surface and Coatings Technology, 182 (2004) 175-183.

[33] B. Saruhan, U. Schulz, R. Vassen et al., Ceramic Engineering and Science Proceedings, (CESP), 25(4), 2004, 363-373.

[34] D. Wortman, 1999, "Multilayer TBC," U.S. Pats. No. 5,942,334 and 5792521.

[35] T. Krell, U. Schulz, M. Peters, W.A. Kaysser, "Graded EB-PVD alumina-zirconia thermal barrier coatings- an experimental approach," in: FGM 98, ed. W.A. Kaysser, Trans Tech Publications LTD, 396-401 (1999).

[36] M. Peters, C. Leyens, U. Schulz, W.A. Kaysser, Advanced Engineering Materials, 3 (2001) 193-204.

[37] M. Maloney, "Method for producing ceramic coatings with layered porosity," U.S. Pat. No. 6365236, 2002

Key Engineering Materials Vol. 333 (2007) pp 147-154
online at http://www.scientific.net
© 2007 Trans Tech Publications, Switzerland

Time-Economic Lifetime Assessment for high Performance Thermal Barrier Coating Systems

M. Bartsch[1,a], B. Baufeld[2,b], S. Dalkilic[3], I. Mircea[1], K. Lambrinou[2], T. Leist[4],
J. Yan[5], A.M. Karlsson[5,c]

[1]Institute of Materials Research, German Aerospace Center, Linder Hoehe, D-51147 Cologne,
Germany

[2]Department of Metallurgy and Materials Engineering, Katholieke Universiteit Leuven, Kasteelpark
Arenberg 44, 3001 Heverlee, Belgium

[3]College of Aviation, Anadolu University, 26470 Eskisehir, Turkey

[4]Material- and Geo-Sciences, Darmstadt University of Technology, Petersenstr. 23, D-64287
Darmstadt, Germany

[5]Department of Mechanical Engineering, University of Delaware, 19716 Newark, DE, USA

[a]marion.bartsch@dlr.de, [b]bernd.baufeld@mtm.kuleuven.be, [c]karlsson@udel.edu

Keywords: Lifetime Assessment, Thermal Barrier Coating, Accelerated Testing, Damage Parameter.

Abstract. Strategies for time-economic lifetime assessment of thermal barrier coatings (TBC) in service are described and discussed on the basis of experimental results, achieved on material systems with coatings applied by electron beam physical vapour deposition. Service cycles for gas turbine blades have been simulated on specimens in thermo-mechanical fatigue tests, accelerating the fatigue processes by an increase of load frequency. Time dependent changes in the material system were imposed by a separate ageing, where the samples were pre-oxidized prior to the fatigue test. Results of thermo-mechanical fatigue tests on pre-aged and as-coated specimens gave evidence of interaction between fatigue and ageing processes. An alternative approach is used, which is focused on the evolution of a failure relevant damage parameter in the TBC system. The interfacial fracture toughness was selected as a damage parameter, since one important failure mode of TBCs is the spallation near the interface between the metal and the ceramic. Fracture mechanical experiments based on indentation methods have been evaluated for monitoring the evolution of the interfacial fracture toughness as a function of ageing time. It was found that the test results were influenced by both changes of the interface (which is critical in service) and changes in the surrounding material.

Introduction

Thermal barrier coating (TBC) systems are used in the hot gas path of gas turbines, e.g. on turbine blades. Components like turbine blades, especially for aero engines, have to sustain thermal and mechanical cycling but also up to 10,000 h at high temperature. Thus, a lifetime assessment of TBC systems has to consider changes of the material properties during service due to time and temperature depending processes, along with damage accumulation due to fatigue. Considering the long time of service, realistic 'real time' testing is not economical. In the case of pure (mechanical) fatigue, tests can be accelerated to cover many cycles by increasing the load frequency. Moreover, it is an established method to accelerate processes depending on time-at-high-temperature by increasing the test temperature. However, at higher temperatures different mechanisms may be triggered, e.g. other oxidation products may form. Nevertheless, limited but reasonable reduction of test time can be achieved. For the case when the fatigue and time dependent processes are dependent on each other (i.e. formation and growth of fatigue damages are influenced by the status achieved by time dependent changes in the material), acceleration of the laboratory tests may give misleading results. An alternative strategy to obtain accelerated testing is to monitor a failure-relevant damage parameter in realistic cyclic experiments and extrapolating the evolution of the damage parameter from interrupted experiments long time before failure occurs.

We will here review two promising test methods for developing time-economic lifetime assessments for high performance TBC systems.

Accelerated close to reality testing - Thermal Gradient Mechanical Fatigue

Specimens. Hollow, dog bone shaped, coated specimens with an inner diameter of 4 mm and an outer diameter of 8.6 mm were used for the thermo-mechanical fatigue tests. The substrate was a nickel-based super-alloy IN 100 DS, which was directionally solidified (DS) with the <100> direction approximately in the axial specimen direction, in order to simulate the elastic behaviour of single crystal materials used in turbine blades. The elastic modulus in the axial direction of IN100 DS was measured under tensile load with a high temperature extensometer and displays a distribution with values between 117 and 138 GPa at room temperature and between 72 and 83 GPa at 950°C. The coating system comprises a metallic oxidation protection layer, the so-called bond coat (BC), and a ceramic top coat. Both coatings were applied by electron beam physical vapour deposition (EB-PVD). The BC is a NiCoCrAlY with the standard composition in wt %: 20Co, 21Cr, 12Al, 0.15Y, balance Ni, and the top coat is partially stabilized zirconia with 7-8 wt % yttria (PYSZ). The thickness of the BC is about 120 μm and of the ceramic top coat about 220 μm. During the coating procedure, a 0.3 μm thick alumina scale, the so-called thermally grown oxide (TGO), forms between the BC and the ceramic top coat. The TGO growth entails the formation of an Al-depleted zone in the adjacent BC. All materials were processed at the German Aerospace Center in Cologne. Part of the specimens was pre-aged in a separate furnace in air before thermo-mechanical testing, in order to economically accumulate time at high temperature. The heat treatment was cyclic, with each cycle for about 24 h at high temperature, accumulating 250 and 500 h at high temperature, respectively. Cooling down to ambient temperature was achieved by removing the specimen from the furnace.

Test procedure. The specimens were subjected to simultaneous thermal and mechanical cycling. The mechanical load was applied by a servo-hydraulic testing machine, which allowed very fast changes of the load level during the cycling. The thermal load was imposed using a radiation furnace powered by quartz lamps. High cooling rates were achieved with an active air cooling from vents in a shutter, which was introduced into the furnace by a pneumatic device. The high heating and cooling rates made it possible to simulate the thermo-mechanical fatigue load of an entire flight of a turbine blade in an aero engine within 3 minutes. A detailed description of the testing set up is given in [1]. Thermal cycles were between 100 and 1000°C, the mechanical load was tensile and in phase. A typical test cycle is displayed in Fig. 1.

During fatigue testing, the specimen was internally cooled by a constant airflow of 45 norm litres per minute. The inlet temperature of the internal cooling air was about 270°C. Internal cooling and external heating and cooling, respectively, generated thermal gradients over the cross section of the specimen. The temperature difference between the outer and the inner surface was measured at a calibration specimen with sheet thermocouples. Under the quasi-stationary conditions during the high temperature sequence of the test cycle, a temperature difference between the inner and outer surface of about 170°C was measured. Because of the thermal gradient, the test is called thermal gradient mechanical fatigue (TGMF), in contrast to conventional thermo-mechanical fatigue (TMF).

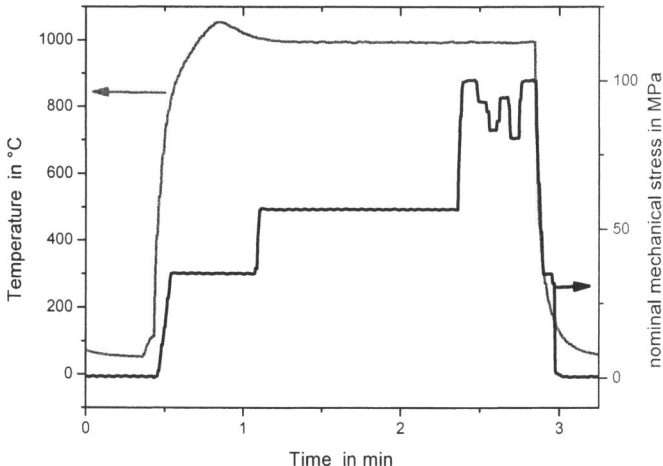

Fig. 1. Typical TGMF cycle with the temperature and the nominal mechanical load profile related to the cross section

Results. Coated specimens, which were pre-aged for 250 h or longer at 1000°C, showed final failure of the TBC system by spallation of the top coat after about 1000 cycles with a nominal maximal mechanical stress of 100 MPa during TGMF. Specimens in the "as-coated" condition have been cycled for even longer times and with higher mechanical loads, but did not show spallation of the ceramic coating [2]. The spallation has been associated to fatigue cracks, which propagated underneath the TGO, parallel to the surface in the axial and circumferential directions [3]. The cracks, which resemble in length sections in their mature state a 'smiley face', always display a crack in the TGO perpendicular to the mechanical load. Fig. 2 gives a schematic of the 'smiley crack' feature and Fig. 3 shows the scanning electron microscope (SEM) image of the crack, which motivated the name 'smiley crack'.

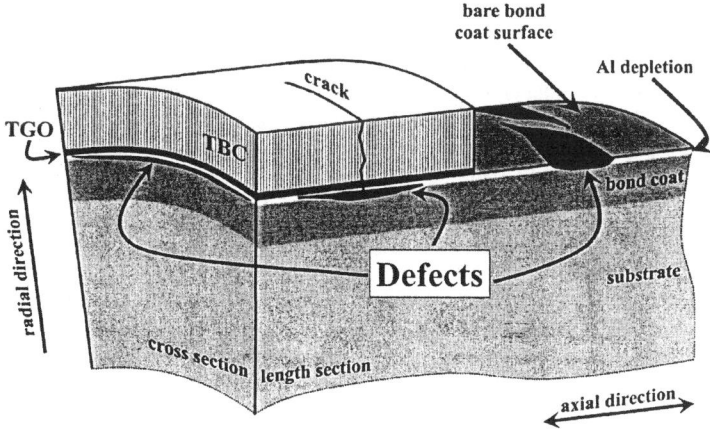

Fig. 2. Sketch, showing the main features of the observed 'smiley cracks' in relation to the specimen geometry

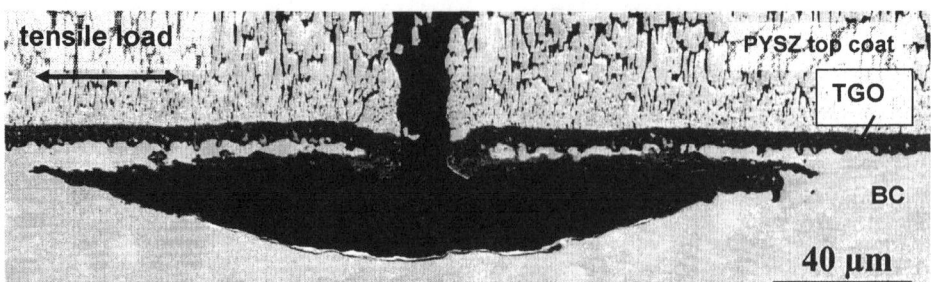

Fig. 3. SEM image of a typical 'smiley crack' prepared by length sectioning

Discussion. Key to the understanding of the 'smiley crack' evolution is the stress situation in the TGO during TGMF-cycling. The SEM images indicate that (in at least one sequence of the TGMF cycle) the tensile stresses in the TGO must have exceeded the TGO strength otherwise a crack through the TGO perpendicular to the mechanical load would not appear. Once the TGO is cracked, oxygen can access the metallic BC and weaken the material by local oxidation and an oxidation assisted fatigue crack is likely to grow. However, linear elastic calculations of the stress distribution across the wall of the coated specimen indicate compressive stresses in the TGO for the case of a nominal maximal mechanical stress of 100 MPa [2]. Looking for inelastic phenomena, which may entail tensile stresses in the TGO, we found (through SEM analysis) that the IN 100 DS substrate shows significant rafting [3]. However, assuming linear elastic TGO in the numerical simulations, the accumulation of inelastic strain in the substrate would develop tensile stresses, not only in the pre-aged but in the as-coated specimens as well. Since the as-coated specimen was loaded during part of the experiment with even higher maximum tensile forces (2182 cycles with maximal nominal tensile stress of 280 MPa), TGO cracks should be more likely to occur in these specimens but were not observed. Thus, the pre-ageing, which results in top coat sintering, TGO growth, and diffusion processes, changes the response of the system. As described by several authors [4, 5] and observed here (Fig. 5), the TGO in the as-coated condition has an intermixed (mainly Al_2O_3 and ZrO_2), very fine grained morphology with grain sizes of less than 100 nm. During high temperature exposure the TGO grows and forms a dense zone (mainly α-Al_2O_3) with grain sizes of more than 1 μm. Following calculations by Rösler et al. [6], the TGO can relax its stresses at high temperature due to creep processes that depend on the TGO grain size, resulting in relaxation times for the as-coated TGO at 1000°C of less than 1 second and for the dense TGO of more than 10 seconds. In the investigated TGMF cycle, both the 'mixed zone' and the dense TGO, should be able to relax most of the compressive stresses (induced due to the combination of thermal gradient, growth stress and thermal property mismatch) during the first 2 minutes of the cycle (before a fast mechanical load step follows). The incremental mechanical load step imposed in the end of the load cycle takes about 5 seconds. Therefore, there may be sufficient time for the fine-grained 'mixed zone' TGO to relax but not for the evolved 'dense zone' TGO, introducing higher tensile stress in the aged samples. This is currently being investigated and will be published at a later state [7].

Extrapolation of damage parameters - determination of interfacial fracture toughness

Test method. Among several proposed methods for determining the interfacial fracture toughness of EB-PVD TBC systems, the Rockwell indentation test with a conical brale C indenter has been selected. This test method has some advantages: in principle it can be applied on specimens and components of any geometry, requiring only a small quantity of material. Indentation of the coating perpendicular to the interface using a brale C Rockwell indenter was analyzed by Drory and

Hutchinson [8] and applied on EB-PVD TBCs by Vasinonta and Beuth [9]. In this test method, the indenter penetrates the coating perpendicular to the surface and generates plastic and elastic deformations in the substrate, which are a driving force for the formation of delamination cracks at the interface between coating and substrate. Experiments on systems with a 100 μm thick TBC resulted in circular, concentric debonding and buckling of the coating behind the crack front [9]. The interfacial fracture toughness is estimated from the calculated in-plane deformation of the substrate and the measured radii of both the indent and the delamination crack. So far, the mechanics of the Rockwell indentation test is analyzed for flat specimens, presuming isotropic material properties and neglecting the interaction between ceramic topcoat and the indenter during the indentation process.

Experimental Procedure. Rockwell indentation tests were performed on flat specimens comprising a 4 mm thick isotropic IN 625 substrate, coated by EB-PVD first with a 100 μm NiCoCrAlY bond coat and subsequently with a 280 μm thick zirconia layer with 7-8 wt% yttria (YSZ). The chemical composition of the coating layers was the same as for the fatigue specimens described above. The material system was investigated in as-coated condition and after heat treatments of 50, 100, 200, and 400 h at 1000°C in air. The indentation was performed using an electromechanical testing machine. The displacement of the indenter was recorded by means of an inductive displacement transducer; the load was recorded by means of a load cell. Test series with indentation loads from 50 to 1000 N were conducted on the specimens. After the indentation, cross sections of the tested specimens at the imprint of the indenter were prepared and investigated by optical microscopy and SEM. The length of the generated crack system at the interface or parallel to the interface was measured, and the crack paths have been evaluated.

Results. The length of the crack systems is a function of the applied load, i.e. higher loads give longer cracks. Comparing as-coated and pre-aged specimens, the same load resulted in longer crack systems for the as-coated specimens (Fig. 4).

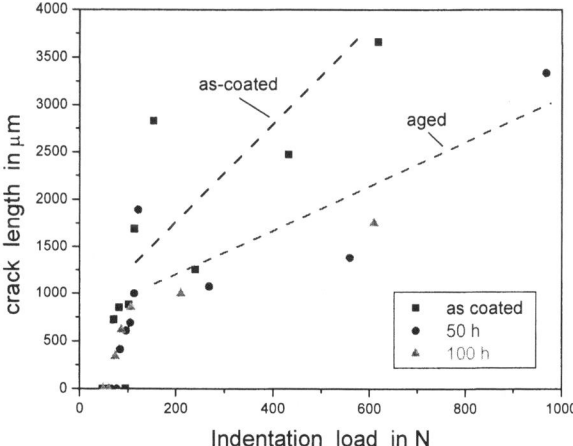

Fig. 4. Crack length as function of indentation load for TBC systems in both as-coated and aged condition

Specimens aged for 400 h showed delayed but spontaneous spallation of the ceramic top coat after removing from the furnace. This phenomenon has been often observed, and it is identified as stress corrosion cracking [10]; it is also nicknamed 'desk-top effect' since it often happens when the specimen is laid on the 'desk top' after the completion of thermal exposure experiments. Thus, it was not possible to perform indentation tests on 400 h aged specimens. Moreover, after indentation of the 200 h aged sample, delayed spallation of the top coat occurred starting from the free edges of the flat specimens. However, it was possible to prepare cross sections of some of these indented specimens for further microscopic investigations.

Analysis of the crack path revealed that in the as-coated condition the crack system propagated mainly at the interface between TGO and BC. After 50 h of thermal treatment the crack path was partially along the interface between TGO and BC, but mainly parallel to this interface within the TGO and near to this interface within the top coat. With increasing ageing time, the crack systems remained within both the top coat and the TGO, never penetrating the dense zone of the TGO. The characteristics of the crack path are illustrated in Fig. 5.

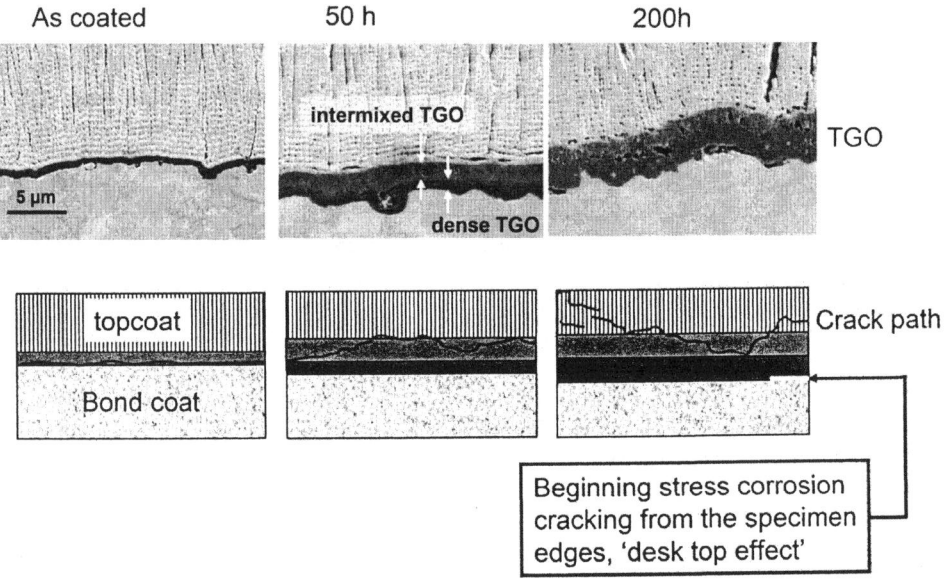

Fig. 5. SEM images displaying the changes of the interfacial region due to ageing, especially the growth of the TGO with a light grey intermixed zone next to the top coat and a darker dense zone, separated from the intermixed zone by a line of pores. The schematics show the respective crack path after indentation.

Discussion. Rockwell indentation tests performed on the TBC surface resulted in crack systems propagating at or parallel to the interface between the TGO and the BC. Given a constant indentation load, the length of these crack systems is shorter after ageing compared to the as-coated condition. Applying the equations given in [9], the experiments would result in an *increase* of the interfacial fracture toughness of the coating system due to ageing at high temperatures, which is in contradiction to the observed coating behaviour in thermal and thermo-mechanical tests as well as in service. Usually, failure of the TBC occurs by spallation of the top coat at the interface between the BC and the TBC. Long-term aged specimens of this test series showed delayed, spontaneous spallation at the TGO-BC interface as well. Examination of the crack paths reveals that the crack systems propagate along different paths in the as-coated and the aged condition. In the aged condition the indentation-induced cracks did not penetrate the TGO but were either trapped in the ceramic top coat or within the TGO at a porous line that forms between the mixed zone and the dense zone during long-term ageing. Thus, in aged specimens the indentation-initiated cracks do not reach the weakest interface. In particular, it appears that the dense zone of the TGO is strong enough, compared to the mixed zone and the top coat, and that it shields the weaker interface between the BC and the dense zone of the TGO. Since the top coat material has gained in hardness and strength during the time at high temperature due to sintering processes, the cracks needed more energy for propagation. Thus, the crack systems in aged specimens are shorter than in the as-coated condition, and they are not a measure of the fracture toughness of the decisive interface.

Concluding remarks

Strategies are needed to reduce the test time for reliable lifetime assessment of components that have to survive long term service. For example, it is impractical to perform realistic thermo-mechanical fatigue tests for turbine blade materials in real time until failure, if the blade or the TBC system on the blade is supposed to function 5,000 to 10,000 flights of 1 to 10 h. In this paper, we discuss two test methods designed to achieve this goal.

Strategies for test-acceleration have been applied on EB-PVD TBC systems, i.e. thermo-mechanical fatigue has been accelerated by high loading rates as well as high heating and cooling rates. Operating time in the fatigue testing facility has been reduced by pre-ageing the specimens separately. The resulting damage features in the pre-aged specimens were significantly different to those in the as-coated specimens. A specific type of fatigue cracks, the 'smiley cracks,' with a TGO crack perpendicular to the mechanical tensile load, evolved only in pre-aged specimens. The pre-aged samples had formed a coarse-grained, dense TGO, which was not present in the 'as-coated' samples. The analysis of the results suggests that the loading rate was too high to allow relaxation processes in the dense TGO of pre-aged specimens but allowed relaxation in the thin fine-grained TGO of the as-coated specimens. Thus, tensile stresses develop only in the pre-aged TGO, entailing fracture of the TGO. For lifetime assessment of the TBC system in service, this result showed how important it is to keep the acceleration of tests within limits, which ensure that damage mechanisms in testing are the same as in service.

Experiments have been performed pursuing an alternative acceleration-strategy, monitoring a failure-relevant damage parameter and extrapolating its evolution from interrupted experiments, long time before failure occurs. In this study, the interfacial fracture toughness of the TBC system was selected and the experiment involved the Rockwell brale C indentation of the surface of both as-coated and aged specimens. It was found that the results were not governed by changes of the interface between BC and TGO (which is critical in service) but by the sintering of the top coat and the growth of a dense TGO, which shielded the weakest interface from indentation-induced cracks. This example shows how important it is to capture with the selected experiment the evolution of the critical feature, which determines the lifetime of the material system in service.

Acknowledgement

The work is supported in part by the European Community's Human Potential Program under contract HPRN-CT-2002-00203, [SICMAC]. I. Mircea and K. Lambrinou acknowledge the financial support provided through the European Community's Human Potential Program under contract HPRN-CT-2002-00203, [SICMAC]. J. Yan and A.M. Karlsson thank the National Science Foundation for financial support under contract DMR-0346664

References

[1] M. Bartsch, G. Marci, K. Mull and C. Sick: Advanced Engineering Materials 2 (1999), p. 127.

[2] B. Baufeld, M. Bartsch, S. Dalkilic and M. Heinzelmann: Surface & Coatings Technology. 200 (2005), p. 1282.

[3] M. Bartsch, B. Baufeld, S. Dalkilic and M. Heinzelmann: International Journal of Fatigue (2006) in press.

[4] U. Schulz, M. Menzebach, C. Leyens and Y. Q. Yang: Surface & Coatings Technology 146-147 (2001), p. 117.

[5] A.M. Karlsson, J.W. Hutchinson and A.G. Evans: Materials Science & Engineering. A351 (2003), p. 244.

[6] J. Rösler, M. Bäker and M. Volgmann: Acta materialia 49 (2001), p. 3659

[7] M. Hernandez, M. Bartsch and A.M. Karlsson, in progress

[8] M.D. Drory and J.W. Hutchinson: Proceeding Of the Royal Society London A 452 (1996), p. 2319.

[9] A. Vasinonta and J. Beuth: Engineering Fracture Mechanics 68 (2001), p. 843.

[10] V. Sergo and D. Clarke, Journal of the American Ceramic Society 81 [12] (1998), p. 3237.

Key Engineering Materials Vol. 333 (2007) pp 155-166
online at http://www.scientific.net
© 2007 Trans Tech Publications, Switzerland

Modeling failures of Thermal Barrier Coatings

Anette M. Karlsson

University of Delaware, Dept. Mechanical Engineering, Newark, DE, 19716 USA

karlsson@udel.edu

Keywords: Thermal Barrier Coatings, Finite Element Analysis, Mechanics Based Modeling.

Abstract. Thermal barrier coatings are commonly used in high temperature parts of gas turbines, to protect the underlying metal substrate from deterioration during high temperature exposure. Unfortunately, the coatings fail prematurely, preventing the design engineers to fully utilize their implementation. Due to the complexity of the coatings, there are many challenges involved with developing failure hypotheses for the failures. This paper reviews some aspects of the current state-of-the-art on modeling failures of thermal barrier coatings, focusing on mechanics based models (such as finite element simulations) where the material physics is incorporated (such as oxidation and diffusion).

Introduction

Thermal Barrier Coatings (TBCs) are commonly used to protect high temperature components in gas turbines from the harsh environment associated with elevated temperatures. The coating reduces the degradation rate of the superalloy and allows for higher operating temperatures, thus making the gas turbine both more economical and fuel efficient. Unfortunately, the full potential of TBCs is not utilized due to premature failures (i.e., in spallation of the coating), leaving the superalloy exposed to the high temperatures. The evolution of the failures is not completely understood, limiting the abilities to redesign and improve the coatings. The lack of understanding of failures in TBCs is primarily due to its complex structure. In order to understand the failures, it is important to correlate experiments and theoretical evolutions by conducting careful experiments and develop mechanics based model. The mechanics based models must incorporate the relevant physics governing the evolution of the failures. [1-6].

In this paper, we will discuss some strategies on how to model failures of TBCs. To this end, we will first discuss the most common TBCs and then discuss some selected failure modes and how these are most successfully modeled according to the current understanding. The discussion is limited to mechanics based simulations, assuming the basic theories of continuum mechanics holds.

Thermal Barrier Coatings

It is pertinent to understand the behavior of the components that constitute the TBC, and how the components interact and evolve during use, to achieve a successful failure model of a TBC. In this chapter, we will summarize some of the key constituents of the coating-system.

A TBC is a multilayered system where a *bond coat* is deposited on the superalloy, along with a ceramic *top coat*, see Fig. 1. The ceramic top coat provides thermal protection by having a low thermal conductivity. A thermal gradient over the top coat, of up to $150°C$, is achieved from active cooling of the super alloy [1-6]. Due to the high temperatures the top coat is subjected to, the top coat typically sinters, resulting in increasing elastic modulus and potentially a transformation strain associated with the shrinkage due to sintering [7]. The metallic bond coat provides oxidation protection to the superalloy by providing aluminum to form an alpha-alumina scale (α-Al_2O_3) between the bond coat and the top coat (Fig. 1). Thus, the bond coat sacrifices itself, resulting in decreasing aluminum content and thus changing properties of the bond coat. The chemical content of the materials are important to control, since even small amounts of critical trace elements can enhance or reduce the interfacial fracture toughness of the structure.

The Ceramic Top Coat. The predominant material currently used for the top coat is 7-8wt.% yttria (Y_2O_3) stabilized zirconia (ZrO_2). There are two major processing methods for the top coat: Depositing the top coat through air-plasma spray (APS) or through electron-beam, physical-vapor-deposition (EB-PVD) [3]. Both methods are developed so that the top coat is porous, thus introducing a "strain tolerance." The strain tolerance allows the top coat to be more compliant, reducing the risk of instantaneous spallation due to thermal mismatch and thermal chock. The processes result in distinct structures and interface morphologies between the top coat and bond coat, Fig 1. In APS top coats, "splats" parallel to the interfaces are formed, and microcracks associated with these splats that governs the strain tolerance, Fig 1A. In the EB-PVD system, the strain tolerance derives from a columnar structure and a micro-porosity (Fig. 1B). Due to the different morphologies, the failure mechanisms tend to be different between the two, as will be addressed below.

Bond Coat Materials: Two classes of intermetallic materials, MCrAlY (M = Ni and/or Co) or Pt-modified Aluminide (Pt-Al), are currently predominantly used as bond coats. The MCrAlYs are normally deposited by plasma spray or EB-PVD where the former case tends to give a rougher surface. Pt-Al made thought a diffusion process where a platinum layer is first placed on top of the substrate and is then aluminized by a thermal process, allowing for diffusion of the constituents, forming the bond coat [4]. The Pt-Al bond coat tends to results in more uniform TGO than the MCrAlYs. Infact, the MCrAlYs can form "pegs" which are localized very thick imperfection of TGO that penetrate the bond coat and may serve as nucleation sites for cracks [6], even though some models suggest that they might serve as crack stoppers [8]. To develop failure models of TBCs it is important to incorporate the evolution of the bond coat during the use: the microstructure can change significantly as the it is depleted from aluminum [9-13].

General Approaches for Modeling

We will here discuss some general modeling guidelines for the various layers in a thermal barrier system. As mentioned previously, the discussion is focused on "mechanics based" simulations, where a continuum mechanics approach is used. Thus, we will focus the discussion towards finite element analyses (FEA), which is currently the most popular approach. Fig. 2 shows a typical finite element model of a TBC, where the imperfection in the interface of the bond coat and top coat is studied [14].

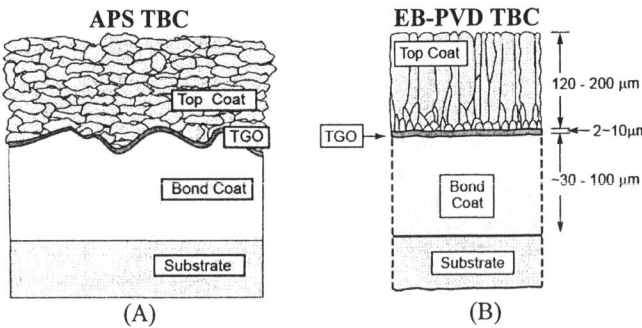

(A)   (B)

Figure 1. Schematics of thermal barrier coatings: (A) Air plasma sprayed top coat and (B) Electron-beam, physical deposited top coat.

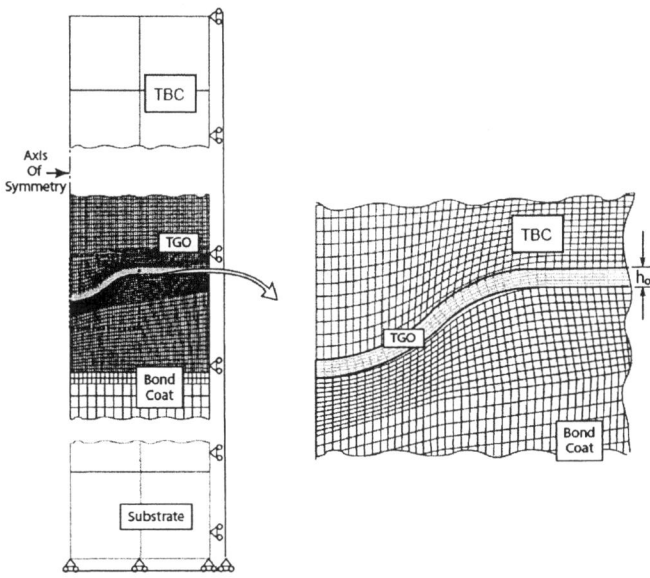

Figure 2. Example of finite element model of a TBC, using an axisymmetric model with continuous boundary conditions. [14]

Initial State of Stress. When starting the simulations, it is important to establish the true state of stress for the conditions considered. The most direct way is to establish the initial stress-free state and then perform the simulations. With few (if any) exceptions, the initial stress-free state is at the deposition temperature. For EB-PVD it can be assumed to be around 950°C and for APS around 350°C. However, process techniques varies, and it is important to make sure the correct temperature is used. A typical simulation will start with a stress-free state at this temperature, followed by a computation step that brings the system to ambient. This will result in residual stress. Thereafter the thermal exposure can be simulated, whether it is isothermal or cyclic condition. The cool down step can be ignored if only elastic properties are assumed, but the correct stress free conditions must be used. (A purely elastic simulation is unlikely to result in the correct material response.)

Thermal cycling. Thermal cycling (after the initial cooling step from processing conditions, described above) is typically simulated by three general steps: (i) heating (ii) thermal exposure and (iii) cooling, repeated to make up the number of cycles to be simulated. Isothermal conditions are simulated with heating, thermal exposure, and cooling. If the material properties are not changing with time (even though they may change with temperature), cyclic conditions will result in an initial evolution of stresses and strains, but these values will reach shakedown after a few cycles, that is, the stress and strains will cycle between two boundary values corresponding to the conditions at high and at low temperature [15, 16]. If all properties are elastic, shake down is reach immediately. Significant long term changes are *numerically* induced if there is no change imposed in the structure. By imposing a change in properties with time, such as TGO growth or top coat sintering, significant evolution can be obtained.

Top coat. The ceramic top coat is by design porous, Fig. 1, making it strain tolerant so to absorb some of the thermal mismatch strains introduced during cooling. Depending on the failure mode investigated, one can either (i) model the top as a homogeneous solid with low elastic modulus or (ii) introduce the overall shape of the splats (APS) or columns (EB-PVD). It is pertinent to keep the model as small as possible, in order to keep the computational effort reasonable and to retain a tractable numerical scheme. It may be sufficient to model the top coat homogeneous, see for example

[14, 17, 18]. The evolution of the top coat properties due to sintering can readily be introduced as a time dependent elastic modulus and an eigenstrain [7]. The details of the microstructures only need to be modeled if the internal morphology is crucial to the response, for example for foreign object damage or erosion. In most cases, the top coat can be assumed elastic and linear fracture mechanics theories can be utilized.

Thermally Grown Oxide. The TGO is the component that drives most of the failures: (i) the TGO grows due to oxidation at thermal exposure and (ii) the TGO has significantly higher elastic modulus than the companion materials in the coating system. The compressive stresses are caused by a combination of the thermal expansion mismatch between the TGO and the rest of the system during the cooling/heating sequence, and by the growth stress, σ_G, that is associated with of new alumina. The compressive stresses can reach several GPa at ambient conditions [15, 19, 20]. Localized tensile stresses can also appear in the TGO, due to the morphology, see for example [11, 15, 21]. An estimate of the stress in the TGO, σ_{TGO}, due to thermal mismatch and growth strain can be obtained by [16, 22]

$$\sigma_{TGO} = \frac{E_{TGO}}{1-v_{TGO}} \Delta\alpha\Delta T + \sigma_G \tag{1}$$

where E_{TGO} and v_{TGO} are the elastic modulus and Poisson's ratio for the TGO; $\Delta\alpha$ is the difference in thermal expansion coefficient between the substrate and the TGO, and ΔT is the change in temperature. Typical properties as given by table 1 and growth stress ranges from 300 MPa to 1 GPa. Fig. 3 show some typical results from FEA of the stresses in the TGO [15, 21] for two possible imperfection shapes of the bond coat – TBC interface. The results are based on a model without the top coat (the bond coat and substrate are not shown for simplicity).

Thus, the TGO is associated with locally high stresses and strain energy, which has potential to drive failures in the vicinity of the TGO. It follows that it is important to simulate the TGO in a realistic manner. What is "realistic" must be determined from case to case, but in general, the thickening of the TGO and the growth strain has to be incorporated. Furthermore, the TGO is prone to creep at high temperatures and is elastic at ambient. Some aspects of this will be elaborated upon in the section that follows.

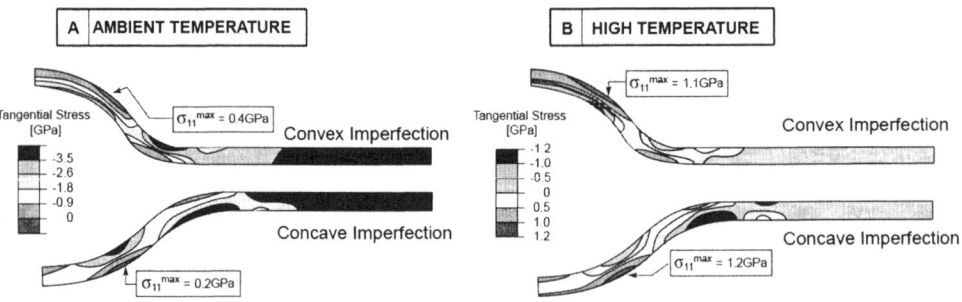

Figure 3. Typical stress distribution in the TGO for two different shapes of the interface between the bond coat and the TGO (convex vs. concave imperfection) after thermal cycling, for (A) ambient and (B) high (operating temperature). Note the tensile stresses in the TGO. [15, 21] (Stresses tangential to the bond coat interface.)

Sample Material Properties

Property	Value at ambient	Value at operating temperature
Top Coat		
Elastic modulus	40-80 GPa	40-80 GPa
Poisson's ratio	0.1-0.2	0.2-0.3
Thermal Expansion	11 $\mu/^{\circ}C^{-1}$	12 $\mu/^{\circ}C^{-1}$
Bond Coat		
Elastic modulus	180-200 GPa	70 GPa
Poisson's ratio	0.3	0.3
Thermal Expansion	10-14 $\mu/^{\circ}C^{-1}$	15-16 $\mu/^{\circ}C^{-1}$
Yield/Creep strength	500-800 MPa	20-100 MPa
TGO		
Elastic modulus	380 GPa	320 GPa
Poisson's ratio	0.2-0.25	0.2-0.25
Thermal Expansion	8 $\mu/^{\circ}C^{-1}$	8 $\mu/^{\circ}C^{-1}$
Growth stress		300-1000 GPa

Table 1: Typical properties of the components of a thermal barrier coating. Note: these are approximate values, typically used in generic simulations. Care must be taken when selecting properties.

Bond Coat. The bond coat is normally responding elastic at ambient, but yield or creeps at the higher temperatures. As for the TGO, the highest stresses are found at ambient. However, the strength is significantly lower at high temperature, resulting in the inelastic behavior even though the stresses are lower. Many simulations predict overall yielding of the bond coat, not only around the stress risers associated with imperfections [11, 23, 24]. The bond coat might have the most interesting evolution of properties. It provides aluminum to form the TGO, thus slowly changing its composition. This frequently results in phase transformations in some or all of the grains making up the bond coat. This can either be modeled as individual grains that undergo phase transformations, or allow the average properties of the bond coat to change, e.g. [10, 11].

Superalloy. The superalloy is frequently modeled as an elastic material with temperature dependent properties. If a single crystal superalloy is considered, the anisotropic nature of the elastic modulus might be important to incorporate.

Interdiffusion zone. As the system is exposed to elevated temperatures, the distinct boundary between the superalloy and the bond coat is many times smeared by an interdiffusion zone, eliminating the distinct boundary between them. This zone is may be ignored, unless the failure mechanisms are seen to be related to its presence.

Material Properties. Sample material properties are shown in Table 1. The properties are temperature depended and varies with time. The properties are extremely hard to measure, see for example [12, 13] and careful judgment must be used when assigning properties to the model being developed. When the properties are not well known, it is customary to conduct several analyses, studying the sensitivity for some critical material parameters.

Examples of Modeling Selected Failure Modes

We will here discuss some predominant failure modes and current model approaches to capture these failures.

Morphological Instabilities. The evolution of the TGO morphology that occurs due to cyclic thermo-mechanical loading, but not during isothermal conditions, are referred to as morphological instabilities (or sometimes ratcheting), schematically shown in Fig. 4. Surface roughening that occurs during isothermal conditions does not belong to this class of problems. This phenomena has received significant attention over the last few years, where Pl-Aluminide have seen to be most prone to exhibit this behavior [10, 11, 14-16, 21-23, 25], but it has also been seen in NiCoCrAlY-alloys in laboratory environments [24, 26] and on coatings in service or extreme temperatures (~1150°C) [27, 28].

The morphological instabilities evolve due to the compressive stresses that develop in the TGO during thermal exposure combined with thermal cycling, as was discussed previously. The TGO strive to relax the compressive stress by deforming out of its plane, deforming the bond coat. The mechanism is controlled by a combination of three non-linear constitutive behaviors: (1) high temperature inelasticity in the TGO; (2) growth strain in the TGO; (3) cyclic inelasticity in the bond coat. The growth strain is driving the system [15, 16, 22]. If the lengthening component of the aluminum formation is removed, the morphological instabilities cannot occur [15, 23].

The growth strain is induced due to the oxidation process when the new alumina is formed [6, 19, 29, 30]. Most of the TGO growth occurs as thickening, but a small part is distributed in the grains of the TGO leading to a lengthening component. The high temperature inelastic strength of the TGO is often referred to as the "growth stress." The growth strain is limited by the growth stress, and once the TGO stress reaches the level of the growth stress, the lengthening strain is reallocated into thickening strain. In Fig. 4, the cases of cyclic versus isothermal scenario are compared, as a function of cumulative growth strain, showing the amplitude increase of an existing imperfection, and the tangential stress in the TGO (i.e., stresses parallel to the interface of the bond coat and TGO). The position designated **Y** refers to the onset of TGO inelasticity, e.g., the growth stress is reached. In the isothermal scenario, the amplitude change essentially stops, whereas under cyclic conditions, displacements and stresses continue to change after the point of TGO inelasticity is reached. This is so, since the stress in the TGO is relaxed during each cycle, allowing for additional accumulation of lengthening strain in the TGO at high temperatures. Morphological instabilities are governed by the state of stress in the bond coat in combination with the inelastic strain of the bond coat at high temperatures, since the inelasticity in the bond coat allows the TGO to relax and deform out of plane – into the bond coat – for each thermal cycle. The growth stress differentiates between isothermal and thermal cyclic exposure: if the growth stress is assumed infinite (i.e., the TGO is fully elastic) isothermal conditions will result in large amplitude changes as well. [16]

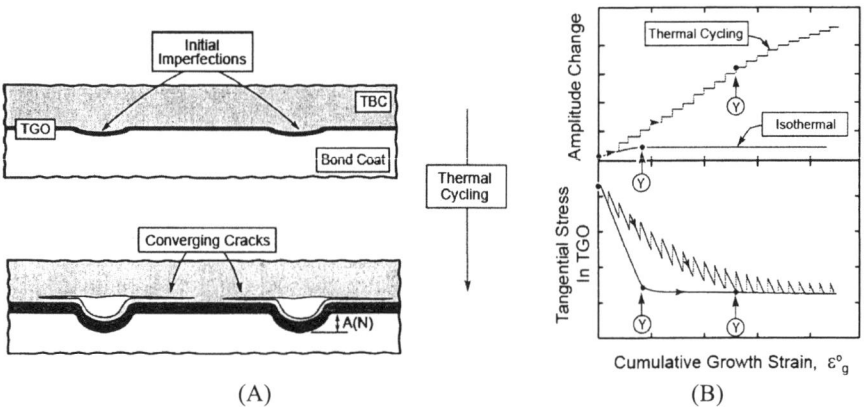

(A)   (B)

Figure 4. (A) Schematic of the development of morphological instabilities. (B) Amplitude change and stress in the TGO tangential to the bond coat surface.

Some factors can suppress morphological instabilities, for example, a full adherence of the ceramic top coat to the TGO/bond coat [21] or high yield strength of the bond coat [16]. Severalfactors that can enhance the rate of instability growth, including thermal mismatch with the bond coat [11] or thermo-mechanical loading [24], bond coat swelling [31], martensitic phase transformation [11, 23], or due to phase transformations between grains [10]. It is important to note that in all these cases, the rate of amplitude growth changes, but if the lengthening growth strain is not present, morphological instabilities do not occur.

The scenario and understanding of the morphological instabilities, as outlined above, could only be established by combining and coordinating experimental and theoretical investigations. Experimental investigations first indicated the formation of the instabilities and a hypothesis was formulated on how they evolved. The hypothesis was quantified in a numerical simulation (FEA) where some of the critical parts to the hypotheses was included (e.g. growth strain and plasticity in the bond coat). These simulations draw some particular conclusions which were verified in model experiments, e.g. [31].

To model the development of morphological instabilities, the volume change associated with the TGO formation must be incorporated. The *lengthening* component has been seen to be the most critical parameter driving the system. A convenient way of incorporating this component is to impose a stress free eigenstrain during the high temperature part of the cycle [15]. In this case, the *total* strain tensor, ε_{tot}, can be written as

$$\varepsilon_{tot} = \varepsilon_E + \varepsilon_P + \varepsilon_T + \varepsilon_G \tag{2}$$

where ε_E is the elastic strain, ε_P is the inelastic strain, ε_T is the thermal strain, and ε_G is the growth strain. The elastic strain and the thermal strain are conservative whereas the inelastic strain and the growth strain (caused by the oxide growth) are non-conservative (i.e., path depended and after unloading still present). Furthermore, the thermal strain and the growth strain are stress-free strains (i.e., if there where no constraints, no stress would be induced).

The implementation of growth strain into a finite element program will depend on the code used. For example, in ABAQUS it is convenient to use the user subroutine UEXPAN that allows the user to prescribe an anisotropic eigenstrain [15, 32]. Even though the thickening of the TGO is – surprisingly – *not* critical for driving the morphology change, it should be included. The concept of eigenstrains has been used successfully to model the thickening as well, e.g.,[11, 15, 24, 33, 34]. As an alternative, the bond coat elements closest to the TGO can be transformed to TGO as a function of cycles. In ABAQUS, this has successfully been done using "state variables" [13, 33, 35].

As mentioned previously the TGO and the bond coat creeps at high temperatures. This material behavior is important to incorporate. The TGO growth stress (which is the high temperature creep strength) is a function of exposure temperature, creep strength of the Al_2O_3 and the bond coat. [Here, we differentiate between the creep properties between a material (i.e., Al_2O_3) and the structure (i.e. TGO): the growth stress in the TGO depends on the properties of the surrounding components, such as elastic modulus and creep strength of the bond coat, since these are constraints for the TGO.] It has been shown that the creep strength of the TGO is what differentiates this system between cyclic and isothermal conditions: if the TGO is assumed elastic, the amplitude change is only depending on time-at-temperature [16, 22].

The most obvious approach is to incorporate the creep behavior through built-in routines in the used finite element package intended to simulate such behavior, which has successfully been done, in related problems, e.g., [36, 37]. However, this may not be very time efficient, since simulations involving visco-plastic effects are very time consuming. A more time efficient way is instead to use the time independent yield strength as the ultimate strength, first introduced in ref. [15]. If using the "end of cycle" stress of the TGO and bond coat, a good estimate of the behavior is found. In the case of morphological instabilities, we are mostly interested in the accumulation of plastic strain, (which will be translated into permanent amplitude change), thus it is not critical *how* the accumulation occurred, just that it *did* occur. (This scheme works well for cases where the long term cyclic behavior is investigated, but is not appropriate in all cases, particular when only a few thermal cycles are considered).

In this case, the failure behavior was successfully established through qualitative rather than quantitative simulations. For example, simplified material properties were used, and in most cases, limited numbers of cycles were considered (typically about 24 full cycles). By keeping the number of parameters to a minimum, and then carefully investigate which were important for the instability problem, qualitative statement could be made explaining the failure evolution. For example, it was seen that the lengthening component of the oxidation growth is the most important parameter driving the system and that by increasing the bond coat strength at high temperature it would be possible to suppress or at least reduce the rate of growth. To this end, the parameters were varied over a range of values, to study its influence of the structural behavior. Since material properties are many times unknown, a qualitative rather than quantitative approach is a powerful method to investigate failure modes.

We also note here that the modeling techniques used for simulating the oxide growth and the cyclic behavior can be successfully used for a range of cases where the TGO stresses contributes to the failure evolution.

Plasma Sprayed Coatings. In APS coatings, the surface is significantly rougher than for EB-PVD, as indicated in Fig. 1, and the morphological instabilities as outlined above are not observed. Nevertheless, it is important to incorporate the growth strain in the TGO, including the inelastic properties. A typical idealization used for investigating the effect the roughness has on the TBC stresses is shown in Fig. 5A, with typical stresses after cycling in Fig. 5B.

If the sintering of the top coat is ignored, simulations show that the stresses in the top coat reach a constant value, reaching shake down, e.g. [34]. The change due to the TGO growth is absorbed by bond coat yielding. If the sintering is incorporated, it is expected that the top coat stresses keep evolving. As can be seen from the stress values in Fig. 5B (based or ref. [34]), these stresses are large enough to induce cracks in the top coat. Using linear fracture mechanics, the crack propagation can be simulated and predicted [34, 37].

Crack Growth. Crack growth in a TBC is most frequently seen in the vicinity of the TGO. It is not uncommon that the crack grows in the top coat, as indicated in Fig. 4A (for a Pt-aluminide with morphological instabilities) or that the crack moves through several interfaces, such as from the TGO-bond coat interface, growing through the TGO and propagating in the top coat, as indicated in Fig. 5C (for a typical APS TBC). In both cases, the cracks are driven by the overall mode II caused by the thermal mismatch and the TGO growth stress, where will find the path of least resistance. For Pt-

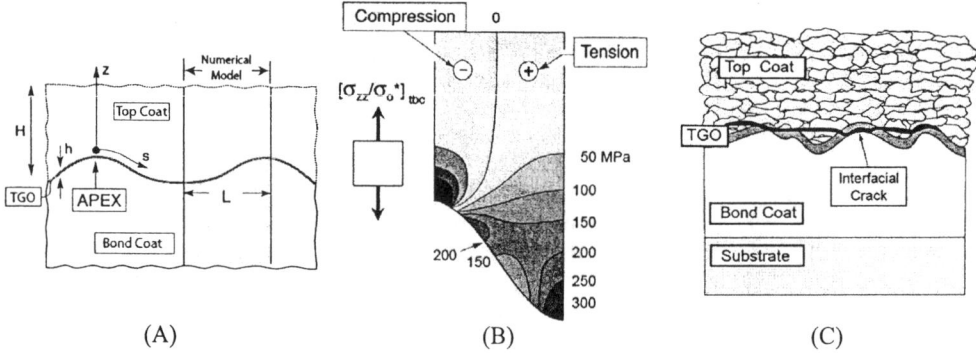

(A)   (B)   (C)

Figure 5. (A) A schematic of a thermal barrier coatings and (B) stresses in the top coat after thermal cycling, (C) Schematic of an interfacial crack in an APS TBC with an overall mode II crack.

Aluminide with EB-PVD coating, the top coat tends to have a very dense layer next to the TGO, where the porosity starts maybe 100 nm above the TGO. The cracks are observed to grow at this distance above the TGO. Based on modeling and experimental results, it is seen that the cracks start at the site of imperfections, and grow parallel to the TGO [14, 17, 38]. For the APS TBC, the crack will find its weakest path in combination of the largest crack driving force. Stresses are tensile in the top coat where the top coat is in a valley (Fig 5B), and the TGO can be tensile at its extreme points [34, 37, 39, 40]. Cracks typically starts as micro cracks which coalesce with time, leading to large scale cracks and final spallation [17, 18, 41].

Externally Caused Damage of the Top Coat. A range of mechanisms may cause damage to the top coat, which are caused by particles present in the gas flow. The smaller particles tend to cause erosion – slowly wearing down the top coat surface – whereas the larger particle can cause "foreign object damage" (FOD) where significant parts of the top coat is instantaneously impacted and removed [42, 43]. To simulate either of these two related damage modes, the individual columns of the EB-PVD top coat must be modeled (or the "splats" of the APS top coat). A random approach, using Monte Carlo simulations to estimate erosion rates, was conducted by Wellman and Nicholls [44]. The impact of larger particles, causing instantaneous critical damage was investigated in [45] where the individual columns and the friction between them where simulated. A large spherical impression was numerically induced and the deformation of the underlying top coat was investigated.

A second class of external damage to the top coat is caused by infiltration of calcium-magnesium-alumina-silicate (CMAS). This substance is commonly present due to sand dust and similar materials (in particular for gas turbines used for propulsion). The substance melts at temperatures of about 1150°C. During shutdown this may cause cold shock and spallation of the top coat [46]. To simulate this behavior, Chen [47] developed a model where the individual columns of the EB-PVD top coat is modeled, along with the volume change – and therefore the additional strain energy – of the CMAS. The energy release rate of assumed cracks in the interface between top coat and the TGO was determined based on the infiltration rate. It was seen, as maybe expected, that the interfacial cracks were more likely to grow as the infiltration rate increased.

Concluding Remarks

In this paper, a brief overview of a range of modeling techniques capturing failure modes in thermal barrier coatings (TBCs) is given. Most failures in TBCs are associated with a combination of mechanical loading and evolution of material properties due to high temperature exposure. The paper outlines some methods to incorporate these evolving properties into mechanics based simulations.

Most simulations discussed in this paper are based on qualitatively approaches rather than quantitatively. In qualitative approaches, simplified material properties and loading conditions are used, and parametric studies are conducted to evaluate the influence particular parameters have on the failure process. This is an excellent and simple approach to establish the overall failure behavior and it can show when one failure mode is dominate over another possible mode. However, if reliable life prediction models are developed, where the true number of cycles before failures is to be predicted, it is important to incorporate the exact material properties into a model, along with how they evolve with time. It is also important to incorporate all the possible failure modes that can occur. To this end, a statistical approach might be needed (which was not discussed in this paper), to randomly distribute the nucleation sites for the failures. A life prediction model will thus require significant experimental work along with a sound numerical scheme. This is still not available.

References

[1] R.A. Miller, Journal of the American Ceramic Society, 67 (1984), p. 517
[2] T.E. Strangman, Thin Solid Films, 127 (1985), p. 93
[3] N.P. Padture, M. Gell, and E.H. Jordan, Science, 296 (2002), p. 280

[4] M. Stiger, N. Yanar, M. Topping, F. Pettit, and G. Meier, Zeitschrift Fur Metallkunde, 90 (1999), p. 1069

[5] P. Wright, Materials Science And Engineering A-Structural Materials Properties Microstructure And Processing, 245 (1998), p. 191

[6] A.G. Evans, D.R. Mumm, J.W. Hutchinson, G.H. Meier, and F.S. Petit, Progress in Materials Science, 46 (2001), p. 505

[7] R. Hutchinson, N. Fleck, and A. Cocks, Acta Materialia, 54 (2006), p. 1297

[8] H. Zhu, N. Fleck, A. Cocks, and A. Evans, Materials Science and Engineering A-Structural Materials Properties Microstructure and Processing, 404 (2005), p. 26

[9] S. Darzens, D.R. Mumm, D.R. Clarke, and A.G. Evans, Metallurgical and Materials Transactions a-Physical Metallurgy and Materials Science, 34 (2003), p. 511

[10] S. Darzens and A.M. Karlsson, Surface & Coatings Technology, 177-178C (2004), p. 108

[11] J. Shi, S. Darzens, and A.M. Karlsson, Materials Science and Engineering A-Structural Materials Properties Microstructure and Processing, 392 (2005), p. 301

[12] M.W. Chen, R. Ott, T.C. Hufnagel, P.K. Wright, and K.J. Hemker, Surface & Coatings Technology, 163 (2003), p. 25

[13] D. Pan, M.W. Chen, P.K. Wright, and K.J. Hemker, Acta Materialia, 51 (2003), p. 2205

[14] A.M. Karlsson, T. Xu, and A.G. Evans, Acta Materialia, 50 (2002), p. 1211

[15] A.M. Karlsson and A.G. Evans, Acta Materialia, 49 (2001), p. 1793

[16] A.M. Karlsson, J.W. Hutchinson, and A.G. Evans, Materials Science and Engineering A-Structural Materials Properties Microstructure and Processing, 351 (2003), p. 244

[17] T. Xu, M.Y. He, and A.G. Evans, Interface Science, 11 (2003), p. 349

[18] X. Chen, J.W. Hutchinson, M.Y. He, and A.G. Evans, Acta Materialia, 51 (2003), p. 2017

[19] V.K. Tolpygo, J.R. Dryden, and D.R. Clarke, Acta Materialia, 46 (1998), p. 927

[20] T. Tomimatsu, S.J. Zhu, and Y. Kagawa, Scripta Materialia, 50 (2004), p. 137

[21] A.M. Karlsson, C.G. Levi, and A.G. Evans, Acta Materialia, 50 (2002), p. 1263

[22] A.M. Karlsson, J.W. Hutchinson, and A.G. Evans, Journal of the Mechanics and Physics of Solids, 50 (2002), p. 1565

[23] A.M. Karlsson, Journal of Engineering Materials and Technology-Transactions of the Asme, 125 (2003), p. 346

[24] J. Shi, A.M. Karlsson, B. Baufeld, and M. Bartsch, Materials Science and Engineering A., Submitted (2006).

[25] M.Y. He, A.G. Evans, and J.W. Hutchinson, Acta Materialia, 48 (2000), p. 2593

[26] M. Bartsch, B. Baufeld, and E.R. Fuller, Ceramic Engineering and Science Proceedings, 24 (2003), p. 497

[27] R. Pennefather and D. Boone, International Journal of Pressure Vessels and Piping, 66 (1996), p. 351

[28] R. Pennefather and D. Boone, Surface & Coatings Technology, 76 (1995), p. 47

[29] D.R. Clarke, Current Opinion in Solid State & Materials Science, 6 (2002), p. 237

[30] D.R. Clarke, Acta Materialia, 51 (2003), p. 1393

[31] A.W. Davis and A.G. Evans, Acta Materialia, 53 (2005), p. 1895

[32] ABAQUS, *ABAQUS 6.5*. 2004, Pawtucket, Rhode Island: ABAQUS Inc.

[33] M.Y. He, J.W. Hutchinson, and A.G. Evans, Acta Materialia, 50 (2002), p. 1063

[34] M. He, J. Hutchinson, and A. Evans, Materials Science and Engineering A-Structural Materials Properties Microstructure And Processing, 345 (2003), p. 172

[35] M. Glynn, M. Chen, K. Ramesh, and K. Hemker, Metallurgical and Materials Transactions A-Physical Metallurgy and Materials Science, 35A (2004), p. 2279

[36] M. Chen, M. Glynn, R. Ott, T. Hufnagel, and K. Hemker, Acta Materialia, 51 (2003), p. 4279

[37] R. Herzog, P. Bednarz, E. Trunova, V. Shemet, R.W. Steinbrech, F. Shubert, and L. Singheiser, ACERs Cocoa Beach Meeting 2006, in press (2006).

[38] T. Xu, M.Y. He, and A.G. Evans, Acta Mater., 51 (2003), p. 3807

[39] K. Schlichting, N. Padture, E. Jordan, and M. Gell, Materials Science and Engineering A-Structural Materials Properties Microstructure And Processing, 342 (2003), p. 120

[40] H. Chai and B.R. Lawn, Acta Materialia, 53 (2005), p. 4237

[41] M. He, D. Mumm, and A. Evans, Surface & Coatings Technology, 185 (2004), p. 184

[42] R. Wellman, M. Deakin, and J. Nicholls, Tribology International, 38 (2005), p. 798

[43] X. Chen, M.Y. He, I. Spitsberg, N.A. Fleck, J.W. Hutchinson, and A.G. Evans, Wear, 256 (2003), p. 735

[44] R. Wellman and J. Nicholls, Wear, 256 (2004), p. 889-899.

[45] X. Chen, J.W. Hutchinson, and A.G. Evans, Acta Materialia, 52 (2004), p. 565

[46] C. Mercer, S. Faulhaber, A. Evans, and R. Darolia, Acta Materialia, 53 (2005), p. 1029

[47] X. Chen, Surface & Coatings Technology, 200 (2006), p. 3418

Key Engineering Materials Vol. 333 (2007) pp 167-176
online at http://www.scientific.net
© 2007 Trans Tech Publications, Switzerland

High Temperature Behavior of Coatings and Layered Ceramics

Ján Dusza

Institute of Materials Research, SAS, Watsonova 47, 043 53 Košice, Slovakia

jdusza@imr.saske

Keywords: Layered ceramics, coatings, thermal shock, creep, creep damage

Abstract. The present contribution summarizes the recent results in the field of high temperature properties of layered ceramics and thermal barrier coatings (TBC), mainly as regards their thermal shock resistance and creep characteristics. The thermal shock and creep behavior of layered ceramics are discussed with the main focus on the influence of layered composition and interlayer boundary on the creep behavior of the composite. In the last part the high temperature deformation and creep of TBC's are discussed.

Introduction

Structural ceramics have a number of excellent properties, however, their wider application is still limited by their brittleness, low flaw tolerance and low reliability. Therefore the production of large load-bearings or rapidly rotating parts, which are subjected to high mechanical loads at high temperatures, is very difficult. Their low flaw tolerance – high brittleness – is strongly connected with the presence of ionic or covalent atomic bonds and with the limited number of independent slip systems, which are cooperative when compared with the number of those necessary for plastic deformation commonly observed in metals and alloys [1,2].

Different approaches have been used during the last two decades with the aim to improve the room temperature properties, reliability, life time, and high temperature properties of structural ceramics. One of the often used approaches – the laminar structure approach [3-6] – improves the structural reliability by designing novel laminar composites which promote crack deflection on interlayer boundaries and/or utilize compressive residual stresses generated during cooling down from the sintering temperature as a result of differences in thermal expansion coefficients between layers with different compositions.

Layered ceramics can be, according to the mechanisms for increasing their mechanical properties, divided into three main groups: composites with constrained transformation zones [7,8], composites with tailored residual stresses [9-11], and composites with weak interfaces. In the layered composites of the fourth category different mechanical responses of the individual layers are combined in order to produce a composite with properties which, for some applications, are superior to those of the constituent ceramics [12-13].

In layered/laminar composites residual stresses can be developed during cooling down from the sintering temperature due to different thermal coefficients of the layers. The sign and magnitude of these stresses can be tailored by composition of the individual layers but also by the thickness of layers [9-11].

Thermal barrier coatings (TBCs) were developed for advanced gas turbines as insulating materials. They need to have sufficient thickness and durability in order to maintain a substantial temperature difference between the internally cooled load bearing alloy and the coating surface. They are usually formed by yttria stabilized zirconia (YSZ) applied by plasma spraying or electron beam physical vapor deposition, but during the last years some new systems have been developed. The challenge in developing new materials (e.g., $Gd_2Zr_2O_7$ or $ZrO2$ with admixed rare earth stabilizers) for TBC's with enhanced overall performance is to understand the balance between improved thermal resistance and diminished durability [14,15]. Applying these coatings prolongs the lifetime of the components and allows application of a higher turbine inlet temperature and therefore results in improved efficiency of the system [16].

The "in service" damage of the TBC coated components usually occur by delamination of the interface between the metal and the coating, or by damage of the coating itself. For lifetime prediction of the parts it is very important to quantify the damage process as a function of loading history [17].

The aim of this contribution is to summarize recent results in the field of high temperature behavior of ceramic multilayers and coatings, mainly as regards their thermal shock resistance and their creep behavior.

Thermal shock and creep behavior of advanced ceramics

Structural ceramics usually are sensitive to thermal shock (only one single severe thermal cycle occurs) and thermal fatigue (number of cycles with moderate stresses occurs). The first investigation of thermal shock behavior of ceramics was made by Kingery [18] and fracture mechanical methods were applied by Hasselman [19]. More detailed analysis were carried out by Pompe et al and Bahr and Weiss [20,21].

An experimental determination of the material sensitivity against thermal shock is usually done by the method proposed by Hasselman [19] when specimens (usually standard bending bars) are heated up to the temperature T_0 and then cooled in an environment to the temperature T_{env}. After this the remaining strength is measured.

Manufacturing of test bars with well-defined shape and surface finish is complicated. In that type of investigation, each test bar can be used only once, implying that separate bars are required for each testing temperature, and in order to improve statistics, several bars are normally tested at each temperature. Using these methods is thus both, time-consuming and costly, which complicates evaluation of the thermal-shock properties. Peterson and Rowcliffe [28,29] applied a new method to determine such properties based on making small initial Vickers cracks on a polished plate of defined furnace and subsequent quenching in a water bath. Then the crack growth is measured and correlated to thickness of the material to be tested. The indented plate is heated in a vertical tube with the temperature difference between the furnace and the water bath. The growth is measured as an average percentage of the original crack length. This method allows to use the same sample throughout a test series of different quenching temperatures, and the statistics are improved by measuring the crack growth for several indentations made on the same sample. Applying the fracture mechanics approach to this method we can defined the further thermal shock resistance parameter as follows, [29]:

$$R_m = \left(\frac{K_c}{\sigma_{th}}\right)^2 = \frac{G_{IC}E}{\sigma_r^2(1-\nu)} = \left[\frac{\chi_r P c_o^{-3/2}(\pi\Omega c)^{1/2}}{\chi_r P\left(c_0^{-3/2}-c^{-3/2}\right)}\right]^2 = \frac{\pi\Omega c}{\left[1-\left(\dfrac{c}{c_0}\right)^{-3/2}\right]^2}, \tag{1}$$

where R_m is the slope of the $(\pi\Omega c)$ versus $[1 - (c/c_0)^{-3/2}]^{-2}$ curve and σ_{th} is the introduced stress.

It must be noted that R_m, can be measured without knowledge of any material property (E, ν, K_c, α, H, χ_r, k) or quenching medium characteristics (ΔT, h). It is derived only from a direct measurement of radial crack lengths before and after thermal shock and depends on the quenching conditions.

The deformation and failure behavior of ceramics at high temperatures is predominantly governed by creep effect. If reasonable creep deformation occurs this itself may lead to a design limit if the function of a component is affected by this deformation. Creep rupture occurs by formation and extension of creep cracks. Inelastic deformation in the polycrystalline ceramics occurs within the grains (motion of dislocations – gliding or climbing or diffusion of vacancies) or at the grain boundaries (viscous flow, diffusion of vacancies, formation and extension of cavities, dissolution and re-precipitation). Damage mechanisms leading to the rupture of the component are the formation of grain boundary cavities, growth of the cavities and their coalescence to cracks and growth of the crack up to their critical size.

Ceramics usually exhibit non-symmetric creep behavior with significant higher creep rates under tension than under compression which effect is significant in the secondary creep range.

Creep behavior of the advanced ceramics was summarized in several reviews, [32-41]. The most up-to-date and extensive review of creep in silicon nitride ceramics was published by Melendéz-Martínez and Domínguez-Rodríguez in May 2004 [42]. Based on the review of hundreds of experimental studies they concluded that it is impossible to determine with accuracy what deformation mechanism is taking place under the fixed experimental conditions, since none of the models accounts for experimental observations in a fully successful manner. Apparently, creep of silicon nitride has to be explained in terms of a series of mechanisms occurring simultaneously. According to their suggestion, the main mechanism occurring during compressive creep is grain boundary sliding (GBS) accommodated by solution-precipitation, whereas tensile creep is greatly influenced by cavitation [43].

The differences in creep behavior of ceramics compared to metals are significantly influenced by differences in the microstructure. The majority of structural ceramics are liquid phase sintered materials and consequently, they contain continuous layers of amorphous secondary phases at the grain boundaries. The presence of residual glass plays fundamental role in the high-temperature creep behavior of ceramics. Amorphous boundary film affects creep in three different ways: firstly, as "lubricant" so that the viscous flow of the boundary phase accommodates sliding of the matrix grains. Secondly, the boundary phase is a fast diffusion pathway, and finally, its presence facilitates nucleation and growth of cavities. Single phase ceramics, e.g., some grades of silicon carbide and alumina, exhibit greater creep resistance compared to the corresponding vitreous bonded materials, but their sintering is more difficult and limited possibilities for microstructure control results in a lower strength and fracture toughness values [5]. The presence of amorphous intergranular phase in vitreous bonded ceramics has crucial importance for the possible operating creep mechanisms.

Analytical solution of the interrelated creep mechanisms is difficult and there is a tendency to use those solutions known from creep of metals. When the minimum creep rate, $\mathrm{d}\bar{\varepsilon}\mathrm{dt}$, is obtained from the steady-state regime under stress σ, the generalized power law (also called Norton-Arrhenius equation [53]) is commonly used to describe creep behavior even in ceramics [35, 38, 44]

$$\mathrm{d}\bar{\varepsilon}\mathrm{dt} = (AEb)\,(b/d)^p\,(\sigma/E)^n\,(D/kT), \tag{2}$$

where A is a dimensionless constant, E is the Young's modulus, b is the Burger's vector, d is the grain size, p is the grain size exponent, n is the stress exponent and k is Boltzmann's constant. D is the effective diffusion coefficient which depends on temperature as $D = D_0 \exp(-Q/RT)$, where D_0 is a constant, Q is the apparent activation energy for diffusion, R and T have their usual meaning.

Thermal shock resistance and creep behavior of layered ceramics

There exist only a limited number of papers dealing with thermal shock resistance and creep behavior of layered ceramics, [54-60]. Bermejo et al [54] investigated the thermal shock behavior of an Al_2O_3-5%tZrO_2/Al_2O_3-30%mZrO_2 multilayer ceramic with tetragonal to monoclinic phase transformation within the Al_2O_3-30%mZrO_2 layers during the cooling down from sintering. Due to this transformation compressive residual stresses were introduced into these layers. A finite element model has been developed to estimate both the thermal strain effects during the sintering process as well as the temperature distribution and stress profile within the laminate during thermal shock testing. Experimental tests on the monoliths and laminates were carried out and compared to the model. According to the results there is present a 103 MPa tensile stress in the ATZ layers and a 688 MPa compressive stress in the thin compressive AMZ layers, in the bulk. It was found that there is a similar ΔT_c for monoliths and laminates having the same geometry, regardless of the residual stresses. This may be explained by the edge cracks in the laminates that grow under thermal shock conditions and might relax the surface stresses, Fig. 1.

As it was observed, transversal cracks originated in the multilayer propagate through the first ATZ layer and then arrest at the interface, Fig. 2. This phenomenon can result in a higher ΔT's in the case of multilayered structure comparing to the monolithic material.

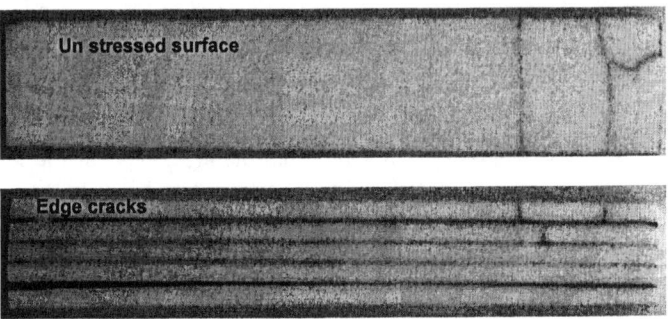

Fig. 1 Specimens after the thermal shock test at ΔT of 300 ^{0}C, Courtesy R. Bermejo

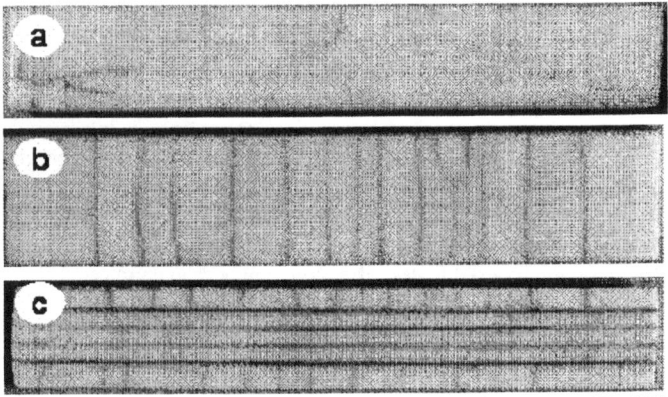

Fig. 2 Monolithic ATZ (a) and multilayer specimens after the test at ΔT of 300 ^{0}C, Courtesy R. Bermejo

Hvizdoš et al [55] studied the thermal shock resistance of particular layers of a functionally graded material (FGM) of $Al_2O_3+10\%ZrO_2$/ $Al_2O_3+30\%ZrO_2$ / $Al_2O_3+10\%ZrO_2$ prepared by electrophoretic deposition and pressureless sintering using indentation-quench method. The results were compared to those obtained from a reference non-layered composite material prepared by the same fabrication technique. The results show a highly enhanced resistance of the FGM against transversal propagation of surface cracks, which in real applications might be formed by impacts, wear or other contact mechanisms during the use, into the bulk. No observable transversal cracks originated from the side surfaces of the specimen were found, even not on the unpolished specimens. This can be contrasted to the crack propagation in the homogeneous material, where no preferential directions for crack growth and no crack arresting were observed.

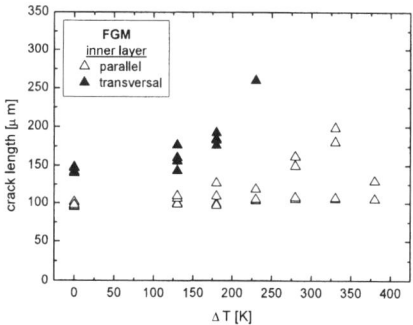
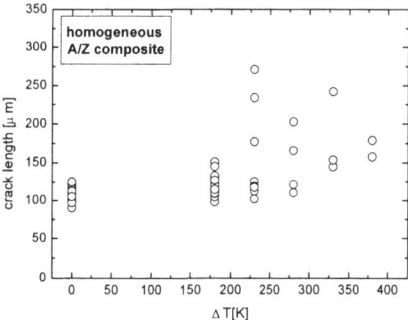

Fig. 3. Influence of the quenching temperature on the crack length in FGM and A/Z composite, [55], Courtesy Hvizdoš

Layered composites of alternate layers of pure Al_2O_3 (thickness of 125 µm) and 85 vol% Al_2O_3–15 vol% ZrO_2 stabilized with 3 mol% Y_2O_3 (thickness of 400 µm) were prepared by sequential slip casting and then fired at either 1550° or 1700°C and tested in constant-strain-rate tests in air at 1400°C at an initial strain rate of 2×10^{-5} s^{-1} [56]. The load axis was applied both parallel and perpendicular to the layer interfaces. Catastrophic failure occurred for the composite that was fired at 1700°C, because of the coalescence of cavities that had developed in grain boundaries of the Al_2O_3 layers. In comparison, the composite that was fired at 1550°C demonstrated ductility of the Al_2O_3+YTZP layer, but at a flow stress level that was determined by the Al_2O_3 layer. Fracture of the composites is dictated in ultimate instance by the failure of the Al_2O_3 layers, although it is delayed by the presence of interfaces. For large-grained composites, the fracture occurs at small strains via growth of the single cracks in the Al_2O_3 layers by coalescence of cavities that are nucleated in two-grain boundaries which are parallel to the stress direction. For materials with interfaces perpendicular to the stress direction, such cracks terminate after extending up to the interfaces. For fine-grained layered composites, the failure occurs by coalescence of the creep damages after attaining large macroscopic strains. The layered composite was found damage tolerant and has the ductility of monolithic ZTA but on the other hand the strength of monolithic Al_2O_3. This indicates that the interface plays a major role in redistributing stress concentrations in the layered material.

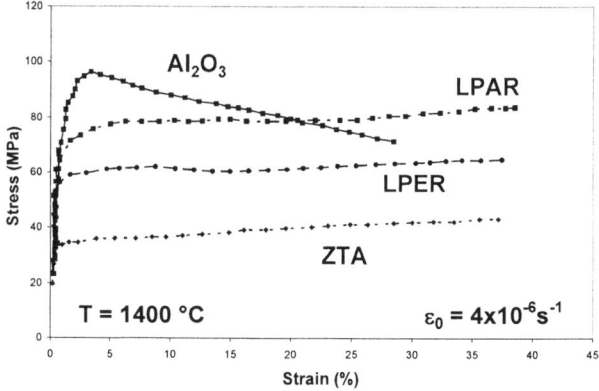

Fig. 4 Sress – strain behavior of laminated composite Al_2O_3/(85 vol.% Al_2O_3+15 vol.% ZrO_2 tested parallel and perpendicular to the layers and monolithic $Al_2O_3$and ZTA, after [57]

At improving strength and ductility of the layered composites, the fracture controlling parameter was found to be the grain size of Al_2O_3 layers.

Jimenez-Mendelo et al [57] investigated mechanical properties under compression at 1400°C of Al_2O_3/(85 vol.% Al_2O_3+15 vol.% ZrO_2 doped with 3 mol% Y_2O_3) layered composites produced by sequential slip casting. The composites were stressed both, parallel and perpendicular to the layer planes. After testing, the layer interfaces maintain their structural integrity. The comparison with monolithic Al_2O_3 and ZTA produced by the same processing technique shows that the laminated composites exhibit simultaneously an enhanced ductility (characteristic of monolithic ZTA) and creep resistance (characteristic of monolithic Al_2O_3), Fig. 4. This behavior cannot be explained by a composite creep model based on individual properties of the two constituent materials. The improvement in mechanical properties is essentially related to the presence of strong interfaces in the laminated composites.

Laminar composites containing layers of 3 mol% yttria-stabilized with tetragonal ZrO_2 (YTZP) and a mixture of Al_2O_3/3YTZP (60/40 vol.%) have been fabricated via a tape casting process that allowed formation of multilayers with strong and sharp interfaces. Compressive tests at constant strain rate and constant load were carried out on these materials in air at temperatures between 1200 and 1400 °C with stress axis both, parallel and perpendicular to the layer interfaces. Microstructure observations of the deformed composites indicated that the deformation is caused primarily by grain boundary sliding. The stress exponent n and the activation energy Q decrease from 2 to 1 and from 650 to 550 kJ/mol, respectively, with increasing stress. Such a trend has already been observed in the monolithic constituents, suggesting that probably the same creep rate controlling mechanisms operate in the composites.

The mechanical properties of the laminates fall in between monolithic YTZP and ZT60A composites tested in two mode, perpendicular and parallel to the applied stress [58]. The stress exponent and the activation energy of layered composites show similarity as it was observed in monolithic YTZP and ZTA ceramics, suggesting that the same rate mechanisms control the strain rate. By using the creep properties of monolithic YTZP and ZT60A, a simple composite creep theory predicts correctly the flow stress in laminated composites with layer inter-faces parallel to the applied stress. However, it was not possible to apply the same theory for the investigation of flow stress in laminates with interface planes perpendicular to the applied stress because of the constraints imposed by the harder phase on the deformation of the softer phase.

Dusza et al [59] studied the creep behavior of a laminar composite consisting of a regularly alternate stacking of layers made from Al_2O_3 and 60 vol% Al_2O_3 + 40 vol% ZrO_2 (ZTA) fabricated by tape casting, followed by worm moulding of different layers, binder burn-out and sintering. The creep tests were carried out in compression. The creep deformation at different temperatures and different loads tested parallel to the layers are illustrated in Fig. 7. The activation energy and stress exponent were found to be 645 kJ/mol and 1.41, respectively.

Dusza et al [60] investigated the high temperature behavior of silicon nitride based multilayers of Si_3N_4/ Si_3N_4+ SiC/ Si_3N_4 and found significantly increased creep and slow crack growth resistance for the layered system with Si_3N_4+ SiC layer as the tensile layer, comparing to the behavior of monolithic silicon nitride. The higher high temperature properties of the layered composite in bending were caused by the improved properties of outer layer, where the intergranularly located SiC nanoparticles hinder the grain growth of the Si_3N_4, changing their shape and the shape and chemistry of the grain boundaries and grain boundary phases. Moreover, the intergranularly located SiC nanoparticles interlock two neighbouring Si_3N_4 grains with a "puzzle" character and therefore prevents the grain boundary sliding.

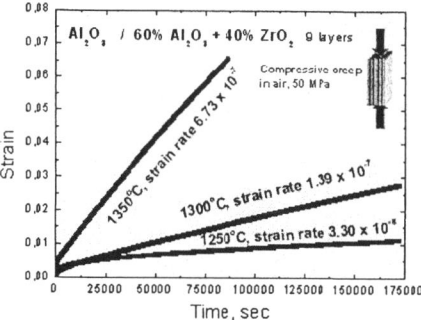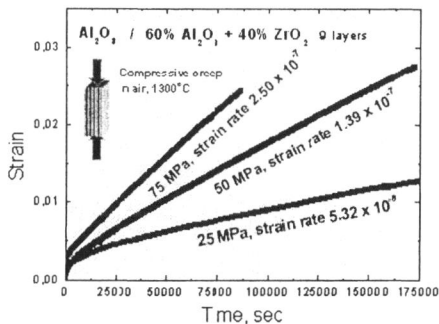

Fig. 7 Creep curves of Al₂O₃ /60 vol% Al₂O₃ + 40 vol% ZrO₂ (ZTA) layered composite tested at different temperatures (a) and different loads (b)

Creep behavior of thermal barrier coatings

The creep of plasma-sprayed TBC's has been investigated by different methods during last decades. The results of Firestone et al. [64] show that creep is a thermally activated process and the main creep mechanism is splat sliding. It was reported, that the TBC's can sinter and creep significantly under compressive stresses at relatively low temperatures. The influence of dopants on the creep mechanisms in the plasma-sprayed zirconia and hafnia based TBC's has been investigated using dilatometry by Zhu and Miller [65]. Close relationship has been found between the observed sintering/creep behavior and chemical phase stability of TBC's. It has been found that materials with higher chemical and therefore also phase stability show improved sintering/creep resistance. It seems so, that the low-stress creep characteristics of the ceramic coating materials, determined by the dilatometer technique, are similar to the creep behavior of plasmasprayed TBC's obtained from high temperature mechanical creep tests [66,67] and the laser sintering/creep test [68]. The low creep activation energy has been attributed to mechanical sliding, surface and grain boundary diffusion, and temperature and stress enhanced transport.

The creep rates of the plasma-sprayed ZrO_2-Y_2O_3 TBC measured by laser sintering technique are in a very good agreement with the creep rates of EB-PVD TBC measured by indentation test during this investigation. The creep rates of plasma –sprayed TBC at 1200°C and 50 MPa is approximately 2×10^{-8} sec^{-1}; this value is well comparable with what was found in our experiment for EB-PVD TBC.

Compressive creep test was used on the temperature interval from 900°C to 1300°C for characterization of the creep behavior of plasma-sprayed ZrO_2 TBC's at stresses between 1.8 and 80 MPa [69]. Power law was used for description the of creep behavior. Beside the creep deformation, shrinkage of the material was observed leading to the change of its elastic properties. Recently Wanatabe et al. [70] and Chen et al. [71] investigated the plastic deformation of thermal barrier oxides at high temperatures using impression technique which was used in the past for the characterization of the creep behaviour of different materials [61-63]. This involves deposition of the coating onto an Al_2O_3 substrate, impressing with a sapphire sphere and simultaneously measuring the loads and displacements. According to the results the penetration depth at the temperature of 1137°C for the 7wt%Y_2O_3+ZrO_2 material at the applied stress of approximately 30 MPa is approximately 60 μm. At the temperature of 1150°C and load 50 MPa, the penetration depths of the cylinder were approximately 50 μm. They also found a zone of deformation and densification below the indent at 1137°C, with extreme plastic bending of crystals in the dense layer.

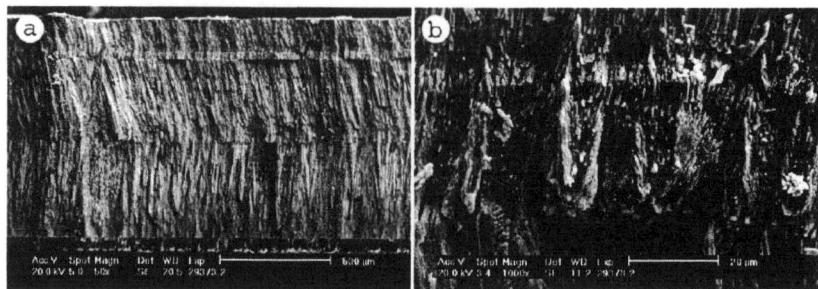

Fig. 8. As recieved specimen of the free standing EB-TBC with several layers (a) and detail of the microstructure of the layers (b)

The indentation creep behavior of a free standing thermal barrier coating (TBC) prepared by electron beam physical vapor deposition (EB-PVD), Fig. 8, has been investigated, [72]. Composition of the experimental material was 7 wt% Y-ZrO$_2$, and the specimens had the form of discs with 5 mm in diameter and thickness 1.2 mm. A flat cylindrical indenter (hot pressed SiC) has been used for indentation creep experiments in the temperature range from 1100 °C to 1300 °C at applied stresses from 30 to 50 MPa, in air. The strain-time relationship was recorded, and the creep exponents and activation energies of the creep have been calculated. The creep micromechanisms have been studied by observation and comparison of the microstructure of the as-received and crept specimens in scanning electron microscopy (SEM). The obtained creep stress exponent varied from 0.05 to 1.12 and the obtained activation energy in the interval from 215 to 329 kJ/mol. The main creep mechanisms are densification of the sublayers with fine columnar crystals, bending of crystals and leaning down of a set of crystals.

Acknowledgments

The work was partly supported by the European Community's Human Potential Programme under contract HPRN-CT-2002-00203, [SICMAC], by NANOSMART, Centrum of Excellence of SAS, Slovak Grant Agency for Science via grant No. 2/4173/04 and by the Science and Technology Assistance Agency under the contract No. APVT – 51– 049702.
Jan Dusza acknowledges the support of the Alexander von Humboldt Foundation.

References

[1] R.F. Cook and G.M. Pharr: in *Materials Science and Technology, 11, Mechanical properties of ceramics*, (ed. M.V. Swain), 339-407 (1994) Weinheim, VCH

[2] P.F. Becher: in *Materials Science and Technology, 11, Mechanical properties of ceramics*, (ed. M.V. Swain), 339-407 (1994) Weinheim, VCH

[3] M. P. Harmer, H.M. Chan, and G.A. Miller: J. Am. Ceram. Soc., 75, (1992), p. 1715

[4] W.J. Clegg, K. Kendall, N.McN. Alford, T.W. Button and J.D. Birchall: Nature, 347 (1990), p.445

[5] F. Toschi, C. Melandri, P. Pinasco, E. Roncari, S. Guicciardi, G. De Portu.: J. Am. Ceram. Soc., 86 [9] (2003), p. 1547

[6] C. Baudín, A. Sayir and M.H. Berger: Key Eng. Mat. (2005), p. 208

[7] D.B. Marshall, J.J. Ratto and F.F. Lange: J. Am. Ceram. Soc. 74 (1991), p. 2979

[8] D.B. Marshall: Ceram. Bull. 71 (1992), p. 969

[9] T. Chartier, D. Merle and J.L. Besson: J. Eur. Ceram. Soc., 15 (1995), p. 101

[10] P. Šajgalík, Z. Lences and J. Dusza: J. Mater. Sci., 31 (1996), p. 4837

[11] M.F. Amateu and G.L. Messing: in International Encyclopedia of Composites, (ed. S.M. Lee), Vol.3, 11-16;1900, Weinheim, VCH Verlag

[12] A.V.Virkar, J.L.Huang and R.A.Cutler: J. Am. Ceram. Soc., 70 (1987), p. 164

[13] C.J. Russo, M.P. Harmer, H.M. Chan and G.A. Miller: J.Am.Ceram. Soc. 75 (1992), p.3396

[14] R.A. Miller: In: Thermal Barrier Coatings Workshop, NASA C.P. 3312, (1995), 17-34

[15] X. Chen, J. W.Hutchinson, A.G. Evans:Acta Materialia 52 (2004), p. 565

[16] M.J.Stiger, N.M.Yanar, M.G.Topping, F.S.Pettit, G.H.Meier.: Z.Metallknd. 90 (1999) 12, 1069-1078

[17] A.G.Evans, D.R.Mumm, J.W.Huntchinson, G.H.Meier and F.S.Pettit: Progress in Materials Science,46 (2001), p. 505

[18] W.D.Kingery: J.Am.Ceram.Soc, 38 (1955), p. 3

[19] D.P.H.Hasselman: J.A.Ceram.Soc. 46 (1963), p. 535

[20] W. Pompe, H.A. Bahr, G. Hille, W. Kreher, B. Schultrich and H.J. Weiss: Current Topics in Material Science, 12 (1985), p. 205

[21] H.A. Bahr and H.J. Weiss: Theor.Appl. Fract.Mech., 6 (1985), p. 57

[22] D. Munz and T. Fett: Ceramics – Mechanical properties, Failure Behaviour, Materials Selection.Springer Series in Materials Science, p.291.

[23] B.E. Bertsch, D.R. Larson and D.P.H. Hasselman: J.Am.Ceram.Soc., 57 (1974), p. 235

[24] C. Chiu, and E.D. Case: J.Mater.Sci., 27, (1992), p. 6707

[25] K. Koumoto, H. Shimizu, W.S. Seo, C.H. Pai and H. Yanagida: *Thermal shock Resistance of porous SiC ceramics*, Internal Report, Univ. of Tokyo, 90, (1991), 32-33.

[26] M.V. Swain: J.Am.Ceram.Soc., 73 (1990), p. 621

[27] D.P.H.Hasselman: J.Am.Ceram.Soc., 52 (1969), p. 600

[28] P. Petterson, M. Johnsson and Z.Shen: J. Europ. Ceram. Soc., 22 (2002), p. 1883

[29] M.Collin and D.Rowcliffe: Acta mater, 48 (2000), p. 1655

[30] F.Tancret, F. Osterstock: Scripta materialia, 37 (1997), 443-447.

[31] D.F.Carroll and S.M.Wiederhorn: *High Temperature Creep Testing of Ceramics*, eds. B.F.Dyson, R.D. Lohr, and R.Morrell, 135-149, (Elsevier Appl.Sci., London, 1989).

[32] D.R.Messier and M.M.Murphy: *Silicon nitride for structural applications – an annotated bibliography*. Final Report AMMRC MS 75-1, (Army Materials and Mechanics Research Center, 1975).

[33] G.D.Quinn: Ceram. Eng. Sci. Proc., 3 [1-2] (1982) p. 77.

[34] W.R. Cannon and T.G. Langdon: J. Mater. Sci., 18, (1983) p. 1

[35] W.R. Cannon and T.G. Langdon: J. Mater. Sci., 23, (1988) p. 1

[36] B.I.Davis and C.H.Carter, Jr.: in *Advanced Ceramics*. Edited by S. Saito, (Oxford Univ. Press, 1988), 95-125.

[37] E.F.Hockings, E.F.Krimmel W.Kurtz and P.Popper: *Gmelin Handbook of Inorganic and Organometallic Chemistry*, B5, (Springer Verlag, Berlin, 1994).

[38] A.Hynes and R.Doremus: Crit. Rev. Sol. State Mat. Sci., 21 [2] (1996), p. 129

[39] D.S.Wilkinson: J. Am. Ceram. Soc., 81 [2] (1998), p. 275

[40] W.E.Luecke: *Review: High temperature creep of Si_3N_4*, unpublished work, 1999.

[41] W.E.Luecke and S.M.Wiederhorn: J. Am. Ceram. Soc., 82 [10] (1999), p. 2769

[42] J.J. Meléndéz-Martínez, A.Domínguez-Rodrígue: Progress Mater. Sci., 49 [1] (2004), p. 19

[43] S.M. Wiederhorn: Z. Metallkd., 9 (2000), p. 1053

[44] J.P. Poirier: *Creep of crystals* (Cambridge Univ. Press, Cambridge, UK, 1983).

[45] T R.W. Evans, B.Murakami and W.Wilshire: Br. Ceram. Trans., 87 [2] (1988), p. 54

[46] D.S. Wilkinson: in *Tailoring of mechanical properties of Si_3N_4 ceramics*, (Kluwer Acad. Publ., The Netherlands, 1994), p. 327

[47] S.M.Wiederhorn, B.J.Hockey, D.C.Cranmer and R.Y.Yeckley: J.Mater.Sci. 28 (1993), p.445

[48] F. Lofaj, S.M. Wiederhorn, G.G. Long and P.R. Jemian: Ceram. Eng. and Sci. Proc., 22 (2001), p. 167

[49] T. Ohji: Ceram. Eng. Sci. Proc., 22 [3] (2001) p.159

[50] K.C. Liu, H.T. Lin, C.O. Stevens and C.R. Brinkman: Key Eng. Mat. 132-136 (1997), p.583

[51] T.-J. Chuang: J. Mat. Sci., 21 (1988), p. 467

[52] T. Fett, K. Keller and D. Munz: J. Mat. Sci., 23 (1986), p. 165

[53] B.I. Davis and C.H. Carter, Jr.: in *Advanced Ceramics.* Ed. by S. Saito, (Oxford Univ. Press, 1988), p. 95

[54] R.Bermejo. L.Llanes, M.Anglada, P.Supancic and T.Lube: Key Eng. Mat. 290 (2005), p. 199

[55] P.Hvizdoš, D.Jonsson, M.Anglada, G. Anné and O.van der Biest: Mechanical properties and thermal shock behaviour of an alumina/zirconia functionally graded material prepared by electrophoretic deposition. Accepted for the J.of the Europ.Ceramic.Soc.

[56] M. Jiménez - Melendo, F. Guitiérez - Mora and A. Domínguez - Rodríguez: Acta Mater., 48 (2000), p. 4715

[57] M. Jiménez - Melendo, C. Clauss, A. Domínguez - Rodríguez, G. De Portu, E. Roncari and P. Pinasco: Acta Mater., 46 (1998), p. 3995

[58] M. Jiménéz-Melendo, C. Clauss, A. Domínguez-Rodríguez, A. J. Sánchez-Herencia and J.S. Moya: J. Am. Ceram. Soc., 80, (1997), p. 2126

[59] J. Dusza, F. Dorčáková, J. Vít and T. Köves: unpublished work.

[60] P. Šajgalík, J. Dusza and Z. Lenčéš: Layered Si_3N_4 based composites, Ceramic Processing Science and Technology, Publ.American Ceramic Society, 603-607.

[61] G. Cseh, J. Bar, H.-J. Gudladt, J. Lendvai and A. Juhasz: Mat. Sci. Eng. A272 (1999), p. 145

[62] J.C.M. Li: Mat. Science and Engineering A322 (2002), 23-42

[63] A.A. Tchizhik, L.B. Getsov, A.I. Rybnikov and I.S. Malashenko: Thin Solid Films 270 (1995), p. 243

[64] R.F. Firestone, W.R. Logan, J.W. Adams and R.C.J. Bill: in: *The 6th Annual Conference on Composites and Advanced Ceramic Materials*, Eds. J.D. Buckley, C.M. Packer, J.J. Gebhardt (The American Ceramic Society, Columbus, OH 43214, 1982), p. 758.

[65] D. Zhu and R.A. Miller: Surface and Coatings Technology 108–109 (1998) p. 114

[66] M.J. Stiger, N.M. Yanar, M.G. Topping, F.S. Pettit and G.H. Meier: Z.Metallknd. 90 (1999) 12, p 1069

[67] F. Wakai: Acta Metall. Mater., 42 (1994), p. 1163

[68] S. Ahmaniemi, P. Vuoristo and T. Mantyla: Materials Science an Engineering A366 (2004), p. 175

[69] G. Thurn, G.A. Schneider and F Aldiger: Mat. Sci. Eng. A, 233(1997), p. 176

[70] M. Watanabe, C. Mercer, C.G. Levi and A.G. Evans:Acta Materialia 52 (2004), p. 1479

[71] X. Chen, J.W. Hutchinson and A.G. Evans: Acta Materialia 52 (2004), p.565

[72] J. Vit, F. Dorčákova and J. Dusza: will be published.

Key Engineering Materials Vol. 333 (2007) pp 177-194
online at http://www.scientific.net
© *2007 Trans Tech Publications, Switzerland*

Ceramic Layer Composites in Advanced Automotive Engineering and Biomedical Applications

Rainer Gadow

Institute for Manufacturing Technologies of Ceramic Components and Composites
University of Stuttgart
Allmandring 7 b, D-70569 Stuttgart, Germany

rainer.gadow@ifkb.uni-stuttgart.de

Keywords: light weight engineering, automotive applications, thermal spraying, cylinder liners, residual stresses, bioceramic coatings, tissue engineering

Abstract

Light weight engineering and composite technologies are key strategies in modern product development in mechanical engineering as well as in biomedical applications, where innovation is driven by novel material concepts and surface functionalities. Designed or customized surface properties by advanced coating technologies are an important discipline in this context. Ceramic, metallurgical and cermet layers can be manufactured in a most appropriate way by high energetic thermokinetic deposition techniques like plasma spraying, electric arc and last not least by supersonic flame spraying (HVOF). These technologies perform high deposition rates, high flexibility to use various materials and their combinations and applications in micro to macro scale products.

The final properties of the coatings and layer composites do not just depend on the properties of the combined materials but, as in the case of ceramic coated light metals, are distinctly affected by the occurring residual stresses and their interaction with operational load stresses. With respect to the complex geometries of most components, their dimensional and positional tolerances a further strong influence of the robot kinematics of the plasma or HVOF torches during coating manufacturing is observed. By combining the expertise in materials and manufacturing engineering coatings and composites with high performance and reliability can be achieved. This is shown in the development of functionally coated cylinder liners and crankcases for ultra light weight engines as well as for ceramic coated bioinert and biodegradable substrates in medical surgery. It will be shown that cast engine block bores can be directly coated by using an automated HVOF process, obtaining improved coating results. The internal coating process by hypersonic flame spraying is a superior technological alternative to the APS process for high quality cylinder liner and engine crankcase applications. The applications of such ceramic and cermet coatings are not limited to automotive and biomedical applications, i. e. for wear and friction properties or biomedical compatibility, but can be used for tailored thermophysical, electrophysical or catalytic properties in various technical systems.

A.   Hypersonic internal coatings for cylinder liners in light weight engines

In advanced mechanical engineering and especially in the automotive industry light weight engineering by materials, by design and processing are key requirements for competitive industries. The reduction of fuel consumption and pollution emission, engine efficiency and comfort improvement, as well as cost reduction in manufacturing are general motivations. Most of these requirements are fulfilled by reducing the total weight of the vehicle and an increasing utilization of light metal components for engine applications. Significant weight savings are obtained by changing the engine block material from cast iron to aluminum. Despite of all advantages the industrial implementation of light metals is often inhibited by their poor surface properties especially concerning wear and tribological behavior. Due to the highly loaded operation conditions a cylinder liner surface reinforcement is necessary.

Nowadays the mass proportion of the engine block on the total vehicle weight is in the range of 10 to 15 %. Therefore full light-metal engine design offers a great potential for a successful mass reduction and finally fuel consumption. Aluminum alloys instead of cast iron or steel are used in a significant and increasing number of engine designs. In the near future also magnesium alloys, because of their low density and excellent damping behavior, are expected to be used in motor applications. For a successful implementation of light metals in heavily loaded tribological systems like cylinder liner and piston rings an interdisciplinary cooperation between material science, engineering and manufacturing technology is indispensable [1]. State of the art in light metal engine design are aluminum crankcases. Demands to reduce the production process complexity, the number of process steps and manufacturing cost on the one side and the improvement of the operation behavior and recycling ability on the other side lead to the investigation of thermally sprayed coatings as protective and functional surfaces on light metal alloys, see Fig. 1.

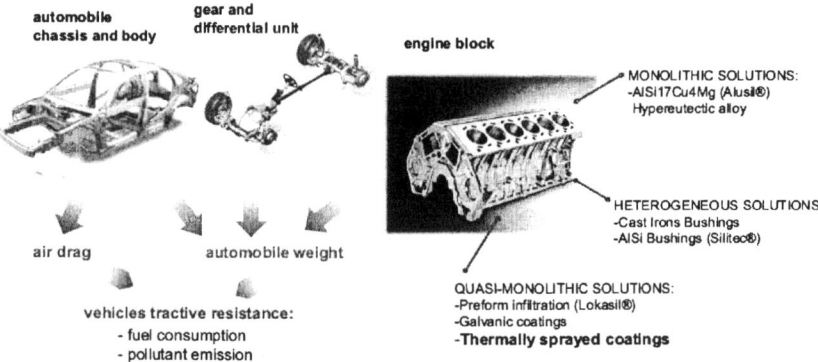

Fig. 1   Light-weight engine design to minimize fuel consumption and pollutant emissions

The thermal spray processes offer the possibility to apply a broad variety of metallurgical, cermetic and ceramic coatings on the liner surface. For the deposition of internal coatings the atmospheric plasma spraying APS as well as the high velocity oxygen fuel spraying HVOF can be used. The HVOF process uses liquid fuels or fuel gases for high energetic combustion with oxygen ($V_{particle} \sim 400$ - 900 m/s; $\vartheta_{gas} \sim 2{,}500$ - $3{,}200$ °C). It leads to extremely dense coatings because of the high kinetic energy of the hot powder loaded expanding gas jet. During the APS plasma spray process, temperatures up to 20,000 °C are obtained; therefore this system is mainly used for refractory materials. During spraying, the powder particles are molten in the plasma or autogeneous flame and accelerated to the substrate material. The average coating thickness ranges from several microns up to a few millimeters. Precedent to the coating process a roughening and degreasing of the liner surface is performed. The roughening of the surface, e.g. with corundum of defined size, improves the mechanical adhesion of the coating to the substrate material. Following to the deposition process a mechanical honing of the coating surface takes place.

State of the art for industrially produced cylinder liner internal coatings are rotating plasma spray devices. The engine block is transported by a transfer line or by a robot system. During the coating process the block is fixed and the bore is coated by a vertically moving rotating plasma torch, e.g. RotaPlasma® system, with a maximum rotation speed of 250 rpm [2]. Other thermally sprayed coating systems are under development. A very promising method is the deposition of internal coatings under rotation of the engine block. The crankcase is fixed on a rotating table, and by means of precision linear motors exactly positioned under rotation. The cylinder bore in the rotation axis is coated by a vertical or elliptical movement of the spray torch due to the higher spraying distance, 150 – 250 mm. The main advantages of internal coatings deposited under rotation of the light metal crankcase are the improved coating microstructure (less porosity) and better coating adhesion, due to an enhanced and more powerful jet propagation and thus higher particle velocity.

The HVOF coating process is adjusted in such a way that the top of the cylinder bores, subjected to maximum wear, are coated perpendicularly, whereas the less intensive loaded bottom regions are coated under a decreasing spraying angle. Using an automated control and regulation system the robot movement and rotation velocity can be flexibly adapted on variable spraying angles (90° - 30°). By adapting the feed rate of the HVOF gun to the spraying angle, a homogeneous coating thickness over the cylinder liner depth can be achieved.

In IMTCCC laboratories, a new automatically centering and rotating device for the coating of complete engine blocks by using HVOF coating techniques was developed and patented [3] see Fig. 2.

Fig. 2   Coating process of a 5 cylinder in line engine at the IMTCCC laboratories

Due to the good adhesion properties of HVOF coatings and their high hardness and wear resistance, a very promising application is in the field of high performance engines. Also a tilting device for the internal coating process of V and W-type engines was developed.

Automated coating system and kinematic analysis and optimization

Reproducibility still displays a problem in the quality control of thermal spray processes. This is especially the case if the specification of the thermal spray coatings requires low tolerances concerning coating structure, hardness, thickness, roughness and porosity. Similar specifications exist with regard to oxide content, coating pre-bonding, etc. [4]. Approaches in the field of process control are needed to achieve a high reproducibility and quality. An automated system was designed to apply internal, HVOF sprayed coatings in light weight cylinder bores. In the automated HVOF system a 6-axis robot is used to handle the HVOF torch. Microprocessor technology enables the control data of various spray parameters to be stored and retrieved when required. The crankcase to be coated is placed on a rotary table, due to the necessity of rotation of the cylinder liner during the coating process. The centered position of the cylinder is achieved by the use of a linear guide system. The connection of the robot via Ethernet with a PC-platform allows the use of CAD programs which facilitate programming tasks. In this way, the software for the robot kinematics can be developed off-line.

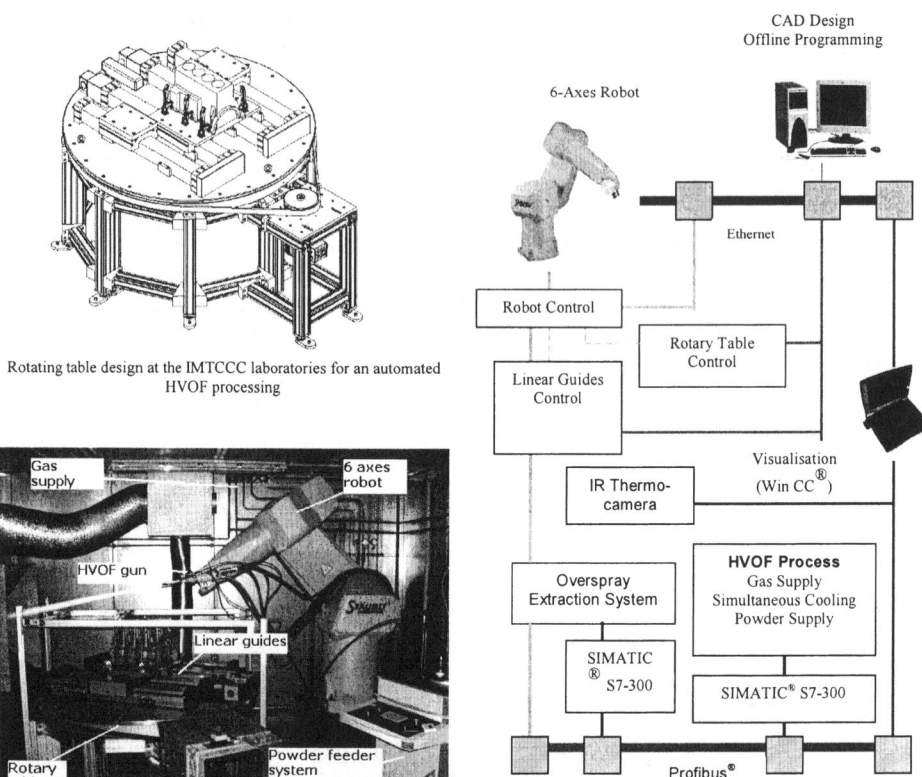

Rotating table design at the IMTCCC laboratories for an automated HVOF processing

Fig. 3a, Facilities in the IMTCCC for the internal coating process of aluminum crankcases

Fig. 3b, Control architecture of the HVOF process

Since coating properties are principally determined by the particle temperature and velocity upon impact, methods have recently been developed to measure these properties on-line. These techniques include Laser-Doppler anemometry, high velocity pyrometry and the use of special high resolution infrared thermo-cameras.

Also a complete kinematic analysis of the moveable components of the system involved in the coating process (spray torch, rotary table and linear guides) is indispensable to reach the desired properties of the coating. Concerning robot kinematics, properties such as coating structure and profile (e.g. coating thickness) are highly influenced by the spraying distance, torch feed drive and deposition angle. Also the rotation velocity of the table is an important parameter which affects the final geometry and properties of the thermally sprayed coating. In order to obtain a homogeneous coating profile the spraying distance has to remain constant. This can be achieved by describing the path of the torch using optimized geometric polynomials. This function has to be defined exclusively for each set of coating material and bore geometry to be coated.

The orientation of the torch at each point of the path depends on the dimensions of the cylinder to be coated. The top areas of the cylinder bores are coated with the torch perpendicularly positioned to the cylinder surface. This angle decreases linearly to a minimum at the final point of the path. Due to the influence of this angle decrement, which contributes to the reduction of the deposition rate over the liner depth, it is impossible to obtain a constant coating thickness when the torch is guided with constant velocity. This problem can be solved by the application of a speed profile in the movement of the torch, a profile which is defined exclusively for each material and set of spraying parameters, see Fig. 4.

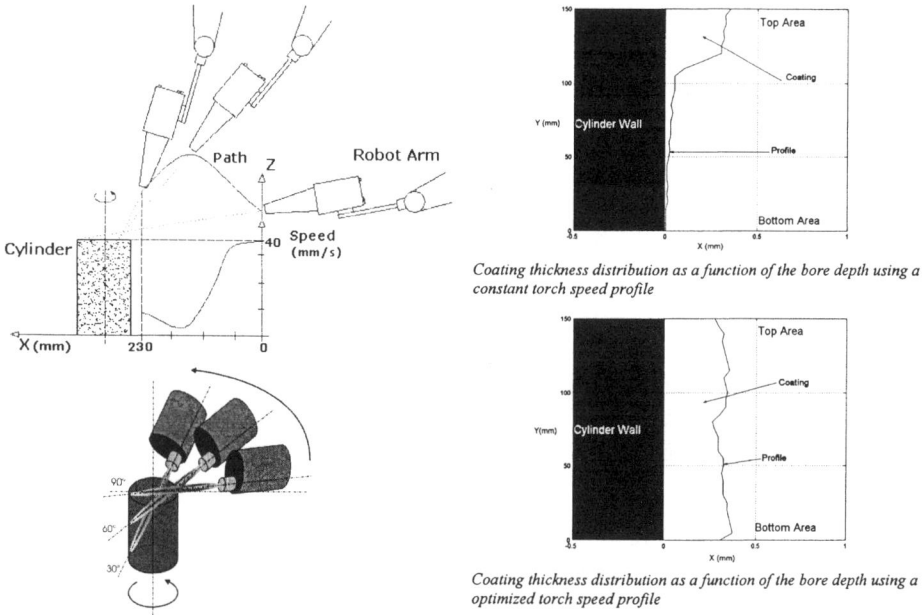

Coating thickness distribution as a function of the bore depth using a constant torch speed profile

Coating thickness distribution as a function of the bore depth using a optimized torch speed profile

Fig. 4, Path and speed profile of a HVOF coating process for cylinder bores and coating thickness optimization for a $Cr_3C_2/NiCr$ coating

As an example the optimization of a HVOF spray process using $Cr_3C_2/NiCr$ 75/25 as spray powder and propane as fuel gas has been shown. Different speed profiles were analyzed and tested. A homogeneous coating thickness of 300 μm in the aluminum liner was required. The coating profile was determined by a 3D-coordinate measurement device.

Material investigation and characterization

The material screening was focused on the development of high-efficient coating systems for tribological applications with regard to low friction and wear coefficients, high corrosion resistance as well as enhanced operation performance. The friction and wear coefficients for the interaction of piston ring and cylinder liner coating under lubrication must be less than $\mu = 0.2$ and $k_v \leq 200 \cdot 10^{-8}$ mm³/Nm. The tribological behavior is influenced by sliding speed, lubricant interaction and wettability, topography and operating conditions (temperature and stresses) as well as by the coating material characteristics and properties. Of special interest were materials with solid lubricant properties or the capability to form lubricious oxides under operational or tribological conditions, like special oxides of titanium and vanadium as well as molybdenum and tungsten. Both types of materials are characterized by low shear strengths due to planar defects in the crystal lattice. For these materials, low friction coefficients and wear rates can be expected under mixed and dry friction conditions [5, 6].

To obtain a general overview of candidate coating materials different ceramics, cermets and metals were investigated. To assess the coating quality and manufacturing reproducibility of the layer composites metallographic investigations were performed. By image analysis the coating porosity, expressed by the relative pore volume content VP [%], is determined. The hardness $HV_{0.05}$ is measured by an automated universal hardness indenter equipment Fischerscope TM HCU with a load of 500 mN.

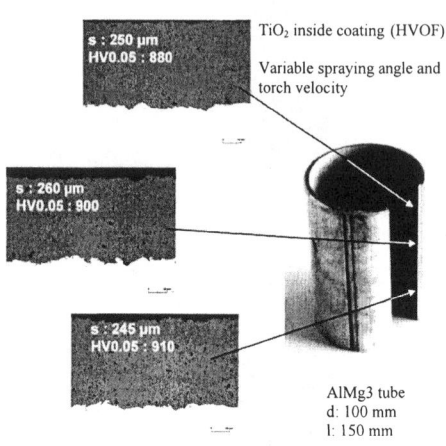

Coating	P [%] APS	P [%] HVOF	HV0.05 APS	HV0.05 HVOF
TiO_2	1 - 12	1 - 4	700 - 1500	1000 - 1300
Al_2O_3/TiO_2-60/40	1 - 4	1 - 5	1200 - 1700	900 - 1400
Cr_2O_3	3 - 10	1 - 4	2200 - 3100	1600 - 2800
Cr_2O_3/TiO_2-60/40	1 - 2	2	1600 - 2300	1200 - 2000
Cr_2O_3/TiO_2-75/25	1 - 2	1 - 2	1700 - 2300	1300 - 2100
MoO_2	16	not measured	1200 - 1400	not measured
Al_2O_3/ZrO_2-60/40	2 - 5	2 - 4	1500 - 1800	1100 - 1400
$Cr_3C_2/NiCr$-80/20	8 - 12	6 - 10	1100 - 1700	1300 - 1900
(Ti,Mo)(C,N)/NiCr-70/30	9 - 12	5 - 8	800 - 1300	800 - 1000
CrB/NiCr-75/25	18	13	1500	1300 - 1600
Mo	5 - 10	3 - 4	400 - 800	1000 - 1200
FeCr17	5 - 10	2 - 5	150 - 300	300 - 450
Mo/NiCrBSi-30/70	6 - 8	5	700 - 1100	1000 - 1200
Mo/FeCr17-70/30	6 - 10	4 - 5	400 - 500	700 - 1100

Fig. 5 Left: Homogeneity of a HVOF sprayed coating microstructure (TiO_2 coating) over the bore depth. Right: Porosity and hardness results of investigated coating systems

The deposition of a reliable and homogeneous coating quality is very important. All the investigated HVOF coating systems can be homogeneously deposited over the bore depth. Fig. 5 shows metallografically prepared cross sections of coated cylinder liners verifying the homogeneity of the coating process. The table shows the measured porosity and hardness characteristics for the investigated coatings obtained by the APS and HVOF spraying processes.

Typical deposition thickness in as sprayed state is in the range of 200 – 300 μm. To complete the process, a special honing operation is necessary, where the coating is finished to a thickness range between 100 – 150 μm with a suitable topography with small cavities which act like oil reservoirs, improving the lubrication process.

Residual stresses and bonding strength

Residual stresses arise during the thermal spray process and significantly influence the coating quality and composite performance. The general reasons for residual stresses in layer coated composites are varying thermal expansions in the different layers caused by temperature gradients and mismatch in the thermo physical properties, as well as nonuniform plastic deformations in metal substrates and coatings due to thermal and mechanical loads.

The knowledge of the occurrence and magnitude of residual stresses is an important factor in the performance of subsequent machining steps and for the determination of operation lifetime, where residual and load stresses interact [7]. To improve the operation and fatigue strength of the composite, compressive stresses are desired. The adhesion between coating and substrate is mainly influenced by residual stresses in the interface. With the circular micro milling method (CMM), the residual stresses in layer components can be experimentally determined. In several micro drilling and milling processes, a circular micro hole is introduced step by step (5 - 10 μm) into the component surface. The residual stresses are relieved by material removal, and are measured as relaxed strains ε_{z0} at the component surface using high precision strain gauges. In combination with calibration curves, Young's Modulus E and Poisson's Ratio μ, the measured relaxed strains at the surface are transformed into nominal strains εz at the bottom of the drilled hole. Out of the nominal strains, the in plane stresses αx, αy are incrementally determined by Hooke's law [7].

The residual stresses of APS and HVOF sprayed TiO_2 cylinder coatings, deposited with different processes and cooling rates, were measured up to a drilling depth of 0.4 mm, as shown in Fig. 6.

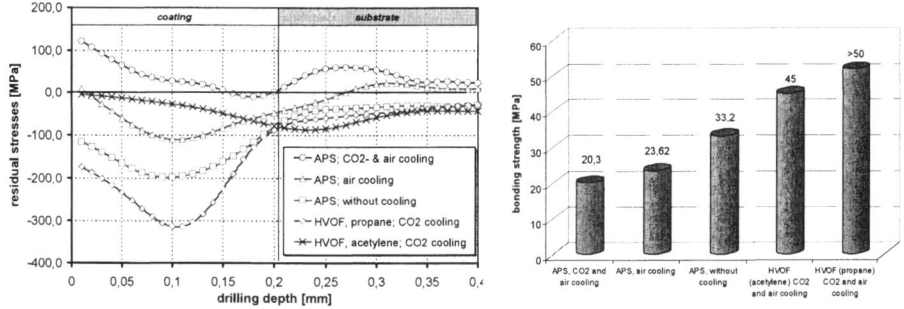

Fig. 6 Measured residual stress depth profiles and bonding strengths in TiO_2 coated cylinder bores, variation of the spraying process and cooling intensity

Due to the relatively low particle velocities (~ 50 - 150 m/s) and the large degree of totally melted particles the resulting residual stresses of APS coatings are mainly influenced by the transient temperature distribution and the different thermophysical material properties of the composite materials. The HVOF process can be characterized by high particle velocities and moderate particle fusion. Therefore thermal stresses are superimposed by additional compressive stresses induced by high particle impact energies. It should be noted that the kinetic particle energy can be controlled via the total gas flow rate as a result of the oxygen/fuel gas ratio. For APS sprayed composites tensile stresses in coating and interface are measured with increasing cooling rates. Decreasing cooling rates result in compressive stresses. Decreasing cooling rates effect an increasing uniform temperature level in the composite. Due to the mismatch in the thermal expansion coefficients of coating α_C and substrate α_S ($\alpha_S \gg \alpha_C$), compressive thermal stresses arise in the coating and the interface.

For the HVOF sprayed composites, compressive stresses in coating and interface are measured. Due to the higher gas flow rates of the propane/oxygen process compared to the acetylene/oxygen process (~ factor 1.6), and the less degree of particle fusion, higher compressive stresses are measured for the HVOF propane process.

The investigations of the coating's adhesion were performed on a Zwick Z100 universal mechanical testing machine. A steel tension rod was glued on the coating surface and the tensile load was continuously increased. As soon as delamination of the coating occurred, the tensile load was measured and the bonding strength was calculated. To ensure a secure and reliable operation process, a bonding strength of at least 30 MPa must be reached. In general it can be said that the bonding strength of the coating increases with increasing compressive stresses in the composite interface. For the HVOF propane/oxygen sprayed TiO_2 coating systems the bonding strength can not be measured because of the lower yield strength of the glue. It is observed that the bonding strength of the HVOF propane/oxygen sprayed TiO_2 coatings is at least factor 2 higher compared to APS TiO_2 coatings.

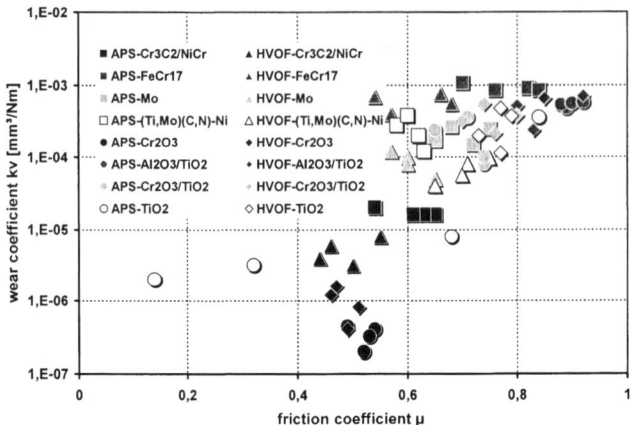

Fig. 7 Measured dry friction coefficients μ versus wear coefficient k_v, Al_2O_3 counterpart

Tribological investigations

The intention of the material screening was the selection of tribological systems with low friction coefficients as well as low wear rates on coating and counterpart (piston ring). Low friction coefficients are required to reduce the energy dissipation during operation. A decrease in the wear rate during operation increases lifetime and quality of the tribological system, and reduces the pollution of the lubricating oil with wear particles. To obtain a general overview of the wear and friction behavior of the different coating systems, dry running oscillating pin on disc tests were performed. The coatings were finished to a surface roughness of Ra ~ 0.05 µm. As a counterpart, Al_2O_3 (2,800 HV0.05) and 100Cr6 balls (1,200 HV0.05) with a diameter of 5 mm were used. The selected number of oscillating strokes was 50,000, sliding velocity 70 mm/s, length of strokes 5 mm with an imposed measurement load of 10 N. For all thermally sprayed coatings dry friction coefficients μ below 0.9 were measured, see Fig. 7.

Extremely low dry friction coefficients between 0.1 and 0.2 were measured for some TiO_2 APS sprayed coatings. Additionally, a very interesting correlation between hardness and friction coefficient was detected in the TiO_2 coating systems. Coatings with low hardness values and friction coefficients were sprayed under an increased percentage of hydrogen by optimized CO_2 and air jet cooling. Therefore, it can be assumed that oxygen sub-stoichiometric Ti_nO_{2n-1} phases with special crystallographic shear abilities are responsible for this excellent friction behavior. The wear investigations show similar advantageous results.

Additional investigations under lubricated conditions (Shell Helix Plus) were carried out, with the two different counterparts Al_2O_3 and 100Cr6. The results are summarized in Fig. 8. The measured friction and wear coefficients particularly of the cermets and ceramic coatings show superior tribological properties compared to the metallurgical coatings and uncoated substrate materials.

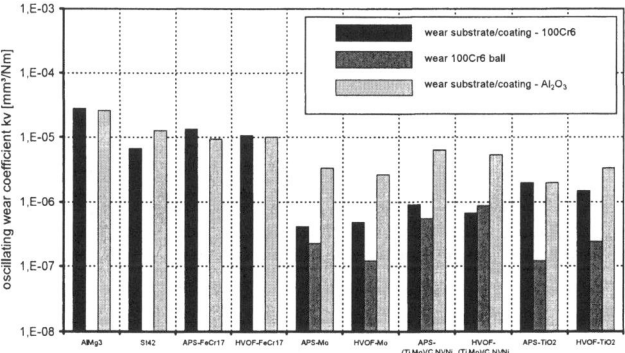

Fig. 8   Wear rates of different substrate and coating materials under lubricated conditions

Series coating process in engine manufacturing

In a series coating process of cylinder bores, after cleaning and roughening, the light metal crankcases are transported on a transfer line into the coating cabin. Due to the high degree of acoustic noise and unhealthy process conditions the coating cabin is isolated from the surrounding. By using a handling robot system the engine blocks are lifted from the line to the rotation device and fixed automatically. Under rotation the bores to be coated are exactly positioned in the rotation axis for each individual bore. The movement of the spraying gun is performed by a multi axes robot system. When the coating process is finished, the engine block is lifted back to the transfer line and transported to the final machining station. For reliability reasons the fuel gas and powder mass flows as well as the positioning coordinates and velocities are automatically steered and recorded by the control units. Fig. 9 shows a schematic drawing of a series production plant for the deposition of HVOF sprayed coatings in cylinder bores under rotation of the light metal crankcase.

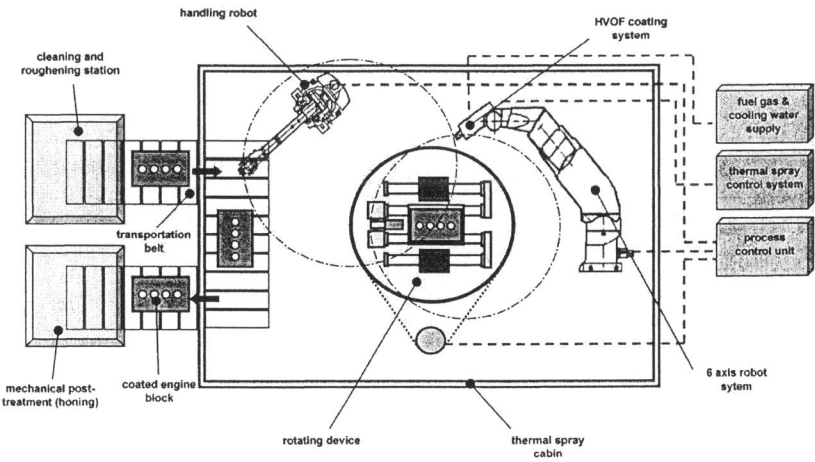

Fig. 9   Schematic drawing of the series process plant for internally coated crankcases

Conclusions

Thermally sprayed coatings offer a broad material variety as well as flexible and cost effective manufacturing processes for refined crankcases. The poor tribological operation behavior of light metal surfaces can be improved by protective coating systems. By APS and HVOF spraying, wear resistant internal coatings were deposited on light metal cylinder bores. During the material screening different ceramic, cermetic and metallurgical coating systems were deposited and investigated. Residual stress measurements have shown that the stresses in thermally coated layer composites strongly depend on process temperatures, as well as on the particle impact velocity. Tribological investigations show friction coefficients lower than 0.9 for all coatings under dry conditions. In the case of plasma sprayed TiO_2 coatings, excellent low friction coefficients between 0.1 and 0.2 were measured under dry running conditions. The investigations during the material screening and first bench tests reflect positive results for most of the coating systems. Oxygen sub-stoichiometric TiO_x phases obtained under reduced APS spraying conditions showed the most promising tribological behavior under dry sliding conditions with a high potential for applications under critical or mixed lubrication conditions in high performance thermal engines. A detailed coating selection and material development will be done after further combustion bench tests in stationary reciprocating engines. By HVOF processing, engine blocks can be coated with improved coating qualities compared to APS and other conventional thermal spray coating techniques. Therefore the HVOF internal coating process is a technological alternative to the APS process with rotating internal plasma torch for high quality cylinder liner and engine block applications. The adhesion to the various liner substrates, the hardness and wear resistance of the HVOF deposited coatings, their homogeneity of density, microstructure and friction behavior over the height of the bore are superior to any other coating deposition. This technology will continue to become more important in the near future in passenger and sport car motor engineering but also in light weight engines for small aircrafts and sport boats.

B.   Bioceramic coatings and composites for implants and prostheses in modern surgery

Bioresorbable polymer implants are a promising research field in maxillofacial surgery, since their use eliminates the need for a secondary operation to remove metal implants. Mechanical properties and biocompatibility of these implants are nevertheless not completely satisfactory, and new composite devices are demanded. Thermally sprayed tricalcium phosphate (TCP) coatings may significantly increase the biocompatibility of polymer implants and contribute to match the resorption rate of the device with the bone healing rate, leading to a correct mechanical stress transfer implant/tissue and therefore to successful fracture fixation even in load conditions. The manufacturing process of suitable β-TCP powders for the Atmospheric Plasma Spraying (APS) via spray-drying granulation and the thermal spray coating process on bioresorbable poly(D,L)lactide devices are reported and discussed.

Today the mostly applied and studied bioresorbable materials are tricalcium phosphate [TCP, $Ca_3(PO_4)_2$] and hydroxyl apatite [HAP, $Ca_{10}(PO_4)_6(OH)_2$] among ceramic materials [8, 9] and polyglycolic (PGA) or polylactid (PLA) acids among polymers [10, 11]. Calcium phosphates are generally brittle materials, which are applied only in form of powder (TCP) as bone filler material or as coating on metal implants (HAP). Calcium phosphates are the most important inorganic constituents of biological hard tissues, such as bones. In particular up to 70 % (wt.) of bones are composed of natural HAP, and up to 99 % (wt.) in case of teeth [12]. Synthetic HAP is a bioactive osteoconductive material, able to create a strong chemical bond with the surrounding tissue, and when implanted in the human body it remains stable and chemically unchanged over years. On the other hand TCP is a bioresorbable, osteoinductive ceramic, which is gradually resorbed and replaced by new bone substance as follows:

$$4Ca_3(PO_4)_2 + 2H_2O \rightarrow Ca_{10}(PO_4)_6(OH)_2 + 2Ca^{2+} + 2(OH)^{2-}$$

TCP is known in two polymorphs, the β phase (rhombohedral) stable up to 1120 °C and the α -phase (monoclinic) stable in the range 1120 °C ÷ 1820 °C. The α-phase can be found at room temperature only as metastable phase. Previous studies pointed out that α-TCP has a much higher resorption rate in physiological environment than β-TCP [13].

Polymer fracture fixation devices present some advantages compared to metal implants in particular regarding their mechanical properties, because they match the natural mechanical properties of bones much better than metal fixation devices. Polymers are less stiff than metals and this prevents the possible atrophy of bones due to stress protection by the rigid metal osteosynthesis plate. Bioresorbable polymer implants prevent the secondary operative procedure to remove the metal implant. Another advantage of bioabsorbable polymer devices is their ability to be shaped and bent to fit bone geometry *in vivo* by heating and forming them above the glass transition temperature (T_g). An ideal bioresorbable fixation implant should be degraded during the bone healing process at the same rate at which the bone repairing occurs [14, 15]. In maxillofacial surgery one of the most recently applied polymer is amorphous poly(D,L)lactide (PDLLA). This material generally doesn't cause any foreign body reaction during resorption and it is employed to produce plates and screws for fracture fixation of mechanically unloaded bony fragments [16]. The application in loaded zones (e.g. in jaw bone fractures) may nevertheless lead to failure of the implant itself. This is due to the resorption rate of the material, which occurs faster than the healing process of the fracture. The mismatch between resorption rate and healing rate leads therefore to an incorrect stress transfer. Long-term aim is to produce new ceramic/polymer composite bioresorbable devices for internal fracture fixation of loaded bony fragments in maxillo-facial surgery with higher bioactive surface, in order to stimulate the bone repairing process and to provide at the same time an optimized stress transfer between implant and healed bones.

Experimental procedures

β-TCP raw powders were processed to be spray-dried in a co-current flow spray-drier in order to get suitable granulates for the Atmospheric Plasma Spraying (APS) technique. The aim is to obtain round granulates with a monomodal and narrow grain size distribution with mean grain size D_{50} in the range 30÷120 μm. These characteristics are requested to get the best spraying conditions, i.e. uniform melting of the powder in plasma stream, better flowability and constant feeding by means of conventional powder feeders. The manufacturing process and powders characterization are reported in previous papers [17, 18]. The powders grain sizes are given in table 1.

Table 1 Grain sizes of the spray-dried powders

Powder	P1	P2	P3
D_{10} (μm)	12	33	32
D_{50} (μm)	82	119	115
D_{90} (μm)	194	340	234

Table 2 Plasma spray parameters

Substrate	PDLLA
Spraying distance (mm)	100 ÷ 120
Gas flow Ar (l/min)	35 ÷ 50
Gas flow H_2 (l/min)	6 ÷ 8
Power (kW)	30 ÷ 37

The coating experiments were carried out by means of a 6-axis computer controlled robot equipped with a GTV plasma torch MF-P-1000 under the conditions listed in table 2. An outstanding advantage of APS with simultaneous cooling is the relatively low thermal load on the substrate during the coating process, so that it is possible to coat thermally sensitive substrates in an appropriate mechanical assembly with robot controlled plasma torches.

Compressed air was used for convective cooling. Thermal spraying on polymers may lead in the worst cases to complete or partial melting of the substrate, destroying it or at least modifying its properties. In particular in case of medical implants it is of paramount importance to avoid as much as possible surface modifications in order to maintain the original resorption properties of the polymer. The combination of fast torch movement and high cooling rate led to very satisfactory results (Fig. 10). The partial melting of a sub-micron thick polymer layer at the interface substrate/coating is nevertheless unavoidable (Fig. 11). Besides, this mechanism leads to mechanical adhesion of the coating to the substrate, which occurs as usual in thermal spraying by shrinkage of the molten particles at the impact with the substrate's surface.

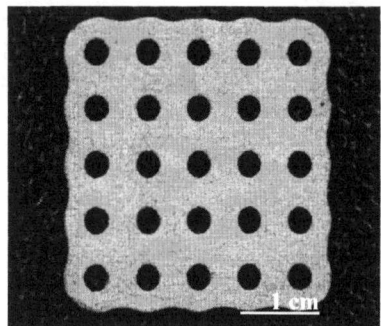

Fig. 10 Picture of a PDLLA implant plate coated with TCP by means of APS

Fig. 11 Light microscope shot of micro- and macropores in the TCP coating

The amount of porosity in the coatings, as well as the coatings' roughness, is influenced by the amount of unmolten particles, i.e. from the combined effect of spraying parameters, powders grain size and grain size distribution. P1 is actually the powder with the lower mean grain size and besides it was sprayed with higher plasma heat content, leading to uniform and complete melting of the particles in plasma and therefore to lower porosity and roughness. P2 and P3 on the other hand have a higher mean grain size and they were sprayed with lower plasma heat content. This led to incomplete melting and therefore to higher porosity and roughness. Roughness and porosity in relation to the powder type are given in table 3. A scanning electron microscope analysis of the coatings surface (Fig. 12) also showed the irregular and incomplete melting of the particles (splats).

Table 3 Coatings porosity and roughness

Powder	Porosity (vol. %)	R_a (µm)	R_z (µm)	R_{max} (µm)
P1	12.6	9.1	54,3	60,3
P2	41,7	13,7	78,5	92,9
P3	31,6	13,9	79,2	95,8

A qualitative bending and shaping test was carried out at 70 °C. At this temperature it is still possible to easily modify the implant's shape without damaging or detaching of the ceramic coating (Fig. 13), allowing to match the implant to the bone morphology. The mechanism of this behavior might be formation of microcracks in the coating, which allow the coating itself to be bended and shaped together with the substrates. Being the coatings highly porous, occurring microcracks are frequently interrupted and are not able to grow all through the coating, avoiding thereby delamination and/or failure.

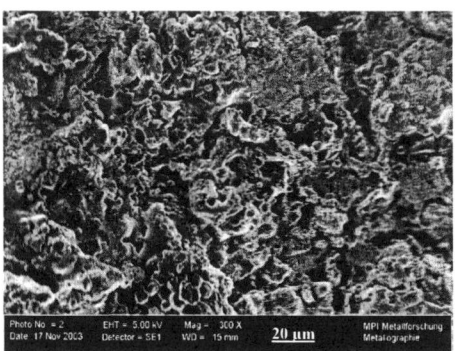

Fig. 12 SEM of the TCP coating's surface.

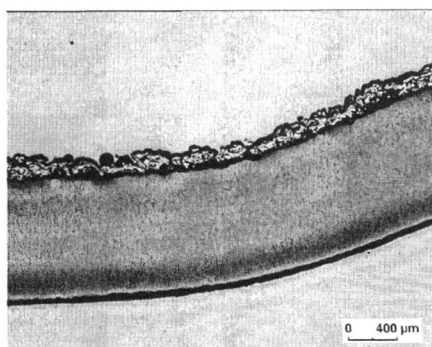

Fig. 13 Composite PDLLA/TCP after manual shaping and bending at 70 °C

The melting process during APS coating deposition involves phase transformations which usually are not reversible. After being molten in the plasma jet, the particles impact on the substrate where they rapidly cool down (quenching). In particular regarding to TCP coatings is it of crucial importance to know whether α- or β-phase is present, or a combination of the two, because of their different solubilities in the human body. The β-TCP particles in plasma undergo the phase transition β to α, which occurs at temperatures around 1120 °C. This transition may occur either completely or only partially, depending upon plasma parameters and powder properties. By means of XRD is it possible to determine the phase composition of each coating (Fig. 14).

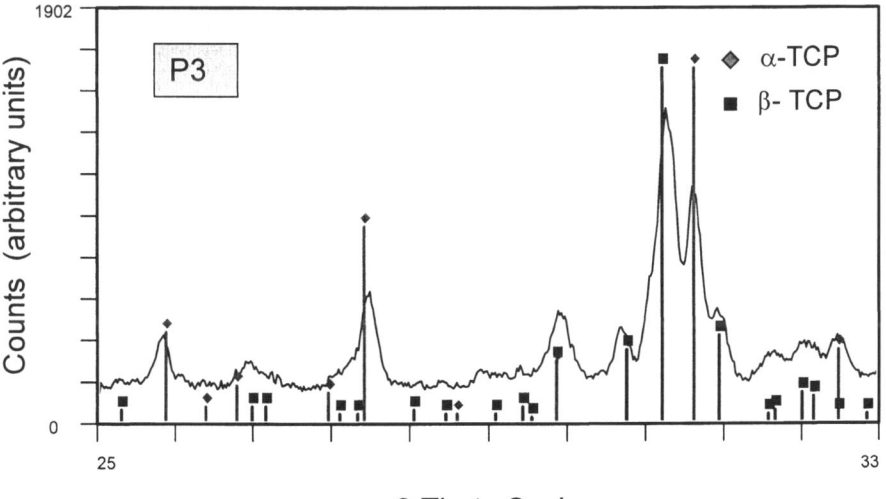

2 Theta Scale

Fig. 14 XRD pattern of α- and β-TCP mixture coating obtained by spraying P3.

In case of a two-phase mixture the following equation (1) permits to calculate the weight fraction w_β for the amount of β-TCP and consequently the amount of α-phase in each coating [19]:

$$\frac{I_\beta}{I_{\beta p}} = W_\beta . \qquad (1)$$

Depending upon spraying parameters and powder properties, the calculated β-phase weight percent W_β resulted to be comprised between 11 and 26 %. It is evident that much higher β-TCP contents cannot be achieved, because the powder particles must be mostly molten in the plasma jet to get sufficient adhesion to the substrate. Non molten particles simply bounce off the substrate at the moment of the impact without creating any coating.

Tissue Engineering by Ceramics and Coatings

Oxide ceramics are promising candidates for bioreactive substrates in selective cell growth. They interact in a nanoscopic range by their various surface functionalities and furthermore by their mesoscopic morphology, in example hollow spheres and cavities (Fig. 15).

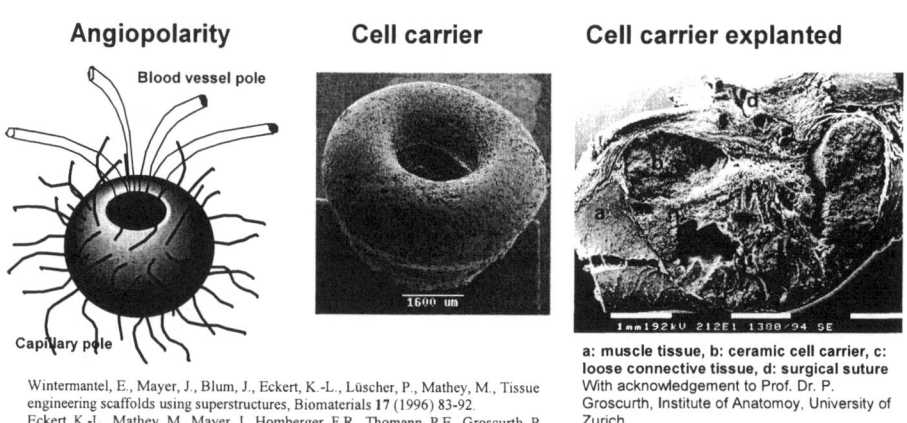

Wintermantel, E., Mayer, J., Blum, J., Eckert, K.-L., Lüscher, P., Mathey, M., Tissue engineering scaffolds using superstructures, Biomaterials **17** (1996) 83-92.
Eckert, K.-L., Mathey, M., Mayer, J., Homberger, F.R., Thomann, P.E., Groscurth, P., Wintermantel, E. Preparation and in vivo testing of porous alumina ceramics for cell carrier applications, Biomaterials **21** (2000) 63-69

a: muscle tissue, b: ceramic cell carrier, c: loose connective tissue, d: surgical suture
With acknowledgement to Prof. Dr. P. Groscurth, Institute of Anatomoy, University of Zurich

Fig. 15 Implantable ceramic hollow sphere cell carriers

With respect to space efficiency of implants in the human body (abdominal cavity) a more advanced strategy is to apply ceramic coatings. To exploit the effect of angiopolarity of micro blood vessels a designed surface morphology with artificial cavities is required. Laser machining is the most suitable method to produce a dense network of microscopic cavities in titania coatings (Fig 16). Thereby flat and thin ceramic structures can be produced with the potential of multilayer assembly with high selectivity and space efficiency.
A novel approach is the deposition of ceramic layers and coatings on flexible fibre structures and woven fabrics [20] (Fig. 17).

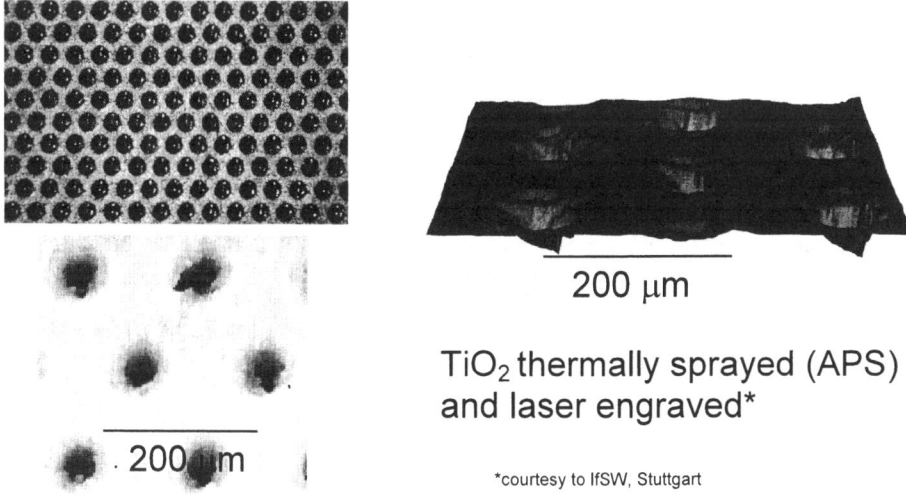

200 µm

TiO₂ thermally sprayed (APS)
and laser engraved*

*courtesy to IfSW, Stuttgart

Fig. 16 Thermally sprayed titania coatings with subsequent laser machining and designed morphology

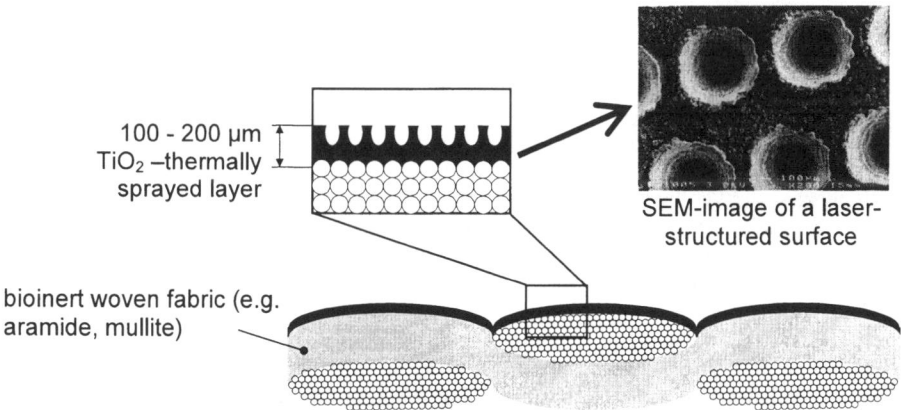

100 - 200 µm
TiO₂ –thermally
sprayed layer

SEM-image of a laser-
structured surface

bioinert woven fabric (e.g.
aramide, mullite)

Fig. 17 laser engraved ceramic coating on flexible fiber structure as carrier system

Conclusions

The manufacturing of composite materials which couple the properties of bioceramics and polymers or fiber woven fabric structures is a promising way to solve some of the problems that these materials meet when they are implanted in the human body. To produce TCP ceramic coatings on polymers the thermal spraying process (APS) was chosen. This process requires appropriate powders. The powders were produced via spray-drying and optimized for the requirements of thermal spraying with automatic powder-feed equipment. Different slurries were produced and afterwards spray-dried. Three spray-dried powders possess the required properties, i.e. monomodal grain size distribution, mean grain size in the range $80 \div 120$ µm, good flowability and round shaped grains. The spray drying conditions can be now adapted from the pilot plant data to small series production. The as-sprayed TCP coatings present a much higher bioactive surface in relation to normal PDLLA implants. This is certified by the high roughness values and also by the high porosity (12.6 ÷ 41.7 vol. %), with pore sizes ranging from few µm up to 100 µm and more (light microscope analysis). The coatings were also analyzed by means of XRD in order to determine phase composition and crystallinity. Depending on the combination of spraying parameters and powder type, the coatings are made of pure α-TCP or of a mixture of β- and α-phase. The β-phase content ranges between 11 and 26 %. The new composite TCP/PDLLA maintains one of the most important characteristics of polymer implants, i.e. the possibility to heat it up over the glass transition temperature in order to successfully bend it and shape it. This way the surgical implant can be adapted to the from patient to patient different shape of the fractured bones.

Future works in the manufacturing process include a complete mechanical characterization of the composite. In particular residual stresses in the substrate should be reduced to the minimum, in order to avoid a possible worsening of the implants mechanical properties. Afterwards the implants will be tested both *in vitro* and *in vivo*, in order to verify the advantages given by the ceramic coating.

Summary

The improvement of tribological operation behavior of light metal surfaces by protective coating systems is a key to light weight design and engineering. Thermally sprayed coatings offer a broad material variety as well as flexible and cost effective manufacturing processes for refined light metal crankcases. By HVOF processing, engine blocks can be coated with improved coating qualities compared to APS and other conventional thermal spray coating techniques. Therefore the HVOF internal coating process is a technological alternative to the APS process with rotating internal plasma torch for high quality cylinder liner and engine block applications. The adhesion to the various liner substrates, the hardness and wear resistance of the HVOF deposited coatings, their homogeneity of density, microstructure and friction behavior over the height of the bore are superior to any other coating deposition.

The manufacturing of composite materials which couple the properties of bioceramics and polymers is a promising way to solve some of the problems that these materials meet when they are implanted in the human body. The TCP ceramic coatings on the polymer were applied with a thermal spray process. The powders were produced via spray-drying and optimized for the requirements of thermal spraying with automatic powder-feed equipment. The as-sprayed TCP coatings present a much higher bioactive surface in relation to normal PDLLA implants. The new composite TCP/PDLLA maintains one of the most important characteristics of polymer implants, i.e. the possibility to heat it up over the glass transition temperature in order to successfully bend it and shape it. This way the surgical implant can be adapted to the from patient to patient different shape of the fractured bones.

REFERENCES

[1]   Hinz R., Schwaderlapp M.: "Potential zur Massenreduktion am Beispiel eines 4-Zylinder-Reihenmotors - Potential of mass reduction, e.g. for a 4 cylinder in line engine"; Leichtbau im Antriebsstrang (1996), pp. 162 – 173, Expert Verlag, ISBN 3-8169-1336-9

[2]   Barbezat G., Keller S., Wuest G.: "The advantages of the plasma spray process for the coating of cylinder bores on AlSi cast alloy in the automotive industry"; Conference Proceedings, United Thermal Spray Conference 1999, pp. 10 - 14, Düsseldorf, ISBN 3-87155-653-X

[3]   Buchmann, M.; Gadow, R.; Killinger, A.; Lopez, D.: "Verfahren und Vorrichtung zur Innenbeschichtung von Hohlräumen durch thermisches Spritzen" Deutsches Patent Nr. DE 102 30 847, AT: 04.07.2002, Erteilt am 05.02.2004

[4]   Nassenstein K., Rickerby D., Gent J.: "GTV/CTA-Universal Spray System to control HVOF Guns for the development of coatings used in the aerospace industry", 5th HVOF Colloquium High Velocity Oxy-Fuel Flame Spraying, Erding, Germany, 2000

[5]   Woydt M.: "Materials-based concepts for an oil-free engine"; New Directions in Tribology (1997), pp. 459 - 468

[6]   Woydt M., Skopp A., Dörfel I., Witke K.: "Wear Engineering Oxides / Antiwear Oxides"; Tribology Transactions, Volume 42 (1999), pp. 21 - 31

[7]   Buchmann M., Gadow R.: "High Speed Circular Microhole Milling Method for the Determination of Residual Stresses in Coatings and Composites"; Ceramic Engineering and Science Proceedings Volume 21, Issue 3, pp.: 109 - 116, ISSN 0196-6219, 2000

[8]   de Groot K., Koch B., Wolke J.G.C., (1990): "X-Ray Diffraction Studies on Plasma-Sprayed Calcium Phosphate-Coated Implants", Journal of Biomedical Materials Research, Vol. 24, 655-667

[9]   Hench L.L., (1998): "Bioceramics", J. Am. Ceram. Soc., 81 [7] 1705-28

[10] Bessho K, Iizuka T, Murakami K, (1997): "A bioabsorbable poly-L-lactide miniplate and screw system for osteosynthesis in oral and maxillofacial surgery", J Oral Maxillofac Surg 55: 941-945

[11] Bos R, Rozema F, Boering G, (1990): "Bioresorbable osteosynthesis in maxillofacial surgery", Oral and Maxillofacial Clinics of North America 2: 745-750

[12] Wintermantel E., Ha S.W., (1996): "Biokompatible Werkstoffe und Bauweisen. Implantate für Medizine und Umwelt", ed. Springer

[13] LeGeros J.P., LeGeros R.Z., (1993): "Dense Hydroxylapatite", in "An Introduction to Bioceramics", Advanced Series in Ceramics – Vol. 1, Editors Hench and Wilson, World Scientific

[14] Epple M., Eufinger H., Rasche C., Schiller C., Weihe S., Wehmöller M., (2001): "Ein optimierter biodegradierbarer Werkstoff für die Behandlung grossflächiger Schädeldefekte", Biomedizinische Technik, Band 46, Ergänzungsband 1

[15] Beckmann F., Epple M., Eufinger H., Rasche C., Schiller C., Weihe S., Wehmöller M., (2003): "Geometrically structured implants for cranial reconstruction made of biodegradable polyesters and calcium phosphate/calcium carbonate", Biomaterials, article in press

[16] Adam C, Hoffman J, Troitzsch D, Zerfowski M, Reinert S, (2003): "Bioresorbable polymer implants in maxillofacial trauma surgery", Eur Surg Res 35: 312-313

[17] Baccalaro, M , v. Niessen, K., Gadow, R.: "Manufacturing of Thermally Sprayed Tricalcium Phosphate (TCP) Coatings for Biomedical Applications". In: *Abstracts of 8th European Interregional Conference on Ceramics, CIEC8*, 03.-05. September 2002, Lyon

[18] Baccalaro, M.; Gadow, R.; v. Niessen, K.: "Manufacturing of thermally sprayed Tricalcium Phosphate TCP Coatings for Biomedical Applications". Symposium 2 on Bioceramics: Materials and Applications: a Symposium to honor Larry Hench, 105th American Ceramic Society Annual Meeting, April 2003, Nashville, Tenn., USA. AcerS Ceramic Transaction Vol. 147, ISBN: 1-57498-202-8

[19] Cullity B.D., (1978): "Elements of X-Ray Diffraction", 2nd Ed., Addison-Wesley ed., London

[20] Gadow, R.; v. Niessen, K.; „Ceramic Coatings on Fiber woven Fabrics" Ceramic Engineering and Science Proceedings 23 [3], eds. Lin, H.-T.;Singh, S., The American Ceramic Society (2002), Westerville, Ohio, ISSN 0196-6219, pp. 277 – 285

Key Engineering Materials Vol. 333 (2007) pp 195-214
online at http://www.scientific.net
© 2007 Trans Tech Publications, Switzerland

Analysis of Free-layer Damping Coatings

Peter J. Torvik

Professor Emeritus, Air Force Institute of Technology, USA

Torvik@att.net

Keywords: Free-layer damping, coatings, damping, Öberst beam

Abstract.   A free layer damping treatment is formed by applying a coating of a high damping material to one or both sides of a structure. If the coating is perfectly bonded to the structure, the coating may be taken as being subjected to the strain at the interface between structure and coating. Consequently, the dominant component of strain is tensile.   Free layer treatments may be symmetric (identical coatings on both sides of the structure), or asymmetric (one side only), and applied to beams, plates or shells in bending or in tension. Two classes of treatments are of interest, those employing thin coatings of a relative stiff damping material, and those employing thick coatings of a relatively soft material. As many materials useful as hard coatings are nominally linear but have inherently nonlinear damping characteristics, provision for this must be made.

A Simplified Analysis

It is useful to consider first the symmetric coating as shown in Fig. 1. We take the structure to be a beam of uniform thickness, h, and Young's modulus E.

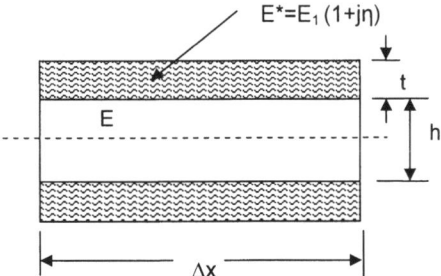

Fig. 1. Symmetric Free-Layer (Extensional) Coating.

We consider first the case of a coating of uniform thickness, t, much less than h, applied to both sides and over the entire width and length of the beam, and having dissipative properties that may be adequately represented by a complex modulus of the form

$$E^* = E_1 + jE_2 = E_1(1 + j\eta)$$

$$(1)$$

so that the ratio of the loss modulus, E_2 , to the storage modulus, E_1 is the loss factor of the *material,*

$$\eta = E_2 / E_1 \qquad (2)$$

In general, E_1, E_2 and η may be expected to be functions of temperature and frequency, but independent of location. If the material is linear, then E_1, E_2 and η are all independent of strain. The energy dissipated in the incremental length of the coating is

$$D = \int_{Vol} \pi E_2 \varepsilon(x,z)^2 \, d\,vol \tag{3}$$

where $\varepsilon(x, z)$ is the local strain. The strain distribution throughout the beam and the coating varies linearly with distance from the neutral axis, which is located at the beam centerline for a symmetric treatment, so

$$\varepsilon(z) = z\frac{\partial^2 w(x)}{\partial^2 x} = \varepsilon_0(x)\frac{2z}{h} \tag{4}$$

where $\varepsilon_0(x)$ is the strain at the beam–coating interface and z is measured from the neutral axis of the beam. The energy dissipated in both layers of the coatings, for a width W and the length Δx, is

$$D = 2\pi E_2 W\Delta x\varepsilon_0^2(x) \int_{h/2}^{h/2+t} (\frac{2z}{h})^2 \, dz = 2\pi E_2 W\Delta x\varepsilon_0^2(x)\frac{h}{2}\frac{1}{3}[(1+\frac{2t}{h})^3 - 1] \cong 2\pi E_2 t\varepsilon_0^2(x)W\Delta x \tag{5}$$

when t/h << 1. But the strain energy stored in the coating is

$$U_C = \int_{Vol} E_1 \frac{\varepsilon(x,z)^2}{2} \, d\,vol \tag{6}$$

For a linear material, $\eta = E_2 / E_1$ is independent of strain and therefore of position, so

$$U_C = \frac{E_2/\eta}{2} \int_{Vol} \varepsilon(z)^2 \, d\,vol = \frac{D}{2\pi\eta} \tag{7}$$

The strain energy in the beam, for the same width and incremental length is

$$U_B = \int_{Vol} E \frac{\varepsilon(z)^2}{2} \, d\,vol = \frac{E}{2}\varepsilon_0^2 \int_{-h/2}^{h/2} (\frac{2z}{h})^2 \, dz = \frac{E\varepsilon_0^2}{6}W\Delta x h \tag{8}$$

The *system* loss factor is defined to be

$$\eta_{SYS} = \frac{D_{TOTAL}}{2\pi U_{TOTAL}} \tag{9}$$

After integrating the energies dissipated and stored over the total length of the beam and substituting Eqs. 5b, 7, and 8 into Eq. 9, and noting that the integrals factor and $E_2 = \eta E_1$, we have:

$$\eta_{SYS} = 6\eta\frac{E_1 t}{Eh}/[1+\frac{6E_1 t}{Eh}] \tag{10}$$

If t/h is not $\ll 1$, the factor of t in numerator and denominator should be replaced by the relationship from Eq. 5a, i.e. $h[(1 + 2t/h)^3 - 1]/6$. But if the stiffness t E_1 is also $\ll$ h E, Eq. 10 can be further simplified: the system loss factor is simply six times the product of the material loss factor and the stiffness ratio. The factor of six arises as follows—2 from the coating on both sides, and three since the mean value of the strain energy density in the bream is only 1/3 of the peak value due to the linear variation in strain. In the case of the same free layer treatment applied to both sides of a bar in tension, of a shell with membrane stresses predominant, the loss factor is only twice the product of the coating loss factor and the stiffness ratio. It is important to note that that the longitudinal distribution of strain appeared in both numerator and denominator of Eq. 10a as the integral of the square, and therefore cancelled. Thus, the system loss factor is the same for all beam boundary conditions. Further, there is no inherent dependence on the mode number. However, if the properties of the damping material vary with either amplitude or frequency, the system loss factor will differ between modes.

Although a number of simplifying assumptions have been made in arriving at this point, e.g. uniform thickness, complete coverage of both sides, and linearity of the coating material, the simple relationship

$$\eta_{SYS} = 6\eta \frac{E_1 t}{Eh} = 6(\frac{E_2}{E})(\frac{t}{h}) \tag{11}$$

captures the essential features of free layer damping. The first thing to be noted is that the loss factor and storage modulus of the coating material as individual properties are irrelevant. All that matters is the product, $\eta E_1 \equiv E_2$, which is the loss modulus. Further, the system loss factor may be made large only through two means. Increasing the thickness ratio, t/h, or increasing the ratio of modulii E_2/E. Taking the structure to be given, the most effective treatment has the thickest possible layer of the material having the greatest possible loss modulus. In this respect, the free layer treatment is very different from the constrained layer treatment. In that case, the modulus of the damping material appeared as $|G^*|/t$, and a low modulus can be offset by a very low thickness. In the case of the free-layer treatment, the modulus appears as the product, $Im\{t E^*\}$, and large values of both are desirable.

Examples (Simplified Analysis)

Before examining in greater detail the consequence of these simplifying assumptions, it is of interest to examine the system loss factors that can be achieved with examples from each of the four classes of materials.

Example 1. "Soft" Viscoelastic Material. The long-chain polymers are characterized by a strong frequency dependence of the storage modulus that typically varies by three or four orders of magnitude between the rubbery and glassy regions. The peak value of the loss factor, typically about unity at a particular combination of temperature and frequency that is unique to each material, occurs at frequencies in the lower part of the transition region, and peak values of loss modulus in the upper. The modulli and loss factors of such materials are also strongly temperature dependent.

The loss modulus as a function of true frequency at several temperatures is given in Fig. 2 for a typical material as might be used in a constrained layer treatment for room-temperature application. The data as plotted was developed from curve fits to the storage modulus and loss factor as given in [1]. Values of shear modulus were converted to extensional modulus by assuming a value of ½ for Poisson's ratio so that E = 3G.

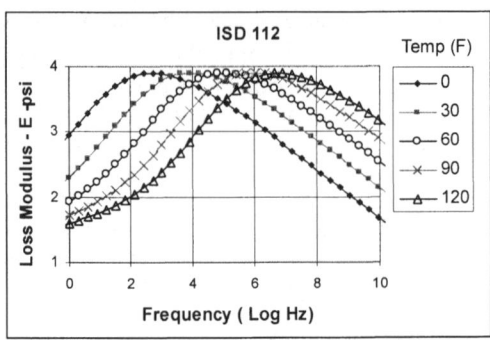

**Fig. 2. Frequency Dependence of Loss Modulus (log scales)
of a Low Temperature Damping Material.**

Assuming that the frequencies of interest will fall in the range of 100-10,000 Hz, (logarithms of 2-4), we see that values of loss modulus much in excess of 1000 psi (logarithm of 3) are largely confined to temperatures somewhat below room temperature. With a loss modulus of 9000 psi, applied to a steel structure at a volume fraction of 20% (t/h) = 0.1, we have from Eq. 11 that the contribution of the coating to the system loss factor would be about 0.00018, which is likely to be less than the inherent damping of the beam, and much less than that due to support and air damping (~0.001) losses. Thus the addition of the coating can be expected to provide little improvement. However, polymers with higher loss modulii are available, and with sufficient coating thicknesses do find application as free layer treatments. Beyond the increase in dissipation due to the additional volume, a thick coating can be said to provide its own strain amplification. If the coating thickness is equal to that of the substrate, the average strain in the coating is 50% greater than the maximum strain in the substrate, and the dissipation is (for a linear material) proportional to the square of the strain. The evaluation of the loss factors for very thick coatings will be considered in a later section.

Example 2. Vitreous Enamel. The maximum damping capacity of a linear anelastic solid has been shown by Zener [2] to be directly proportional to the difference between the isothermal (low frequency) modulus, and the adiabatic (high-frequency) modulus. Thus, any material which undergoes a transition associated with a significant change in modulus, i.e. rubbery to glassy, can be expected to have significant damping capacity within that transition region. In the case of glass cooled from a melt, the high viscosity precludes the assumption of a crystalline state. Rather, the glass remains as a super cooled liquid until further cooled to a glass transition temperature. The high viscosity in the super cooled state serves as the damping mechanism. Accordingly, in this range, a strong dependence on frequency as well as temperature is to be expected. It has been found that the application of glass as a coating (e.g. porcelains or vitreous enamels) can provide a significant enhancement of damping when applied as a coating on metal structures.

Using curve fits to the loss factor and storage modulus of Corning 7570 glass as reported by Drake (*loc. cit*), the loss modulus of this material as a function of frequency was determined at several temperatures. The results are given in Fig. 3. The abscissa and ordinate are again the logarithm to the base 10. The loss modulus given here is the imaginary part of the complex Young's modulus.

Again assuming the frequencies of interest are in the range of 100 to 10,000 Hz, a loss modulus of more than 100,000 psi will occur only for temperatures between (about) 850 deg F and 950 deg F. At 900 degrees, the loss modulus will remain above 200,000 psi and if applied to a steel structure at 20% volume fraction (t/h = 0.1), the contribution of the coating to the system loss factor (from Eq. 11) can be expected to be at least 0.004 and, if the inherent quality factor of the uncoated system is about 1000, the addition of the coating should lead to a system Q of about 200. Further, at 900 degrees even a high temperature alloy will experience some reduction in modulus. Hastelloy X, for example, has a modulus at 900 degrees which is about 14% lower than at room temperature. As less energy is stored in the beam for a given strain, the loss factor will be (approximately) increased in this proportion.

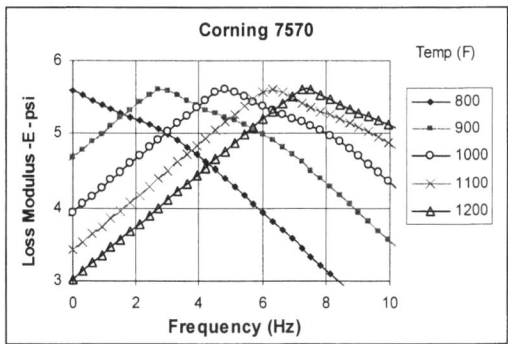

Fig. 3. Loss Modulus for a Representative Glass (log scales).

Example 3. Shape-Memory Alloy.   An example of an alloy family with high inherent damping is the MnCu series, typically with 50-80 % Mn. These materials are sometimes referred to as "shape memory" alloys, and achieve high damping through phase transformations. A relatively strong dependence on strain amplitude is found, but at strains on the order of 1000 microinches/inch, these materials may have loss modulii of several hundred Ksi.   Values as developed from three sets of experimental data are given in Fig. 4.

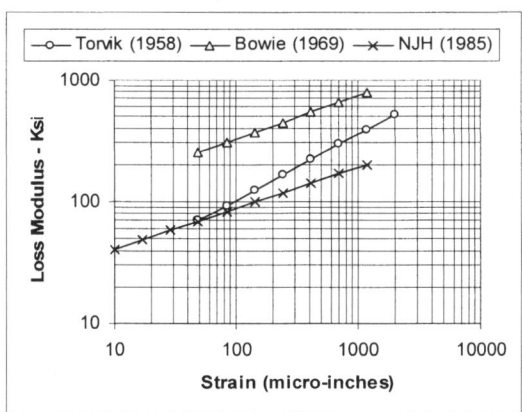

Fig. 4. Loss Factors for various Mn/Cu Alloys at Room Temperature.

The measurements by Torvik [3] are for a 80/20 MnCu alloy developed by the Chicago Development Corporation. Data by Bowie [4] is for a 65/35 MnCu (solution heat treatment for 3hr in air at 1525 deg F, oil quench to RT, age at 825 air for 4 hours, air cool). The data coded as NJH is taken from [5], where data on a Sonoston© alloy, developed for application on marine propellers was reported. Considerable variation among the alloys is evident. Using the results by Bowie, *et. al.*, we find that the loss factor at 10,000 psi to be 0.0712 which, with a storage modulus of 10.3 Mpsi, suggests a loss modulus of 734,000 psi. If such a coating is applied to steel at a 20% volume fraction, i.e., t/h = 0.1, one finds from Eq. 11 that the contribution of the coating to the system loss factor) should be about 0.0147.

If the system Q of the uncoated system is 1000, the addition of the coating should reduce the system quality factor to about 64. While such a system Q is quite good, it should be noted that other alloys and heat treatments of MnCu may only have ½ to 1/3 of this damping capacity. In the latter case, the resulting system quality factor might still be as much as 170.

The influence of temperature on the damping of these alloys was investigated by Bowie, *et. al.* and by Nashif, *et. al.* In both cases, a precipitous drop in damping (more than an order of magnitude) was found to occur at temperatures around 200 deg F. In consequence, these materials appear to be limited to applications near to room temperature. Although the inclusion of other components in the alloys has been found to extend the useful temperature range somewhat, such gains are normally at the expense of the room temperature loss factor. Further, as a consequence of the amplitude dependence, effectiveness at 1 Ksi will be only about 1/3 of that at 10 Ksi.

Example 4, a Ceramic Coating. Components exposed to a high-temperature environment are often provided with a protective coating. Some of these coatings, with sufficient thickness, have been found to provide significant enhancements to the structural damping. One such family of materials are the aggregates of magnesium oxide and alumina, $(MgO.Al_2O_3)$ consisting of 25 to 28% MgO and small amounts of other oxides, e.g. CaO, SiO_2, and $Mg_2 O Fe_2O_3$. The storage modulus for this material has been determined to be about 6 Mpsi at low strain, with an amplitude dependent loss modulus as shown in Fig. 5. The values shown are a composite from measurements [6] in several modes for beams were coated on both sides by an air-plasma-sprayed layer of magnesium aluminate spinel. The plasma-spray process does not lead to a truly homogeneous material, and the damping mechanism is thought to be that of friction between the platelets resulting from the spray process. The strong amplitude dependence below a threshold value (as is seen in Fig. 5) is characteristic of such a mechanism [7].

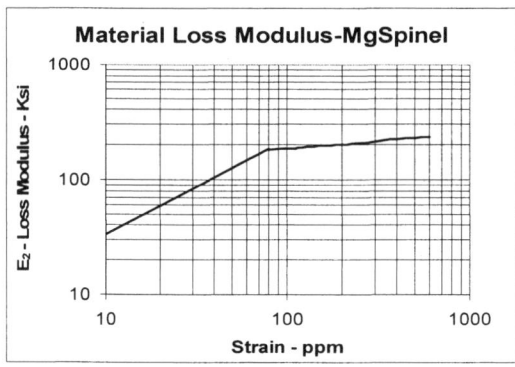

Fig. 5. Loss Modulus of a Hard Coating at Room Temperature.

The loss modulus is nearly constant for strains between 100 and 500 microinches/inch, with an average value of about 200 Ksi. Substitution of parameters in to Eq. 11 shows that, for a coating a volume fraction of 20% (t/h = 0.1) and a steel substrate, the contribution of the coating to the system loss factor should be about 0.004. The actual value will be somewhat less, as portions of the coating near the beam modes will have strains below 100 ppm. If the system Q of the uncoated structure is 1000, the coated system may be expected to show a Q of about 200. However, it should be noted that, if applied in the same thickness ratio to a titanium beam, the contribution to the system loss factor would be approximately doubled, as the modulus of titanium is approximately half that of steel, leading to a system quality factor of 118. These materials have been found to retain their damping capacity to at least 1000 deg F. However, there is some evidence that the use of these hard coatings can reduce the fatigue life of the structure. It is conjectured that cracks originating in the coating material may propagate across the interface. Further, as damping mechanism involves a frictional process, the possibility of changes throughout the service life can not be neglected.

The Asymmetric Free-Layer Treatment

While the assumption that the structure is to be coated on both sides with treatments of thin but equal thickness led to an appealingly simple results, this is not feasible in all applications. When the coating is to be applied on only one side of the structure, the shift in the neutral axis must then be taken into account for all but extremely thin and soft coatings. We retain our other assumptions, namely that the structure is a beam of constant thickness, that a linear coating of uniform thickness extends over the entire length and width, and that elementary beam theory may be applied.

Formulation. A beam with a free-layer damping treatment on one side gives rise to asymmetric configuration shown in Fig. 6. The axis of the beam is that of x, and a transverse (bending) displacement w (x, t) in the z direction is assumed. Let the substrate elastic beam be of width W, of thickness h, and have Young's modulus E_b and the damping layer be of a uniform thickness t. The analysis is first performed by treating the modulus as having a real value, E_C, and then, after taking a Fourier transform so as to eliminate time, replacing the real modulus with a complex value $E^* = E_1 + jE_2$ in accordance with the viscoelastic correspondence principle [8].

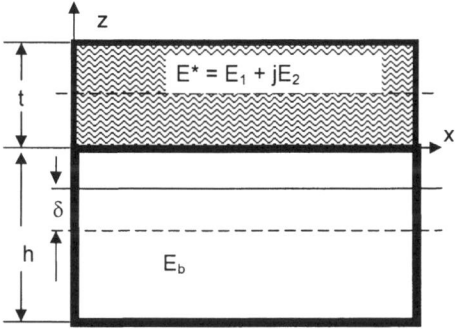

Fig. 6. Asymmetric Free-Layer Damping treatment on a Beam.

If the effect of shear deformation in each layer is ignored, the strain due to bending varies linearly through the thickness. However, the distributions of bending strains are superposed on uniform tensile strains generated by equal and opposite axial forces acting in the two layers. Thus, if z' is measured from the geometric center of the beam, and z" from the geometric center of the coating, the strain distribution is:

In the substrate (-h $\leq$ z $\leq$ 0): $\varepsilon_{bTOTAL} = -z' \dfrac{\partial^2 w(x,t)}{\partial x^2} + \varepsilon_b(x,t)$ (12)

And in the coating (0 $\leq$ z $\leq$ t): $\varepsilon_{cTOTAL} = -z'' \dfrac{\partial^2 w(x,t)}{\partial x^2} + \varepsilon_c(x,t)$ (13)

Assuming a perfect bond, the strain in the substrate and in the coating layer must be the same at the interface, z = 0, i. e., z' = h/2 and z"=-t/2. Equating these, we find the longitudinal strains at the centerlines are related through:

$$\varepsilon_b(x,t) - \varepsilon_c(x,t) = \frac{(h+t)}{2} \frac{\partial^2 w(x,t)}{\partial x^2} \qquad (14)$$

As the resultant axial load on the two-layer beam must be zero, we may integrate the stress distribution over each layer, sum, and set to zero. For the case of linear materials the axial force is:

$$F = WE_b h \varepsilon_b(x,t) + WE_c t \varepsilon_c(x,t) = 0 \qquad (15)$$

This may then be combined with the other relationship between the centerline strains, Eq. 14, and the centerline strains found for each component.

$$\varepsilon_c(x,t) = -\frac{(t+h)}{2} \frac{\partial^2 w(x,t)}{\partial x^2} (1 + \frac{E_c t}{E_b h})^{-1} \qquad \text{and}$$

$$\varepsilon_b(x,t) = \frac{(t+h)}{2} \frac{\partial^2 w(x,t)}{\partial x^2} (1 + \frac{E_b h}{E_c t})^{-1} \qquad (16a,b)$$

Evaluation of the Effective Bending Stiffness. Having found the strain distribution, the stress distribution may be determined. From that, the resulting bending moment about any convenient axis, such as the central plane of the beam, may be found by multiplying the local stress by the appropriate moment arms, and integrating. We again treat the modulus of the free layer as real, with the intention of later replacing it by the complex modulus.

$$M_y(x,t) = WE_b \int_{-h/2}^{h/2} (-z'^2 \frac{\partial^2 w(x,t)}{\partial x^2} + z' \varepsilon_b(x,t)) dz'$$

$$+ WE_c \int_{-t/2}^{t/2} [-z''\{z'' + (t+h)/2\} \frac{\partial^2 w(x,t)}{\partial x^2} + \{z'' + (t+h)/2\} \varepsilon_c(x,t)] dz'' \qquad (17a)$$

The area moments of inertia of each beam about their respective centerlines have been introduced as I_b and I_c. Eliminating the longitudinal strain ε_c by substitution of Eq. 16a leaves

$$-M_y(x,t)/\frac{\partial^2 w(x,t)}{\partial x^2} = (EI)_{EFF} = (I_b E_b + I_c E_c)[1 + \frac{W[(t+h)/2]^2}{I_b E_b + I_c E_c}(\frac{1}{E_c t} + \frac{1}{E_b h})^{-1}] \qquad (17b)$$

An effective stiffness per unit width may be expressed in terms of the bending stiffness per unit width of the two layers taken separately and a dimensionless coupling parameter, Y_0. Let

$$D_1 \equiv (I_b E_b + I_c E_c)/W = (E_b h^3 + E_c t^3)/12 \qquad (18)$$

$$Y_0 \equiv \frac{[(t+h)/2]^2}{D_1}(\frac{1}{E_c t} + \frac{1}{E_b h})^{-1} \qquad (19)$$

Then

$$-M_y(x,t)/\{W\frac{\partial^2 w(x,t)}{\partial x^2}\} = (D_1)_{EFF} = D_1[1 + Y_0] \qquad (20)$$

By introducing two dimensionless parameters, a stiffness ratio and a thickness ratio,

$$S \equiv E_c t / E_b h \quad \text{and} \quad r = t / h \tag{21a,b}$$

the nominal stiffness and coupling parameters may be rewritten as:

$$D_1 = E_b h^3 (1 + Sr^2)/12 \quad \text{and} \quad Y_0 = \frac{3S}{(1+S)} \frac{(1+r)^2}{(1+Sr^2)} \tag{22a,b}$$

from which the effective stiffness is found to be:

$$-M(x,t)/\{W \frac{\partial^2 w(x,t)}{\partial x^2}\} = (D_1)_{EFF} = \frac{E_b h^3}{12}[(1+Sr^2)+3S(1+r)^2(1+S)^{-1}] \tag{23}$$

The term in square braces isolates the proportional increase in bending stiffness due to the presence of the coating layer. This result may be shown to be equivalent to that given by Jones [9] for the Öberst beam [10]. When the coating is described by a complex modulus, S is a complex number.

Evaluation of the Loss Factor. The differential equation of a Bernoulli-Euler beam in free vibration in the nth normal mode [11] is that

$$EI \frac{\partial^4 w(x,t)}{\partial x^4} + \rho W h \frac{\partial^2 w(x,t)}{\partial t^2} = 0 \tag{24}$$

Assuming a separable solution, w(x, t) = f(t) W(x), and taking a Fourier transform so as to eliminate the time variable, and then replacing the nominal stiffness and mass with effective values,

$$\omega_n^2 = \frac{(EI)_{EFF}}{(\rho W h)_{EFF}} \frac{W_n^{IV}(x)}{W_n(x)} = \frac{(D_1)_{EFF}}{(\rho t)_{EFF}} k_n^4 \tag{25}$$

where ω_n and $W_n(x)$ correspond to the frequency and mode shape of the nth mode, $(\rho h)_{EFF}$ is total coated beam mass per unit length and width (a real quantity) and k_n is the dimensional eigenvalue. When the effective stiffness is allowed to take a complex value, the frequency becomes complex, but the eigenvalues and the density are real, and since the loss factor is the ratio of imaginary to real parts of the square of the frequency, the loss factor of the system becomes:

$$\eta_{SYS} = \frac{\Im\{(D_1)_{EFF}\}}{\Re\{(D_1)_{EFF}\}} = \frac{\Im[(1+Sr^2)+3S(1+r)^2(1+S)^{-1}]}{\Re[(1+Sr^2)+3S(1+r)^2(1+S)^{-1}]} \tag{26a}$$

The parameter S (Eq. 21a) now has a complex value. If the material properties are independent of frequency, the loss factors for all modes and boundary conditions are seen to be the same. But if the modulus of the dissipative coating is frequency dependent, then the complex stiffness for each mode must be evaluated at frequency ω_n, leading to a different loss factor for each mode. As the modulus will affect the frequency, this may require iteration.

While this general result is not conveniently written in as a simple expression in terms of the loss factor of the coating, the result is easily evaluated numerically. Eq. 26a may also be expressed as:

$$\eta_{SYS} = \frac{\eta_c S_1 \{3(1+r)^2 + r^2 |1+S|^2\}}{|1+S|^2 + S_1 r^2 |1+S|^2 + 3(1+r)^2 (S_1 + |S|^2)} \tag{26b}$$

which demonstrates that, while the dominant term is the product $3S_1\eta_c$, the material loss factor remains in other terms, and the *system* loss factor can not be simply and explicitly expressed in terms of the *material* loss factor alone unless $|S| \ll 1$.

As it has been assumed in the analysis leading to this result that there are no energy losses other than those due to the dissipative coating, it is more precise to identify the loss factor of Eq. 26 as the contribution of the coating to the system loss factor. Some numerical results are given in Fig. 7 for various values of the ratio E_1/E_b and the thickness ratio, $r = t/h$, for a coating-material loss factor of 0.1. The ordinate of Fig. 7 is the ratio η_s/η_c. The results are therefore indicative of the efficiency of various configurations in converting the dissipative potential of the coating material into a system loss factor.

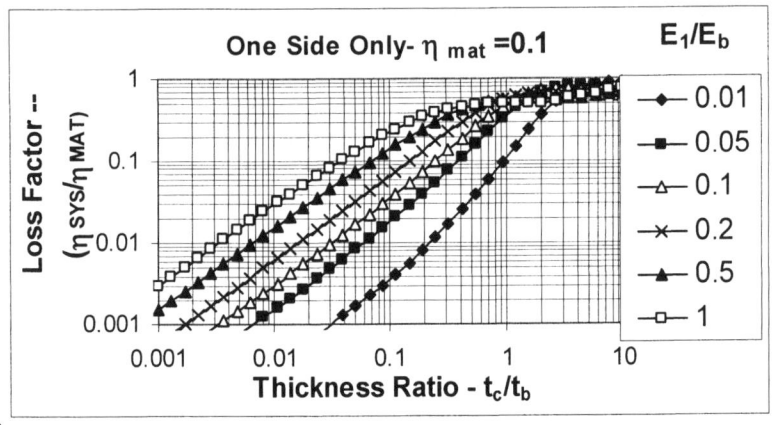

Fig. 7. Loss Factors (Normalized) for Free Layer Treatments - for Various Values of the Modulus Ratio.

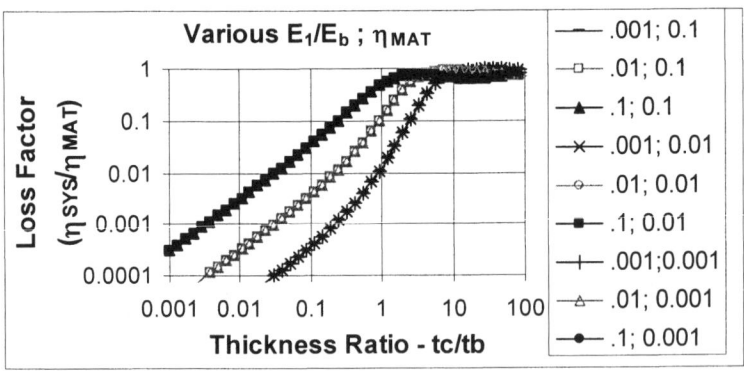

Fig. 8. Normalized System Loss Factors for Various Material Loss Factors.

The appropriateness of such normalization is verified by the results given in Fig. 8. The dependence of the normalized loss factor on the stiffness and thickness ratios are there compared for three stiffness ratios: 0.001, 0.01 (open points), and 0.1(solid points); each with three coating loss factors, 0.1, 0.01, and 0.001. The normalized loss factors (system loss factor over material loss factor) resulting from three different material loss factors are found to be indistinguishable. Thus, the normalized loss factor is essentially independent of material loss factor for system parameters in the range considered.

The left halves of Figs. 7 and 8 correspond to the range of parameters where the assumptions leading to Eq. 11 are applicable. Under these conditions, the loss factor for the beam with asymmetric coating is one-half that of the beam with symmetric coating, or

$$\eta_{SYS} \cong 3\eta_c \frac{E_1 t}{E_b h} = 3\frac{E_2 t}{E_b h} \tag{27}$$

Anomalous Behavior of Thick Asymmetric Coatings. The right halves of Figs. 7 and 8 are somewhat counter-intuitive. Two issues are of a special interest. First, for low ratios of modulii, the slope of loss factor vs. thickness is seen to change from unity at low thickness to about two at thickness values near unity. Second, for higher thickness ratios, the loss factor ratio appears to decrease slightly before increasing asymptotically to the value of unity.

We consider first the case of an asymmetric thick coating of a soft material such that the stiffness ratio, $S = Mr$, is of a small value, where

$$M = M_1 + jM_2 = E_1 / E_b + jE_1 / E_b \tag{28}$$

Neglecting powers of M, the complex factor of Eq. 26a can then be expressed as:

$$Z = \frac{(1 + Mr^3)(1 + Mr) + 3Mr(1 + r)^2}{(1 + Mr)} \cong \frac{1 + Mr + Mr^3 + 3Mr(1 + r)^2}{(1 + Mr)} \tag{29}$$

We may consider thickness ratios of approximately unit value by letting $r = 1 + \lambda$, where λ is a small, real number. Then, expanding in power series and retaining only terms of first order,

$$Z \cong [1 + 14M(1 + 2\lambda)][1 - Mr] \cong 1 + 13Mr^2 \tag{30}$$

The loss factor may then be evaluated from:

$$\eta_{SYS} = \frac{\Im\{(D_1)_{EFF}\}}{\Re\{(D_1)_{EFF}\}} \cong \frac{\Im\{1 + 13Mr^2\}}{\Re\{1 + 13Mr^2\}} = \frac{13(E_2 / E_b)r^2}{1 + 13(E_1 / E_b)r^2} \tag{31}$$

so for (t/h) near unity and $13E_1/E_b \ll 1$.

$$\eta_{SYS} \cong \frac{\Im\{1 + 13Mr^2\}}{\Re\{1 + 13Mr^2\}} = 13(\frac{E_2}{E_b})(\frac{t}{h})^2 \tag{32}$$

A relationship of this form (but lacking an evaluation of the numerical factor of 13) has been given by Kerwin [12] as being applicable when $M_1 \cong 10^{-3}$ and $r \cong 1$ and is in agreement with Eq. 32.

Thus, we have seen that when a small stiffness ratio is achieved by making t small, the asymptotic behavior is linear in t. However, if a small stiffness ratio is achieved by making the ratio of modulii small, with the thickness ratio near unity, a qualitatively different asymptotic behavior is found. The use of the thin coating approximation (Eq. 27) at thickness ratio of 1 and small ratios of modulii will underestimate the damping by more than a factor of four. This may be seen in the "upturning" of the curves in Figs. 7 and 8 for the lowest modulus ratios.

The upper right hand corner of Figs. 7 and 8 shows an apparent behavior anomaly in that the loss factor appears to drop as the stiffness ratio is increased.   The effect is especially dramatic in the case of very thin, but very stiff (high E_1 t) coatings. Because of the much lower thickness, the ratio of coating energy to substrate energy is very low, and, in consequence, so is the system loss factor. With thicker coatings, the effect diminishes, but is still present. Even in a "hard" coating application, the ratio of modulii is not likely to exceed 0.5 with a thickness ratio of no more than 0.2.   Combinations of stiffness ratio and thickness leading to design points to the right of the maxima of Fig. 7 correspond to configurations well outside the range of normal usage of damping coatings. As such, the decrease in loss factor with increasing stiffness does not appear to have a practical significance.

Comparison of Symmetric and Asymmetric Treatments. The analysis as given above (Eqs.12 through 26a) may be repeated for symmetric treatments of arbitrary thickness. The result may be combined with the previous to give a system loss factor applicable in both cases of

$$\eta_{SYS} = \frac{\Im\{(D_1)_{EFF}\}}{\Re\{(D_1)_{EFF}\}} = \frac{\Im\{1 + NSr^2 + 3S(1+r)^2[(1+(2-N)S)]^{-1}\}}{\Re\{1 + NSr^2 + 3S(1+r)^2(1+(2-N)S)^{-1}\}} \tag{33}$$

where N = 1 for the asymmetric application and N = 2 for the symmetric. In the latter case,

$$\eta_{SYS} = \eta_{coat} \frac{(E_1/E_b)[2r^2 + 3(1+r)^2]}{1 + (E_1/E_b)[2r^2 + 3(1+r)^2]} \tag{34}$$

and the loss factor ratio goes to 1 as r =t/h $\Rightarrow \infty$ and to the result of Eq. 32 as r $\Rightarrow$ 0, as expected.

The loss factors achieved through the use of the asymmetric (1 coating layer) and symmetric (2 coating layers) are compared in Fig. 9 for the case of 10% coating thickness and a loss factor of unity. The results obtained through the use of the thin coating approximations of Eqs. 11 and 27 are given for comparison. For low ratio of modulii, the approximations are seen to provide a small underestimate, a consequence of the fact that the strains in the coating are actually greater than the strain at the interface. However, as the storage modulus of the coating approaches that of the beam, the approximations are seen to lead to over predictions. Perhaps surprisingly, the use of a coating on both sides leads to loss factors slightly less than twice the loss factor achieved with the same thickness of coating on only one side.

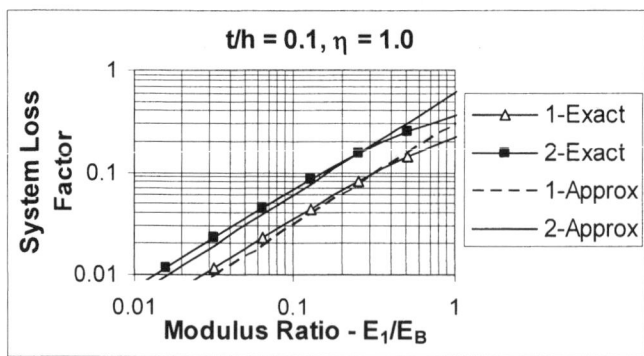

Fig. 9. Asymmetric vs. Symmetric Free Layer Treatments.

Significance of the Complex Coordinates of the Neutral Plane for Asymmetric Coatings. The neutral axis of a beam is customarily defined as that plane where the strain during bending without axial load is zero. To locate this plane, we may write the strain in the beam by using Eqs. 12 and 16b to find the location of the plane of zero strain. In the case of a coating represented by a complex modulus, we find that plane to be at

$$z' = \delta = \frac{(t+h)}{2}(1 + \frac{E_b h}{E^* t})^{-1} \tag{35}$$

which is a complex valued quantity, $\delta = \delta_1 + j\delta_2$. The proper interpretation of this result is not that the neutral axis is at a coordinate of complex value, but rather that, whenever $\delta_2 \neq 0$, there is no real, physical plane where the strain is zero for *all* time. We must consider that the solution obtained through the utilization of a complex stiffness is restricted to harmonic motion, so that

$$\delta_{Physical} = \Re\{(\delta_1 + j\delta_2)\exp(j\omega t)\} = \delta_1 \cos(\omega t) - \delta_2 \sin(\omega t) \tag{36}$$

But the out of phase component of the location, δ_2 is small. It may be shown that the ratio of the out-of-phase component of strain at the constant physical plane $z = |\delta|$ to the maximum strain in the beam is only $\eta_{SYS}/3$ and, as system loss factors will only rarely be greater than 0.01, the out of phase component of strain is negligible.

Alternative Formulations. Other methods of analyzing the free layer damping treatment are available, such as treating the free layer beam as a special case of the Ross, Ungar, Kerwin three layer beam, or by evaluating the energies dissipated and stored from the strain distributions of Eq. 16a and 16b and the fundamental definition of the loss factor as given in Eq. 9. Both lead to the result of Eq. 26b. A computation by the method of Modal Strain Energy, however, will give Eq. 26b with S_1 replacing S, and will lead to a somewhat different result unless $\eta_C \ll 1$. A comparison for $\eta_C = 1$ showed that the difference is not significant for thickness ratios of 1 unless the stiffness ratio is greater than about 0.2 or thickness ratios of 0.01 unless the stiffness ratio is greater than about 0.3. The difference between the results from the two methods is greatest in the region of anomalous behavior, previously seen to be of little practical interest. In general, the modal strain energy method when applied to free-layer coatings will be found to provide a small overestimation of the system loss factor, but in any case where *either* the stiffness ratio or the loss factor is less than (say) 0.1, the error resulting in use of MSE is not of practical significance. Such overestimates of the loss factor, however, are a general characteristic of predictions by MSE [14].

Influence of Coating on the Natural Frequency. The addition of the coating will also change the frequency, which may increase or decrease depending on whether the additional stiffness makes a greater or lesser contribution to the stiffness than the added mass does to the kinetic energy of vibration. The frequency is found by substituting the complex stiffness as in Eq. 33 into Eq. 25, i.e.,

$$\omega_n^2 = \frac{E_b h^3}{12(\rho t)_{EFF}} \frac{(k_n L)^4}{L^4}[(1 + NSr^2) + 3S(1+r)^2\{1 + (2-N)S\}^{-1}] \tag{37}$$

In computing the frequency, it must be noted that the observable frequency is the real part of the complex frequency. Thus, the square root of the right hand side must be taken, i.e.,

$$\omega_n = \sqrt{\frac{E_b h^3}{12(\rho t)_{EFF}} \frac{(k_n L)^2}{L^2} \Re\{[(1 + NSr^2) + 3S(1+r)^2\{1 + (2-N)S\}^{-1}]^{1/2}\}} \tag{38}$$

The processes of extracting the square root and the real part of the complex term are not commutative. Note that $N = 2$ for the symmetric coating, and $N = 1$ for a coating on one side.

Frequency Dependent Materials: Boundary Conditions and Multiple Modes

In the case of a coating materials for which the properties are independent both of frequency and amplitude, the loss factors as computed either by Eq. 11 or Eq. 26 are found to be the same for all modes of a structure. However, none of the classes of materials considered satisfied these idealizations. In the case of metallics and ceramics, amplitude dependence was observed, and this will be considered in a later section. In the case of the polymers and enamels, significant frequency dependence was noted. In consequence, the in-service boundary condition of the structure must be well known so that the natural frequencies may be estimated. Further, the material will have different properties (storage modulus and loss factor or loss modulus) in each mode and, in consequence, the structure to which it is applied will also have a different loss factor in each mode. Designing with such materials must be an iterative process, as the frequency influences the stiffness, and conversely. The iterations must be conducted separately for each boundary condition and for each mode and, because of the interrelation between the temperature dependence and frequency dependence of the properties, for each service temperature of interest.

As an example, consider a beam of Hastelloy X beam (E = 28.5 Mpsi and a density of 8.22 gm/cm^3) 0.095 thick, 8 inches long and 0.75 wide, coated on one side only with 10% of the beam thickness with Corning 7570 (Fig. 3). At 900F, the loss factors for the first three modes are found to be, respectively, 0.00232, 0.00442 and 0.00511. (Q = 431, 226, 196). The loss factors, in this case, increase with mode numbers because all three frequencies are on the left-hand side of the peak in Fig. 3. For the fourth and higher modes, the loss factor will diminish with mode number.

Influence of a Coating Non-Linearity

Two of the classes of coating materials (high damping alloys and ceramics) were observed to have a nonlinear damping relationship, that is, the loss factors (and consequently the loss modulii) that are dependent on the amplitude of strain. In consequence, the energy dissipated is dependent on strain raised to some power greater than two. As the materials are nominally elastic, the stored energy remains dependent on strain raised to the second power. Thus, the ratio of integrals (as in
Eq. 10a) does not cancel, but must be evaluated.

To investigate the consequences of this nonlinearity, we return to the fundamental definition of the system loss factor, considering only the energy dissipated in the coating but including the energy stored in both the coating and the substrate. For simplicity, attention will be restricted to the case where the coating is nominally linear, i.e., the energy stored is proportional to the square of strain, and the substrate structure is a beam, coated on both sides at the same thickness. The system loss factor is then as given in Eq. 9.

It is assumed that Bernoulli-Euler beam theory may be applied so that the strain distribution may be assumed to vary linearly through the thickness of the beam and the coating. A perfect bond is assumed at the interface. The strain distribution everywhere in both beam and coating may therefore be related to the local strain at the interface, denoted e(x). The variation of this strain, as a function of position on the beam, may be determined from the mode shapes of the structure, with the magnitude determined from the mode-shape and an observation of amplitude (displacement, velocity, or acceleration) at any one point on the beam.

If the displacement of the beam is taken as w (x, t) = X(x) cos Ωt, then for a beam of thickness h with coating of thickness t on both sides over the entire (uniform) beam width W and length, the strain anywhere in beam and coating is:

$$\varepsilon(x,z,t) = -z\frac{d^2 X(x)}{dx^2}\cos(\Omega t) = \frac{-2z}{h}e(x)\cos(\Omega t) \qquad (39)$$

The peak strain energy in the beam is:

$$U_b = W \int_0^L \int_{-h/2}^{h/2} \frac{E_b \varepsilon^2}{2} dx dz = \frac{W E_b}{2} \int_{-h/2}^{h/2} \left(\frac{2z}{h}\right)^2 dz \int_0^L e(x)^2 dx = \frac{W h E_b}{6} \int_0^L e(x)^2 dx \tag{40}$$

The peak strain energy for the two symmetrically located coatings of thickness t is:

$$U_C = 2W \int_0^L \int_{h/2}^{h/2+t} \frac{E_1 \varepsilon(x,z)^2}{2} dx dz = t W E_1 T(2,t/h) \int_0^L e(x)^2 dx \tag{41}$$

$$T(N,t/h) = \frac{h}{2t} \frac{1}{N+1} \left[(1+2t/h)^{N+1} - 1\right] \tag{42}$$

For small t/h, T (N, t/h) $\rightarrow$ 1 for all values of N.

In order to evaluate the energy dissipated from the strain distribution, we assume that the nonlinear damping relationship may be represented in the form used by Lazan [15]:

$$D = J_\varepsilon \varepsilon^N \tag{43}$$

where D is the unit damping, that is, the energy dissipated in a unit volume of the coating material in a state of homogeneous strain. Since the strain energy in a unit volume undergoing homogeneous strain is U = $E_1 \varepsilon^2/2$, the *material* loss factor is

$$\eta = \frac{1}{2\pi} \frac{D}{U} = \frac{1}{2\pi} \frac{J_\varepsilon \varepsilon^N}{E_1 \varepsilon^2/2} = (\frac{J_\varepsilon}{\pi E_1}) \varepsilon^{N-2} \tag{44}$$

Clearly, unless N = 2, the material loss factor is amplitude dependent. The material parameter J_ε is seen to be related to the loss modulus by:

$$E_2 = \eta E_1 = (\frac{J_\varepsilon}{\pi}) \varepsilon^{N-2} \tag{45}$$

and is also amplitude dependent, even when the material is nominally linear so that $\sigma \cong E_1 \varepsilon$.

For the two layers of coating material bonded to a beam with strain e(x) at the beam coating interface,

$$D_C = 2W \int_0^L \int_{h/2}^{t+h/2} J_\varepsilon \left(\frac{|-2z|}{h}\right)^N e(x)^N dx dz = 2 W t J_\varepsilon T(N,t/h) \int_0^L e(x)^N dx \tag{46}$$

We are now in a position to evaluate the loss factor for the coated beam when all energy dissipated is dissipated in the coating. Substituting the dissipated energy (Eq. 46) and the stored energies (Eqs. 40 and 41) into Eq. 9 and letting e_n (x) denote the strain distribution at the beam coating interface for the nth mode,

$$\eta_{SC} = \frac{D_C}{2\pi(U_b+U_C)} = \frac{1}{2\pi} \frac{2WtJ_\varepsilon T(N,t/h)\int_0^L e_n(x)^N dx}{\frac{WhE_b}{6}\int_0^L e_n(x)^2 dx + tWE_1 T(2,t/h)\int_0^L e_n(x)^2 dx} \tag{47}$$

The remaining integrals may both be expressed (for the nth mode) as:

$$I(N,n) = \frac{1}{L}\int_0^L \hat{e}_n(x)^N dx \tag{48}$$

where the strain distribution has been normalized by division by the maximum value, $e_n(0) = e_0$, occurring at the root of a cantilever beam, so that $\hat{\varepsilon}_n(x) = \varepsilon_n(x)/\varepsilon_n(0)$. Eq. 47 then becomes:

$$\eta_{SC} = \frac{6tJ_\varepsilon}{\pi E_b h} \frac{T(N,t/h)}{1+6(E_1 t/E_b h)T(2,t/h)} \frac{I(N,n)}{I(2,n)} e_0^{N-2} \tag{49}$$

from which it is seen that if the exponent N in the material loss factor is other than two the system loss factor will also be amplitude dependent. The system loss factor may be rewritten in terms of the amplitude dependent loss modulus of the material as given by Eq. 45.

$$\eta_{SC} = 6\frac{tE_2(e_0)}{E_b h} \frac{T(N,t/h)}{1+6(E_1 t/E_b h)T(2,t/h)} \frac{I(N,n)}{I(2,n)} \tag{50}$$

where e_0 is the peak strain amplitude anywhere in the coating. The first factor is recognized as being the thin coating approximation of Eq. 11, except that the loss modulus is now amplitude dependent. For small t/h, the middle term approaches 1, but the ratio of integrals is independent of the thickness ratios, t/h, and therefore does not have unit value unless N = 2.

    The volume integrals, I (N, n) are readily computed by numerical integration of second derivatives of the mode shapes [11]. Results for representative values of the parameters N are given in Table I. Also given are values of the T function for the specific case where the coating thickness is 10% of the beam thickness. Note that the values of the I-integral are mode specific, but that for N ≥ 3 the variation between the modes n ≥ 2 is less than 3%.

	n = 2	n = 2.2	n = 2.3	n = 2.4	n = 2.6	n = 2.8	n = 3	n = 4	n = 5
1	0.25	0.2334	0.2259	0.2189	0.2124	0.1949	0.1847	0.1468	0.1219
2	0.25	0.2277	0.2176	0.2082	0.1909	0.1755	0.1618	0.1113	0.0801
3	0.25	0.2267	0.2162	0.2063	0.1883	0.1723	0.1580	0.1058	0.0739
4	0.25	0.2261	0.2153	0.2052	0.1867	0.1703	0.1557	0.1023	0.0698
T	1.213	1.238	1.250	1.1.263	1.289	1.315	1.342	1.488	1.655

Table I Representative Values of the Integral I (N, n) and T (N, 0.1)

    As an example, let us consider a coating of magnesium aluminate spinel applied to the same Hastelloy X beam as before with the coating applied on both sides, each being at 10% of the beam thickness. As the bi-linear character of the loss modulus seen in Fig. 5 introduces a considerable complication, necessitating the identification of the regions above and below the break point at each desired level of peak strain, for this example we will consider only strains below (about) 80 microinches/inch. In this case the loss factor is adequately represented by strain to the power 0.8206 for $\varepsilon < 80$ units, with the unit being the micro inch of strain and $E_1 = 5,864,000$ psi.

Values of the thickness function and the ratios of integrals I (2.8, n) / I (2, n) from Table I the system loss factors for the three modes may be found. These are shown graphically in Fig. 10 as the quality factor ($1/\eta_{SYS}$) for the first three modes.

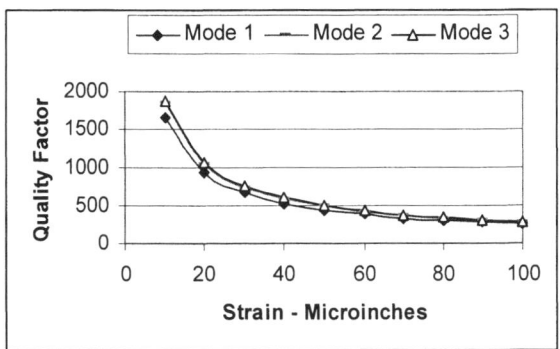

Fig. 10. **Quality Factors for Mg Spinel Beam,**

Symmetric Coating at 10% Thickness

The actual quality factor would be somewhat lower (i.e. higher damping) because of other dissipation from support damping, air damping, and the inherent dissipation of the beam. A slight dependence on mode number is seen, and is a consequence of the non-linearity. The results for modes 2, 3, and higher modes, however, are indistinguishable. This occurs because the mode shapes of a cantilever beam then each become approximately sinusoidal, and therefore more similar to each other.

Influence of Coatings on System Frequencies

It is also of interest to evaluate the influence of the coating on the system natural frequencies. In the case of the symmetric treatments, there is no shift in the neutral axis. This enabled the strain distribution to be written a priori, simplifying the evaluation of stored energies, Eqs. 40 and 41. The peak kinetic energy of the symmetrically coated beam for nth mode is, in terms of displacement $w (n, x, t) = X_n(x) \cos \Omega t$,

$$K = \Omega^2 \frac{\rho_b hw}{2} \int_0^L X_n{}^2 dx + \Omega^2 \frac{2tw\rho_C}{2} \int_0^L X_n{}^2 dx = \Omega^2 \frac{\rho_b hw + 2tw\rho_C}{2} \int_0^L X_n{}^2 dx \qquad (51)$$

where ρ_b and ρ_C are beam and coating densities, respectively. But the ratio of strain to kinetic energies leads to the frequency for the beam-coating system by use of the Rayleigh quotient. Using the strain energies as evaluated in Eqs. 40 and 41,

$$\Omega_S{}^2 = \frac{[1 + 6tE_1 T(2, t/h)/(hE_b)]}{1 + 2t\rho_C/(\rho h} \cdot \frac{E_b}{3\rho_b} \int_0^L e_n(x)^2 dx / \int_0^L X_n{}^2 dx \qquad (52)$$

The frequency is seen to be independent of any damping non-linearity. In the case of a material with significant frequency dependence, it is to be understood that the storage modulus, E_1, must be evaluated at the frequency Ω_S, necessitating iteration. But the last factor of Eq. 52 is the squared frequency of the substrate alone, so, the frequency of the coated beam is related to that of the bare beam through:

$$\frac{\Omega_S^2}{\Omega_{BB}^2} = [1 + \frac{6tE_1}{hE_b}T(2,t/h)]/[1 + \frac{2t\rho_C}{\rho h}] \tag{53}$$

The function T (2, t/h) as defined by Eq. 42 and approaches 1 as t / h $\Rightarrow$ 0. But an approximation has been introduced. Eq. 52 is a Rayleigh quotient. Thus, if an exact mode shape is employed, the frequency is exact. But the true mode shapes will differ slightly from the classical mode shapes due to the influence of fixture deformation and amplitude dependent coating properties, as well as dissipation. However, if these effects are small, Eq. 52 will produce a satisfactory approximation.

While it is evident that a successful free layer treatment must provide a significant increase in system damping, other factors must be considered. The shift in frequency, as given by Eq. 53 may be a critical design constraint. In the case of metallic or ceramic coatings, the addition of the coating will lead to a fractional increase in frequency on the order of 2 t / h or t/h, respectively, for small values of t / h. These estimates are for the case of symmetrically applied coatings. For thin asymmetric coatings, the changes will be approximately one-half of these values. In the case of the soft viscoelastic materials, the increase in stiffness will be nearly negligible, but the added mass of the large thicknesses required for significant damping will lead to reductions in the natural frequencies.

Summary

The equations for predicting the loss factors of beams with a damping coating (the Öberst beam) have been developed. An approximate solution, applicable for thin coatings and linear damping materials was first considered. This approximation is quite satisfactory for the initial screening of candidate coating materials as it requires only the knowledge of the storage modulus and loss factor for the coating material. Even in the case of materials with non-linear damping it proves useful, as long as storage modulii and loss factors are known *at the levels of strain to be encountered* in the application.

The properties of four classes of coating materials: polymers, vitreous enamels, high-damping metallics, and ceramics where reviewed. Damping properties of representative materials were given, and the estimation of the contribution of the coating to the system loss factor was demonstrated through calculation using the result of the simplified analysis.

A general analysis for the asymmetric (one-sided) or symmetric (two-sided) coating (Öberst beam) applicable for arbitrary ratios of the coating to substrate thickness and the coating to substrate modulus was developed and discussed. A complex modulus representation of the coating material was assumed and loss factors were obtained from the resulting complex-valued frequencies. The effectiveness was compared for symmetric and asymmetric coatings. The conditions for and the degree of the inaccuracies to be expected from an analysis by Modal Strain Energy were identified, and an interpretation of the apparent anomaly of a complex valued location for the neutral axis of the asymmetric systems was given.

Two complicating factors, the frequency dependence of viscoelastic materials and the amplitude dependence of metallics and ceramics, were given special attention. Modifications applicable in each case were introduced and discussed, and an illustrative example given. And finally, the methodology for predicting the influence of a coating on system natural frequencies was demonstrated.

While this discussion has been approached from the point of view of designing a free layer treatment with the coating properties presumed to be given, these techniques are also applicable to the inverse problem, that is, of material characterization. By measuring the loss factor of a coated structure and making suitable corrections for such extraneous losses as air damping, support damping and the inherent damping of the substrate, the contribution of the coating to the system loss factor may be estimated. This process is relatively straight-forward in the case of the linear materials, but the presence of a damping nonlinearity introduces significant complications. Deducing the local (material) property from the global (system) property in the presence of an amplitude dependent loss factor requires, in essence, the numerical solution of an integral equation.

If the beam properties and the density of the coating material are known, and if very precise measurements can be made of the natural frequencies of coated and uncoated beams, an estimate of the storage modulus of the coating material can also be made.

In closure, it should be noted that, in addition to the limitations on coating selection brought about by considerations of damping and frequency, other factors must be considered. Some applications may lead to design constraints on the maximum allowable thickness. In some cases, such as auto body panels, a thick coating may be quite tolerable. In others, such as turbine blades, aerodynamic considerations may preclude the use of coatings of a thickness greater than a few thousands of an inch. A successful coating must survive and continue to operate in the environment (temperature, erosion, corrosion, impact, etc.) to be encountered in service. And finally, the coating must not degrade the structure to which it is applied through such mechanisms as a chemical interaction or as a transmitter of cracks in the underlying substrate.

For further reading and reference, two excellent sources including design application of free layer damping treatments and material property data are those by Nashif *et. al.* [5] and Jones [9].

References

[1] Drake, M. L.: *Damping Properties of Various Materials*, AFWAL-TR-88-4248, Materials Laboratory, AFFWAL (1989).

[2] Zener, C.: *Elasticity and Anelasticity of Solids*, University of Chicago Press, Chicago (1948).

[3] Torvik, P. J.: Appendix 72fg, in Status Report 58-4, WADC Contract, B. J. Lazan, University of Minnesota (1958).

[4] Bowie, G. E., J. F. Nachman and A. N. Hammer: ASME Paper 71-Vibr 105, ASME Vibrations Conference, Toronto (1971).

[5] Nashif, A. D., D. I. G. Jones, and J. P. Henderson: *Vibration Damping*, Wiley Interscience, New York (1985).

[6] Torvik, P. J.in: *Proceedings of the 8th National Turbine Engine High Cycle Fatigue Conference*, Monterey, CA (2003).

[7] Shipton, M. and Patsias, S. in: *Proceedings, 8th National Turbine Engine High Cycle Fatigue Conference*, Monterey, CA (2003).

[8] Christensen, Richard M.: *Theory of Viscoelasticity*, 2nd Ed., Academic Press (1982).

[9] Jones, David I. G.: *Handbook of Viscoelastic Vibration Damping*, John Wiley, New York (2001).

[10] Öberst, V, H,: Acustica, Heft 4 (1952), p. 181,

[11] Timoshenko, S.: *Vibration Problems in Engineering*, 3rd Ed., Van Nostrand, Princeton, N. J. (1955).

[12] Kerwin, E., M. in: *Internal Friction, Damping and Cyclic Plasticity*, ASTM STP 378, American Society for Testing and Materials (1965).

[13] Torvik, P. J., S. Patsias, and G. R. Tomlinson in: *Proceedings, 7th National Turbine Engine High Cycle Fatigue (HCF) Conference*, Palm Beach Gardens, FL (2002).

[14] Torvik, P. J. and B. Runyon in: *Proceedings of the 10th National Turbine Engine high Cycle Fatigue (HCF) Conference*, New Orleans (2005).

[15] Lazan, B. J.: *Damping of Materials and Members in Structural Mechanics*, Pergamon Press, Oxford (1968).

Key Engineering Materials Vol. 333 (2007) pp 215-218
online at http://www.scientific.net
© *2007 Trans Tech Publications, Switzerland*

Porous Multilayer PZT Materials Made by Aqueous Tape Casting

Lisa Palmqvist[1, a], Karin Lindqvist[1, b] and Chris Shaw[2, c]

[1]IVF Swedish Ceramic Institute, Argongatan 30, SE-431 53 Mölndal, Sweden

[2]Cranfield University, Cranfield, MK43 0AL, United Kingdom

[a]lisa.palmqvist@ivf.se, [b]karin.lindqvist@ivf.se, [c]c.p.shaw@cranfield.ac.uk

Keywords: Aqueous, ceramic, colloidal processing, composites, laminates, multilayers, piezoelectric, porosity, PZT, starch, tape casting

Abstract. Porous piezoelectric ceramics are of interest for hydrophones and medical imaging because of their enhanced coupling with water or biological tissue due to acoustic impedance matching. Multilayer lead zirconate titanate (PZT) substrates with dense and porous interlayers were produced by tape casting of aqueous PZT slips with high solids contents. The use of latex binders with low viscosity enabled addition of starch as a fugitive additive to create air/ceramic composites with ´3-3´connectivity. Microstructures, piezoelectric and mechanical properties of sintered, poled laminates were evaluated. The relative permittivity, ε_{33}, decreased by 40% for laminates with porous interlayers compared to dense ones, whereas the relative decrease in piezoelectric longitudinal coefficient, d_{33}, was 35%. Laminates with porous interlayers maintained 72% of their bending strength compared to dense ones.

Introduction

Piezoelectric ceramics like lead zirconate titanate (PZT) are widely used today in actuators and sensors. Porous piezoelectric ceramics are of interest for hydrophone and medical imaging applications as well as sensors, due to enhanced coupling with water or biological tissue [1]. The introduction of air filled pores in the piezoelectric material results in a ceramic/air composite with reduced acoustic impedance, $Z = \rho\, v$, where ρ is the density of the material and v is the sound velocity. The difference in acoustic impedance of the piezoelectric and the surrounding medium determines the reflection, R. Increased amount of porosity also leads to a decrease in the transverse piezoelectric coefficient, $-d_{31}$, relative to the longitudinal piezoelectric coefficient d_{33} [2]. This generates higher electrical charges per unit hydrostatic force due to an increase in the hydrostatic strain coefficient, $d_h = d_{33} + 2d_{31}$. Moreover, when air in the pores replaces ceramic material the permittivity of the material, ε_{33}, is reduced. This increases the hydrostatic voltage coefficient, $g_h = d_h/\varepsilon_{33}$, which is a measure of the electric field per unit hydrostatic stress [3].

Various processes to create porosity in piezoelectric ceramics have been reported. Starch has previously been used as a fugitive additive to create pores in layered ceramics [4]. Tape casting is a reliable and cost effective method for multilayer ceramics, and porous PZT tapes made in organic systems have been studied [5]. Aqueous tape casting is an environmentally attractive processing route, which has been explored for making PZT based materials using PVA as binder [6, 7]. High solids tape casting systems including emulsion based latex binders have been developed [8]. However, only one study of aqueous tape casting of PZT using latex binders has so far been reported [9].

The mechanical strength of ceramic laminates with porous interlayers can be improved due to crack deflection, as shown previously [10]. As expected, the piezoelectric characteristics of such a graded PZT substrate were shown to depend strongly on the porosity level in the interlayer [11]. Moreover, the acoustic impedance and the relative permittivity, ε_{33}, of porous PZT samples decreased linearly with increasing pore volume fraction in completely porous substrates [5]. The electrical and mechanical properties of PZT substrates with porous interlayers have not previously been compared to those of dense and completely porous substrates.

In this work, multilayer PZT substrates with dense and porous interlayers were produced by tape casting of aqueous PZT slips with high solids contents. The use of latex binders with low viscosity enabled addition of starch as a fugitive additive. A porous structure with '3-3'connectivity (where both the ceramic phase and the pores are interconnected) was created. Microstructures, piezoelectric and mechanical properties of sintered, poled laminates were compared to the properties of dense substrates as well as completely porous ones.

Experimental

Sample preparation. Lead zirconate titanate (PZT) Navy type VI (EC-76, EDO Ceramics) with T_c = 190 °C, ρ_{powder} = 7.74 g/cm^3, d_{50} = 1.5 µm and BET area = 1.35 m^2/g was used. The PZT powder was dispersed in deionized water with poly(acrylic acid) dispersant (PC75, Zschimmer & Schwarz) and ball milled for 12 hrs with zirconia balls. Latex binders (LDM 7651S, Celanese and Resicel E50, Lamberti) were added followed by equilibration for 3 hrs and sieving through a 31.5 µm sieve. For porous tapes, 40 vol% potato starch (Microlys, Lyckeby) was added. Casting was done in a continuous caster (TC155, Wallace Inc.) at 50°C and a carrier speed of 2 cm/s followed by drying in humidified atmosphere. Tapes were cut, stacked and pressed at 60 MPa in room temperature and sintered in lead rich atmosphere on porous alumina or Pt foil at 1280°C for 2 hrs. Sintered laminates were electroded by screen printing of Ag paste prior to poling in mineral oil at 115°C, 3 kV/mm.

Characterization. Electroforetic ζ-potential measurements (ZetaMeter 3.0) were done in $5\cdot10^{-3}$ M KNO$_3$ solution using HNO$_3$ and NH$_4$OH to adjust the pH. Viscosity was measured with cup and couette in a Stress Tech CS Rheometer (Rheologica Instruments). Porosity was measured by Hg intrusion (Micromeritics III). Microstructures and lead loss were examined by EPMA/EDS (Jeol JXA-8600). Capacitance and tan δ were measured with a Component Analyzer (Wayne Kerr 6425) and a Berlincourtmeter (Take control PM25) was used to measure d$_{33}$. Biaxial bending strength was measured in ring-on-ring setup (Ø12/30 mm, sample size 40x40 mm, Zwick Z050).

Results and Discussion

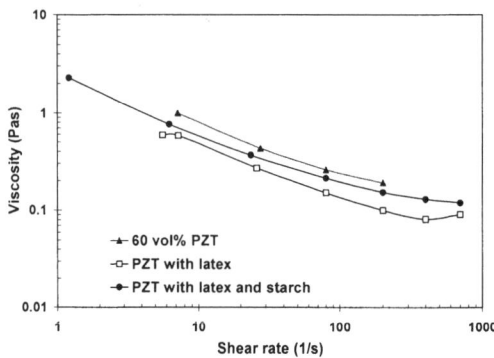

Fig. 1. Viscosity curves of aqueous PZT suspensions

Rheology of PZT suspensions. The dispersant concentration was optimized through rheological evaluation in a previous study [12]. A concentration of 0.12 wt% poly(acrylic acid) (PAA) was found to give the lowest viscosity. Fig. 1 shows viscosity curves for 60 vol% PZT aqueous suspensions stabilized with 0.12 wt% PAA dispersant. PZT slips including latex binders displayed shear thinning behavior as desired for tape casting. A suitable, low viscosity level was also obtained in slips with 40 vol% of pore former (potato starch).

ζ**-potential measurements.** Fig. 2 shows ζ-potential curves for pure PZT powder as well as PZT with 0.12 wt% PAA. The isoelectric point (IEP) decreased from pH 7.9 for pure PZT to about pH 4.4 with adsorbed dispersant, in agreement with other studies [11]. At processing conditions around pH 9-10 the powder with adsorbed dispersant is electrosterically stabilized with a high ζ-potential of approximately -30 mV.

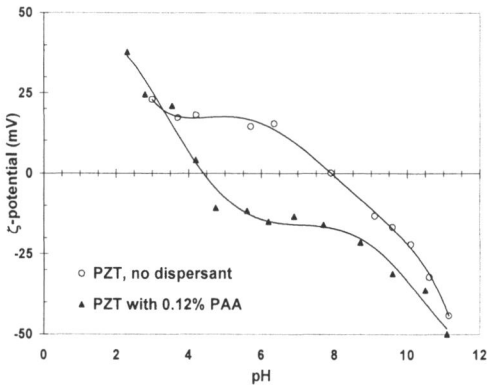

Fig. 2. ζ-potential curves of PZT

Microstructure of sintered laminates. Sintered, electroded laminates as seen in Fig. 3a were examined by Electron Probe Micro Analyser (EPMA). A crack free material can be expected, since no external pressure is applied in the tape casting process and the starch particles are likely to deform plastically during lamination. As shown in a cross section in Fig. 3b, there are no visible cracks present in the sintered substrate with porous interlayers. The total open porosity in porous layers was 25 vol% as determined by Hg intrusion.

PZT ceramics are commonly sintered on Pt, but sintering on zirconia has also been reported [13]. Here, PZT tapes sintered on alumina in lead rich atmosphere showed no lead gradient, as seen from the EDS line profile in Fig. 3c. For comparison, PZT tapes were also sintered on Pt foil. Higher levels of micropores in dense layers were observed in samples sintered on alumina compared to Pt. Despite this, porous alumina was chosen as sintering substrate since PZT samples sintered on Pt displayed more warping and surface defects.

a)

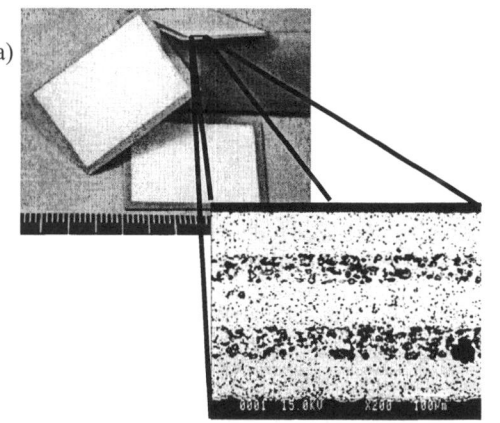

b)

c)

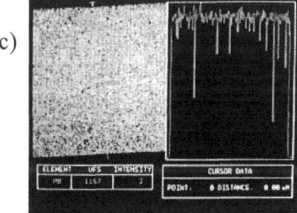

Fig. 3
a) Sintered PZT substrates with Ag electrodes
b) cross section (EPMA)
c) lead profile across dense PZT layer sintered on alumina (EDS)

Piezoelectric and mechanical properties. Table 1 shows the biaxial bending strength of sintered PZT samples and piezoelectric characteristics after poling. The laminates with porous interlayers have a total porosity of only 10%, which can be expected to give a decrease in relative permittivity, ε_{33} of less than 25% according to previous studies [5, 11]. Interestingly, these laminates show a decrease in ε_{33} of almost 40% compared to dense ones. The strong decrease in ε_{33} is most likely due to the localization of pores to interlayers, which creates areas with high degree of porosity in a dense matrix. In the porous substrates, where the total porosity is 25 vol%, ε_{33} decreased with 60% compared to dense samples. Moreover, the longitudinal piezoelectric coefficient, d_{33}, decreased less than ε_{33} for both laminates with porous layers and for porous substrates, by 35% and 45% respectively. Laminates with porous interlayers maintained 72% of their mechanical strength compared to dense ones, whereas porous laminates had a relative strength of 49% of dense samples. The overall strength was low, likely due to insufficient sintering in the dense PZT layers.

Piezoelectric and mechanical properties of poled PZT laminates				
Sample	ε_{33} (1 kHz)	tan δ (1 kHz)	d_{33} $[10^{-12}\,m/V]$	Bending Strength [MPa]
EC-76 bulk	3450	-	583	-
Dense laminate	2908	0.010	424	51
With porous interlayers	1778	0.010	276	37
Porous laminate	1208	0.011	23	25

Table 1. Piezoelectric and mechanical properties of PZT substrates (average of 3 samples).

Conclusions

Aqueous PZT tape casting suspensions with starch as fugitive additive can be processed without increased viscosity levels. This can be used for designing sintered multilayer laminates with porous interlayers. In these air/ceramic composites the relative permittivity, ε_{33}, decreased by 40% for laminates with porous interlayers compared to dense ones, whereas the relative decrease in piezoelectric longitudinal coefficient, d_{33}, was 35%. Laminates with porous interlayers maintained 72% of their bending strength compared to dense ones. By optimization of the sintering process to reduce microporosity in dense layers, further improvement of the mechanical properties is expected.

Acknowledgement

The work was made possible through financial support from the Knowledge Foundation and IRECO. Professor Fred Lange and his group at UCSB are gratefully acknowledged for use of the ζ-potential equipment.

References

[1]   C. R. Bowen, A. Perry et al.: J. Eur. Ceram. Soc. **24** (2004), p. 541
[2]   M. J. Creedon and W. A. Schulze: Ferroelectrics **153** (1994), p. 333
[3]   P. Guillaussier and C. A. Boucher: Ferroelectrics **187** (1996), p. 121
[4]   K. Lindqvist and E. Lidén: J. Eur. Ceram. Soc. **17** (1997), p. 359
[5]   E. Roncari, C. Galassi et al.: J. Eur. Ceram. Soc. **21** (2001), p. 409
[6]   J.-H. Feng and F. Dogan: Materials Science and Engineering **A283** (2000), p. 56
[7]   A. Navarro, J. R. Alcock et al.: J. Eur. Ceram. Soc. **24** (2004), p. 1073
[8]   A. Kristoffersson and E. Carlström: J. Eur. Ceram. Soc. **17** (1997), p. 289
[9]   J. E. Smay and J. A. Lewis: J. Am. Ceram. Soc. **84** (2001), p. 2495
[10]   J. B. Davis, A. Kristoffersson et al.: J. Am. Ceram. Soc. **83** (2000), p. 2369
[11]   A. Navarro, R. W. Whatmore et al.: Journal of Electroceramics **13** (2004), p. 413
[12]   D. Cedercrantz: Diploma Work at the Swedish Ceramic Institute, Chalmers University of Technology, Göteborg (2005)
[13]   D. Piazza, C. Capiani et al.: J. Eur. Ceram. Soc. **25** (2005), p. 3075

Key Engineering Materials Vol. 333 (2007) pp 219-222
online at http://www.scientific.net
© 2007 Trans Tech Publications, Switzerland

Laminated Ceramic Structures within Alumina / YTZP System Obtained by Low Pressure Joining

J. Gurauskis [a], A. J. Sánchez-Herencia [b], C. Baudín [c]

Instituto de Cerámica y Vidrio (CSIC), C/Kelsen 5, 28049 Madrid, Spain

[a] gurauskis@icv.csic.es [b] ajsanchez@icv.csic.es [c] cbaudin@icv.csic.es

Keywords: Al_2O_3, ZrO_2, processing, laminates, residual stresses.

Abstract. The production of multilayer ceramics by laminating stacked green ceramic tapes is one of the most attractive methods to fabricate layered materials. In this work, a new lamination technique was employed to obtain laminated ceramic structures in the alumina-zirconia system with residual stress compression at the outer layers. This reinforcement mechanism would lead to ceramics with changed material properties and R-curve behaviour. The optimization of processing parameters for fabrication of defect free monolithic and laminated structures is described. The residual stresses developed in the laminated structures are discussed in terms of the results obtained from piezo-spectroscopic technique measurements and finite element method calculations.

Introduction

Laminated structures have been investigated due to their capability for the reinforcement of ceramics. One of the reinforcement mechanisms associated with laminated structure is the residual stress compression at the outer layers. This reinforcement mechanism can lead to higher fracture strength and toughness values of the ceramic material, and to activate rising crack growth resistance (R-curve) behaviour [1-3].

One of the most versatile ceramic systems alumina - yttria stabilized zirconia (YTZP) was chosen in this work. In this system, tapes with different compositions can be combined to give rise to the desired residual compressive stresses in the outer layers for macro-structural scale reinforcement [2-4]. In addition, the presence of YTZP might originate reinforcing mechanisms, which operate at the microstructural level [5, 6].

The control of the residual stresses and, consequently, the reinforcing mechanism, is achieved by a strict control of the composition and thickness of the layers that have to be designed for an optimum behavior. In this sense colloidal processing technique based on tape casting is adequate for the fabrication of designed layered structures [7-9].

The lamination of individual tapes into one structure involves the use of elevated pressures and temperatures [7-10]. Such a process is time and energy consuming, furthermore various authors [10, 11] observed the presence of varying mechanical properties for monolithic pieces processed by this route. It has been suggested that this behaviour might be due to the development of some degree of texture in the microstructure related to the implication of elevated pressures and temperatures. Therefore, new lamination technique is proposed in order to simplify this step. The quality of the interfaces between the different tapes and the presence of any texture in the microstructure within sintered specimens obtained via new lamination technique was evaluated by means of the fracture behaviour of the pieces during the ball on three balls test.

No clear data is presented in literature that would indicate the reason of the presence of curve-R behaviour within laminated structures with residual compressive stresses in the outer layer and the influence of magnitude of residual stresses to this phenomenon. Therefore the laminated structures with compressive residual stresses of different magnitude at the outer layer with fixed thickness were fabricated and their values were evaluated by the piezo-spectroscopic technique and finite element method approximation.

Processing of laminated structures

Two compositions within the system alumina - YTZP, A-5 (alumina-5vol.%YTZP) and A-40 (alumina-40vol.%YTZP) were selected for fabrication of ceramic tapes via tape casting procedure of aqueous ceramic slurries. A new processing route for production of layered structures by laminating stacked green ceramic tapes with help of adhesive layer and pressing at room temperature was employed. Full details of tape pre-treatment, gluing agent application and the pressing procedure are given elsewhere [12, 13].

As a first approach, the optimum pressure levels to obtain defect free sintered monolithic materials were investigated. For this purpose, the quality of the interfaces between the tapes and any formation of textures in sintered monolithic specimens A-5 and A-40 were evaluated by means of the fracture behaviour of the pieces during the ball on three balls test [14]. This analysis was chosen because any processing defect will be better revealed under a biaxial stress distribution. Different fracture patterns were found as a function of processing pressure and the characteristics of the tapes, giving the value of optimum pressure of 18 MPa for both compositions. By applying this pressure dense pieces with maximum sizes of the defects of the same level as those of the microstructural ones and without interface effects on mechanical behaviour were obtained via this new processing route employed.

The established processing route was applied for the processing of designed laminated structures, L1 and L2, using the tapes with different compositions, A-5 and A-40. Two different distributions of tapes were selected in a way to obtain two different symmetrical laminated structures with external A-5 layers. In one of them (L1) four A-5 layers were alternated with three A-40 layers of the same thickness ($\approx$430 μm). In the other (L2) three A-5 layers ($\approx$430 μm) were alternated with two A40 layers of double thickness ($\approx$860 μm).

The cross-sections of sintered pieces showed the presence of large transverse cracks though the internal A-40 layers. Smooth and rounded grains at the surface of these cracks suggested that crack formation took place before or during sintering process. Following this observation, the colloidal processing parameters of the tapes, in particular the suspension solid load, were modified to obtain tapes with similar green densities for both compositions. The result was that the initial differential sintering was avoided and defect free laminated structures L1 and L2 were obtained (Fig.1).

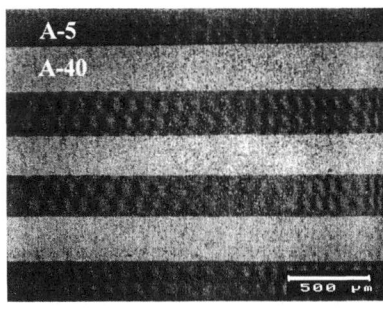
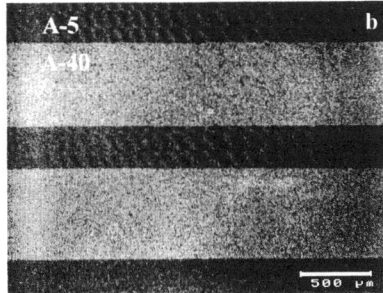

Figure 1. SEM micrographs at the polished cross-section showing the symmetrical laminate structures.
   a)   Laminate L1.
   b)   Laminate L2.

Evaluation of residual stresses

The difference in thermal expansion between the compositions A-5 and A-40, resulted in the development of residual stresses in layered structures L1 and L2. The magnitude and sign of the expected residual stresses were evaluated using the simplified model of a symmetric plate constituted by alternate layers of the same thickness having a uniform biaxial distribution of stresses across each layer [4]. This model gave compressive stresses in the A-5 layers of -182 and -268 MPa for L1 and L2 structure, respectively.

   The magnitude and the distribution of residual stresses in layered structure L1 were assessed by the piezo-spectroscopic technique [15]. As expected, A-5 layers underwent compression, whereas A-40 layers tension. The residual stress within the layers varied with a parabolic trend, with the highest values at the interfaces and a reduction toward the centre of the layers. The maximum values obtained showed a good agreement with the ones evaluated using the simplified model of a symmetric plate. The observed distribution of residual stress values was attributed to the free surface effect, which is recorded by low penetration of fluorescence spectra ($\sim$5-20 μm) [16].

   In order to confirm the effect of the free surface on the distribution of residual stress within the laminate structures L1 and L2, the simulation of residual stresses using finite element method (FEM) was performed. The simulation results for laminated structure L1 near the free surface ($\sim$5 μm) showed the coincident results to the ones observed by the piezo-spectroscopic technique. This fact confirms the correctness of residual stress determination techniques employed and depending on details needed about the residual stresses present in the structure it is possible to use one or other.

   The profiles of residual stress distributions obtained by FEM at the outer layer in the middle of sample (minimum effect of free surface) showed similar trends for both laminated structures (L1 and L2). Showing the minimum values near the surface (L1 = -161 MPa, L2 = -247 MPa) and the maximum values near the interface with A-40 tape (L1 = - 174 MPa, L2 = -268 MPa). This distribution of residual stresses should be taken into account during evaluation of the reinforcement mechanism based on residual compressive stresses at the outer layer of laminated structures.

Summary

Dense multilayer ceramics with strong interfaces and free of defects can be obtained by stacking water based tapes via new processing route and pressureless sintering. By controlling green density of the tapes, development of stresses during initial stages of sintering as well as the associated defects in laminated structures can be avoided. Determination of residual stresses by the piezo-spectroscopic technique and by FEM approximation gives similar results. Even the minimum effect of free surface affects the distribution of residual stresses.

Acknowledgments

Special thanks to Luca Ceseracciu from Departament de Ciencia dels Materials i Enginyeria Metalurgica (Universitat Politécnica de Catalunya) for the help provided with simulation of residual stresses using finite element method .

This work was supported by the projects CICYT MAT 2003-00836 and CAM GR MAT07072004 (Spain).

Work supported in part by the European Community's Human Potential Programme under contract HPRN-CT-2002-00203, [SICMAC].

Jonas Gurauskis acknowledges the financial support provided through the European Community's Human Potential Programme under contract HPRN-CT-2002-00203, [SICMAC].

References:

[1]   C.J. Russo , M.P. Harmer, H.M. Chan and G.A. Miller: J. Am. Ceram. Soc. Vol. 75 [12] (1992), p. 3396.

[2]   O. Sbaizero and E. Lucchini: J. Eur. Ceram. Soc. Vol. 16 [8] (1996), p. 813.

[3]   D.J. Green, P.Z. Cai and G.L. Messing: J. Eur. Ceram. Soc. Vol. 19 [13-14] (1999), p. 2511.

[4]   T. Chartier, D. Merle and J.L. Besson: J. Eur. Ceram. Soc. Vol. 15 [2] (1995), p. 101.
       [5]   F.F. Lange and M.M. Hirlinger: J. Am. Ceram. Soc. Vol. 67 [3] (1984), p. 164.

[6]   W.H. Tuan, R.Z. Chen, T.C. Wang, C.H. Cheng and P.S. Kuo: J. Eur. Ceram. Soc. Vol. 22 [16] (2002), p. 2827.

[7]   B. Schwartz and D.L. Wilcox: Ceram. Age Vol. 83 [6] (1967), p. 40.

[8]   R.E. Mistler: Am. Ceram. Soc. Bull. Vol. 69 (1990), p. 1022.

[9]   J.S. Reed: *Principles of Ceramic Processing* (New York: John Willey & Sons, 1994).

[10]   T. Chartier and T.Rouxel: J. Eur. Ceram. Soc. Vol. 17 [2-3] (1997), p. 299.

[11]   P. Boch, T. Chartier and M. Huttepain: J. Am. Ceram. Soc. Vol. 69 [8] (1986), p. C191.

[12]   J. Gurauskis, A.J. Sanchez-Herencia and C. Baudin: J. Eur. Ceram. Soc. Vol. 25 [15] (2005), p. 3403.

[13]   J. Gurauskis, A.J. Sanchez-Herencia and C. Baudin: J. Eur. Ceram. Soc. Vol. 26 [8] (2006), p. 1489.

[14]   J. Gurauskis, J. Pascual, T. Lube, A.J. Sanchez-Herencia and C. Baudin: Key Eng. Mater. Vol. 290 (2005), p. 203.

[15]   G. de Portu, J. Gurauskis, L. Micele, A.J. Sanchez-Herencia, C. Baudin and G. Pezzotti: J. Mater. Sci. (2006). In press, corrected proof. Available online 10th of April, 2006.

[16]   G. Pezzotti: Compos. Sci. Technol. Vol. 59 [6] (1999), p. 821.

Key Engineering Materials Vol. 333 (2007) pp 223-226
online at http://www.scientific.net
© 2007 Trans Tech Publications, Switzerland

Functional Gradient Alumina Ceramics With Controlled Porosity

Jana Andertová[a], Jiří Havrda, Radek Tláskal

Department of Glass and Ceramics, Institute of Chemical Technology Prague, Technická 5,
Prague 6, Czech Republic, CZ-166 28

[a]andertoj@vscht.cz

Keywords: alumina, porosity, gradient materials, slip casting

Abstract. The work deals with preparation of functional gradient alumina ceramics with controlled porosity by slip casting method of aqueous alumina suspension containing pore-generating agent. The sol-gel transition of AlO(OH) was employed to stabilize pore-generating agent in the suspension. The composite bodies with layers of variable porosity were prepared. Based on dilatometer measurement the admission difference of irreversible dilatation changes $\Delta\alpha_{irr}$ between compounded layers was determined in order to prepare defect free bi-layer bodies. The dependence of physical and mechanical properties of as fired composite bodies on the porosity value was expressed.

Introduction

The alumina ceramics has been applied for ceramics implants preparation for many years. These materials are fully sintered and their density is close to the theoretical density of the used materials. The alumina implants are characterized by high values of mechanical properties. However, their main disadvantage is brittleness and a significant difference in the Young's modulus value and the ceramic implant and bone tissue fracture toughness. One option to draw near the ceramic and bone tissue texture and to improve the implant fixation in the bone tissue (tissue growing through the implant) is utilization of the functional gradient ceramic (FGC). The value of Young's modulus can be controlled through the porosity, pores size, and shape. The porosity gradient brings a positive effect in terms of the ability of Young's ceramics modulus to more closely resemble Young's modulus of bone. The composite materials consisting of layers with increasing porosity will enable to draw near the bio-ceramics and bone tissue mechanical properties. The mechanically stressed functional parts of implant should be prepared from the sintered non-porous ceramics which meets standard properties. The composite materials ending by the porous layer with open pores in the point of implant and the bone and the bone tissue contact allows the bone tissue growing into the open pores.

The aim of this work is preparation of the single-phase functional gradient alumina ceramics with controlled variable porosity by slip casting method into porous mold. The work was divided throuth two time periods; the first was utilized for the single-layer bodies' preparation with controlled constant porosity and the determination admission difference of irreversible dilatation changes $\Delta\alpha_{irr}$ between compounded layers; in the second one was prepared the bilayer bodies with variable porosity of individual layers. All prepared bodies are characterized by the determination of the mechanical and physical-chemical properties and their microstructure [1-4].

Materials and methods.

The work was carried out with the powder α-Al_2O_3 (AKP 15-Sumitomo Chemical Co., Ltd., Japan), d=0.6-0.8 µm. Sokrat 32A (CHZ Sokolov) as a deflocculant and the thin-walled balls were used (Al_2O_3, d=150–190 µm) as a pore-generating agent were used. The boehmit gel AlO(OH) (Dispersal sol P2) was applied for a pore-generating agent stabilization. The method of slip casting of the suspensions with variable pore-generating agent content into plaster mould was used for preparation of the monolayer and bilayer samples with variable porosity. The samples were fired at 1570°C (heating rate 2°C min, dwell at maximum temperature 2 hours) after drying. The physical and mechanical properties (porosity, Young's modulus, fracture toughness and flexural strength) were measured on as-fired bodies [2-4]. Porosity was determined by Archimedes´ method. Bending strength values and fracture toughness values were determined in three-point bending, Young's modulus values were determined by the resonant frequency method using a resonant frequency tester (Erudite, CNC ELECTRONICS, UK) on mono-layered specimens with constant variable porosity; for bilayered specimens the Young's modulus values were calculated. The image analysis (LUCIA, Laboratory Imaging, Czech Republic) was used for the optical microscopic analysis of the bodies' surface [3-4].

Experimental procedure

For the purpose of the rheological measurement it was suggested the optimum composition of the aqueous suspension in the ratios: 76.0 wt% of alumina mixture, 3.0 wt% of AlO(OH), 2.3 wt% of deflocculant and different content pore-generating agent in range of 0-6 wt%. All procedure is detail referred in [2]. The bodies with the variable defined porosity were cast from the prepared suspension with the variable content of the pore-generating agent. Experimentally verified mathematical model of the body creating kinetics has enabled a serious estimation of the body casting time of the required body thickness. On basis of the acquired knowledge the monolayer and bilayer bodies with different content of the pore-generating agent were prepared. The prepared bodies were low and high temperature processed. The values of porosity, bulk density, water absorption and mechanical properties (Young's modulus, fracture toughness and flexural strength) were determined. The determined values for monolayer bodies are presented in Table I. In result have shown that the pore-generating agent content has changed bulk density of monolayer in range of 3.8-2.4 g.cm^{-3}; value of water absorption in range of 0.0- 4.9 % ; porosity value in range of 5.6-39.0 % and bending strength value in range of 377- 82 MPa within the interval of pore-generating agent content 0-7 wt % [1-5]. The average diameter of spherical pores was not exceeded 200 µm.

Table 1 Physical and mechanical properties of bodies with constant growing porosity

Content of pore-generating agent	Bulk density ρ_v	Porosity P	Young's modulus E	Water absorption	Bending strength
[wt.%]	[g.cm^{-3}]	[%]	[GPa]	[%]	[MPa]
0	3.80	5.6	-	0.0	377
1	3.51	12.4	310	0.0	235
2	3.39	15.0	275	0.1	187
3	3.13	21.8	-	0.2	148
5	2.92	27.0	226	0.7	142
6	2.82	29..	-	0.8	139
7	2.40	39.0	187	4.9	82

With increasing volume of the pore-generating agent in the suspension increases the prepared bodies' porosity and their bending strength and Young's modulus decreases. The second step of the work was aimed at the bodies preparation development created by mutually joined layers of different porosity and thus by gradually changing properties. The successfully preparation of the functional gradient ceramics is conditioned particularly by solution of suitable combination of the joined layers of various porosity, e.g. with different length changes. The solution basis was determination of the bodies' length changes curves of different porosity and evaluation of the value of admission difference of irreversible dilatation changes for the most frequently evaluated temperature intervals in ceramic firing. From the deformation of bodies with different porosity value was created the non-linear model of irreversible dilatation changes on porosity value. The admission value of difference irreversible dilatation changes $\Delta\alpha_{irr} = 0.2 \times 10^{-6}$ K^{-1} between joined layers can be considered as the limit value for defect-free FGC preparation. The results concerning this measurement have already been reported [4]. The final stage of the work presented consist in the preparation of the set of bilayers samples with porosity gradient layers (thickness of each layer 1 mm). Consequently, the porosity, the bending strength, Young's modulus, and the fracture toughness values were determined (Table II). Interface character between layers with different porosity and character of pores shape is show in Fig 1.

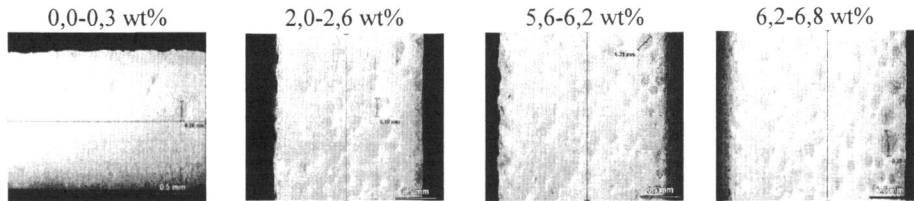

0,0-0,3 wt%   2,0-2,6 wt%   5,6-6,2 wt%   6,2-6,8 wt%

Fig.1. Character of areas in sample of constant porosity; top: content of pore-generating agent

In next step, the second set of bilayer samples (thickness of each layer 1 mm) was prepared; the pore-generating agent content was constant in one layer (0 wt %) and variable increased in second layer (0-7 wt %) in such way that the difference of value $\Delta\alpha_{irr}$ gradually increased in the prepared green bodies in range of $\Delta\alpha_{irr} = 0.2\text{-}2.1 \times 10^{-6} K^{-1}$. The deformation of samples was studied. If the limit value is exceeded, the bilayer bodies' defection increases significantly, as documented in Fig. 2.

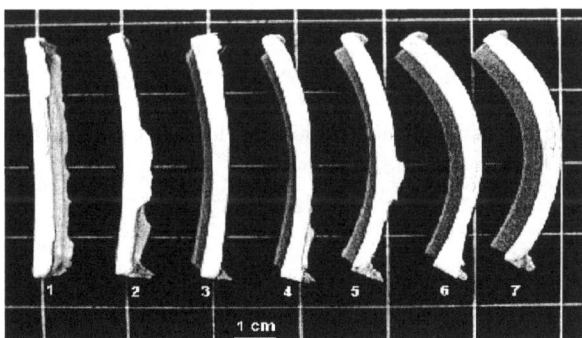

Fig. 2. Character of bilayers bodies deformation with growing pore-generating agent content in second layer, value of $\Delta\alpha_{irr}$: 1– $\Delta\alpha_{irr} = 0.2 \times 10^{-6}$ K^{-1} (0,0-0,3)*; 2– $\Delta\alpha_{irr} = 0.8 \times 10^{-6}$ K^{-1} (0,0-1,2)*; 3– $\Delta\alpha_{irr} = 1.0 \times 10^{-6}$ K^{-1} (0,0-1,7)*; 4– $\Delta\alpha_{irr} = 1.4 \times 10^{-6}$ K^{-1} (0,0-2,7)*; 5– $\Delta\alpha_{irr} = 1.7 \times 10^{-6}$ K^{-1}(0,0-3,8)*; 6– $\Delta\alpha_{irr} = 2,0 \times 10^{-6}$ K^{-1} (0,0-5,7)*;7– $\Delta\alpha_{irr} = 2,1 \times 10^{-6}$ K^{-1} (0,0-6,8)*;
*) number in brackets give the content of pore-generating agent in the second layer (in wt %).

Table II. Properties of bilayers functionally graded ceramic bodies with variable porosity

The pore-generating agent content	Porosity	Bulk density ρ_v	Young's modulus $E_{\parallel}$*)	Young's modulus $E_{\perp}$*)	Bending strength	Fracture toughness **)
[wt.%]	[%]	[g.cm^{-3}]	[GPa]	[GPa]	[MPa]	[MPam$^{1/2}$]
0.0-0.0	4.3	3.83	349.2	349.2	377	3.2
0.0-0.5	5.7	3.77	340.3	339.9	283	3.7
0.5-1.0	9.6	3.62	326.2	326.2	180	2.8
1.0-1.5	11.9	3.52	302.2	300.9	193	2.2
1.5-2.0	13.2	3.47	281.0	281.0	174	2.7
2.0-2.6	15.5	3.38	275.0	274.9	136	1.7
2.6-3.2	18.0	3.28	259.5	259.0	116	2.3
3.2-3.8	20.4	3.19	238.4	238.1	127	1.3
3.8-4.4	22.3	3.11	223.7	223.	118	1.2
4.4-5.0	26.9	2.93	217.0	217.0	114	0.9
5.0-5.6	27.5	2.90	214.7	214.7	108	0.8
5.6-6.2	27.9	2.88	201.2	200.4	94	0.8
6.2-6.8	29.6	2.82	186.7	186.7	81	0.7

*) The values of Young's modules were calculated $E_{\perp}=E_1E_2/(E_2V_1+E_1V_2)$, $E_{\parallel}= E_1V_1/E_2V_2$, where E_1, E_2 are Young's modules of individual layer and V_1, V_2 volume content each phase, $E_{\parallel}, E_{\perp}$ are Young's modules determined for equidistantly a perpendicularly applied stress respectively [5]
**) average values [1]

Discussion and conclusion

The experimental approach of the casting suspension with the pore-generating agent and the porosity effect on the functional graded alumina ceramics properties has been studied in this work. An experimentally verified model of the body generating kinetics has enabled an estimation of the layer casting of the required thickness of the individual bodies of the graded ceramics consisting of bilayers. It is successfully, that preparation of the alumina ceramics consisting of the both porous and non-porous layer with partial unification of the Young's modulus values of the ceramics and the bone while the mechanical strength and fracture toughness comparable with the cortical bone tissue are maintained. After the spherical pores opening, the improved implant fixation due to the tissue growing through these pores is expected. The presented results validity is related to the alumina properties, pore-generating agent properties and measured range of the bodies' porosity.

References:

[1] R. Tláskal: *PhD Thesis*, (in Czech) ICT Prague (2005).

[2] J. Andertová, J. Havrda, and R. Tláskal: Characterization of Porous Solids VII (ed. P. L. Llewellyn, J. Reuquerol, F. Rodriges-Reinoso, N. A. Seaton), submitted Elsevier (2006).

[3] J. Andertová, R.Tláskal, J.Havrda, Proceeding of CHISA, (ed. J. Novosad), Prague 48 (2004).

[4] J. Andertová, R. Tláskal, M. Maryška, J. Havrda, submitted J. Eur. Ceram. Soc. (2001).

[5] J. Andertová, R. Tláskal, J. Havrda, I. Zedníková, Proceeding of Conference on Porous Ceramic Materials, VITO, Belgium (2005).

Key Engineering Materials Vol. 333 (2007) pp 227-230
online at http://www.scientific.net
© 2007 Trans Tech Publications, Switzerland

Glass-alumina Functionally Graded Materials Produced by Plasma Spraying

V. Cannillo[1, a], L. Lusvarghi[1,b], T. Manfredini[1,c], M. Montorsi[1,d], C. Siligardi[1,e], A. Sola[1,f]

[1]Dipartimento di Ingegneria dei Materiali e dell'Ambiente, University of Modena and Reggio Emilia (Italy)

[a]valeria@unimo.it, [b]lusvarghi.luca@unimo.it, [c]manfrediti.tiziano@unimo.it, [d]monia@unimo.it, [e]siligardi@unimo.it, [f]sola.antonella@unimo.it

Keywords: Glass-alumina FGM. Plasma spraying. Mechanical properties.

Abstract. The present work was focused on glass-alumina functionally graded materials. The samples, produced by plasma spraying, were built as multi-layered systems by depositing several layers of slightly different composition, since their alumina and glass content was progressively changed. After fabricating the graded materials, several, proper characterization techniques were set up to investigate the gradient in composition, microstructure and related performances. A particular attention was paid to the observation of the graded cross sections by scanning electron microscopy, which allowed to visualize directly the graded microstructural changes. The scanning electron microscopy (SEM) inspection was integrated with accurate mechanical measurements, such as systematic depth-sensing Vickers microindentation tests performed on the graded cross sections.

Introduction

In recent years, great attention has been devoted to glass-ceramic FGMs, due to their relevant potentialities [1]. For example, if the compositional gradient is properly designed, glass-alumina FGMs may show really interesting properties, such as an enhanced superficial resistance to Hertzian cracking [2] and sliding contact [3]. Since the final performances of glass-alumina FGMs are governed by the compositional profile, the distribution in space of the constituent phases should be carefully designed and realized.

Plasma spraying is an innovative production technique which enables to obtain the wanted graded profile by building it layer by layer [4]. Therefore, due to its flexibility and reliability, plasma spraying was applied to fabricate self-standing glass-alumina FGMs, whose properties gradient was evaluated by means of a systematic Vickers microindentation test.

FGM production

Constituent phases. The *glass* used in the present research belonged to the ternary system $CaO-ZrO_2-SiO_2$ and its composition was purposely formulated in order to minimize the thermo-mechanical mismatch with the alumina [5]. The powder raw materials (calcium carbonate, zirconium silicate, quartz) were mixed and melted; then the glass was fritted in cold water and wet ball milled; the final powder was spray-dried in order to improve its flowability (NIRO Atomizer, Denmark – located at the Centro Sviluppo Materiali S.p.A., Roma) [4].

As regards the *alumina*, a commercial, high purity (> 99.5%) powder (Sulzer Metco 105SFP) was employed.

Fabrication method. The FGMs were produced by plasma spraying; the spray runs were carried out by a F4-MB plasma torch installed in a Controlled Atmosphere Plasma-Spraying plant at Centro Sviluppo Materiali S.p.A. (Roma, Italy).

The samples were designed as multi-layered systems, made up of several strata whose mean composition was progressively changed form 100 vol% alumina to 35 vol% alumina-65 vol% glass (a

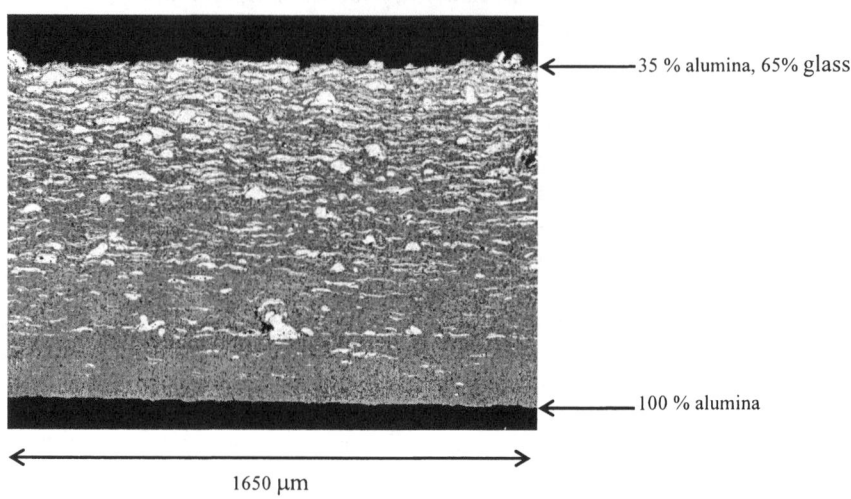

Figure 1 – Cross section of the plasma sprayed FGM.

composition comparable with that usually observed in glass-alumina FGMs obtained by percolation [5]). Moving from a layer to the following one, the alumina content was decreased of 5 vol%, while the glass content was increased of 5 vol%; the thickness of the layers was kept constant (about 25 μm).

The compositional change was not obtained by spraying pre-mixed powders [6]; instead, the glass and alumina powders were stored in two different feeders and separately introduced into the plasma flux. In order to create the compositional gradient, the feeding flow rate of the two powders was progressively changed according to preliminary deposition efficiency tests [4].

After depositing the last layer, the sprayed systems were mechanically removed from the substrate (a commercial alumina bulk, Kéramo ceramiche tecniche, Tavernerio (CO), Italy).

Characterization

The SEM inspection (SEM – Philips XL-30) of the polished cross section (total thickness: about 1.1 mm) immediately confirmed the smooth compositional variation along the spraying direction, as shown in Fig. 1. As a matter of fact, the splats of the constituent phases could be readily distinguished (in Fig. 1, white: glass; grey: alumina). It is worth noting, however, that the interface between the ingredient materials was very regular, since the glass could fit the alumina morphological peculiarities during the splat quenching [4]. Moreover, the SEM investigation revealed that the designed strata could not be identified any more in the resulting system, since the characteristic dimension of the sprayed splats was comparable with the layer thickness, and therefore the final gradient could be considered continuous [7].

In order to evaluate the variation in mechanical properties associated with the compositional gradient, a proper depth-sensing Vickers micro-indentation test (OpenPlatform, C.S.M. Instruments) was performed on the FGM cross section. In fact several indentations were carried out along equally spaced lines parallel to the deposition plane; a maximum load of 1 N was applied for 15 sec. During each indentation the complete loading-unloading cycle was recorded by the instrument and the unloading data were analysed according to the Oliver and Pharr method [8] to deduce the local elastic modulus. In this way, the mean elastic modulus was calculated as a function of depth.

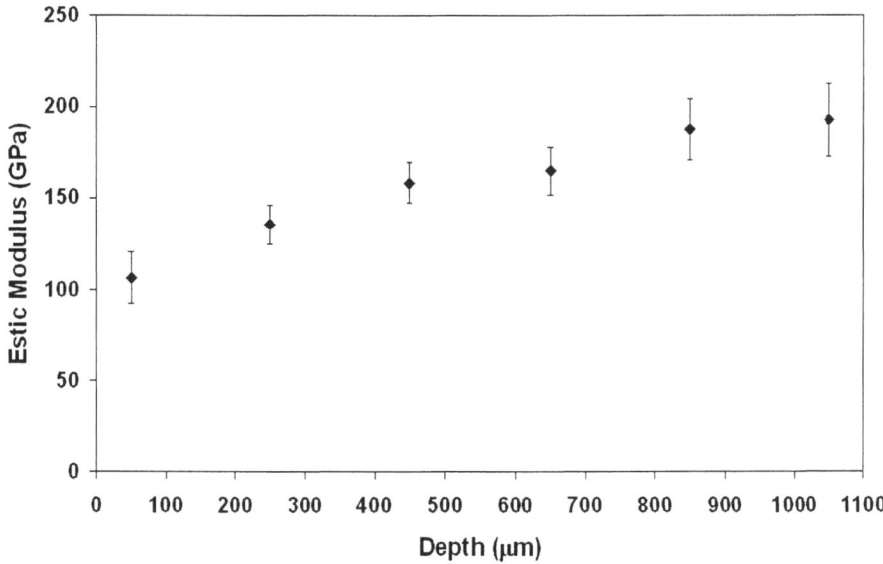

Figure 2 – Variation of the elastic modulus along the spraying direction as determined by depth-sensing microindentation.

As reported in Fig. 2, the elastic modulus progressively changed along the spraying direction, raising from about 100 GPa in the glass rich area to about 200 GPa in the alumina rich area of the cross section. This trend was a consequence of the compositional gradient, since the glass elastic modulus was lower than the alumina one [4], confirming that the graded structure induces spatial variations of the mechanical properties.

This depth-sensing indentation test, therefore, may represent a useful tool to appreciate the spatial change of the elastic properties which arises in FGMs as a consequence of their peculiar constituent phase distribution.

Summary

Plasma spraying proved to be a flexible and reliable production technique to fabricate self standing glass-alumina FGMs; the designed compositional gradient, in fact, could be carefully realized, as clearly proved by the SEM observation.

The change in composition along the spraying direction resulted in a gradual variation of the mechanical properties, such as the elastic modulus, which could be appreciated by performing a proper systematic depth-sensing Vickers microindentation test on the graded cross section.

Acknowledgements

The present research was partially supported by PRRIITT (Regione Emilia Romagna), Net-Lab "Surface & Coatings for Advanced Mechanics and Nanomechanics" (SUP&RMAN).

Centro Sviluppo Materiali (CSM) S.p.A. (Roma, Italy), Surface Engineering Unit, is gratefully acknowledged for the spraying sessions.

Many thanks to Ing. Giovanni Bolelli (Dipartimento di Ingegneria dei Materiali e dell'Ambiente, Università di Modena e Reggio Emilia, Italy), for his precious help in the functionally graded material design.

References

[1]   Y. Miyamoto, W. A. Kaysser, B. H. Rabin, A. Kawasaki and R. G. Ford: *Functionally Graded Materials. Design, Processing and Applications* (Kluwer Academic Publishers, 1999).

[2]   J. Jitcharoen, N.P. Padture, A.E. Giannakopoulos and S. Suresh: Journal of the American Ceramic Society, 81(9) (1998) 2301-2308.

[3]   S. Suresh, M. Olsson, A.E. Giannakopoulos, N.P. Padture and J. Jitcharoen: Acta Materialia, 47(14) (1999) 3915-3926.

[4]   V. Cannillo, L. Lusvarghi, M. Montorsi, C. Siligardi and A. Sola: Characterization of glass-alumina functionally graded coatings obtained by plasma spraying, submitted.

[5]   V. Cannillo, T. Manfredini, M. Montorsi, C. Siligardi and A. Sola: Glass-alumina Functionally Graded Materials preparation and compositional profile evaluation, Journal of the European Ceramic Society, in press.

[6]   K.A. Khor, Y.W. Gu and Z.L. Dong: Journal of Thermal Spray Technology, 9(2) (2000) 245-249.

[7]   A. Mortensen and S. Suresh: International Materials Review, 42(3) (1997) 85-116.

[8]   W.C. Oliver and G.M. Pharr: Journal of Materials Research, 7[6] (1992) 1564-1583.

Key Engineering Materials Vol. 333 (2007) pp 231-234
online at http://www.scientific.net
© 2007 Trans Tech Publications, Switzerland

Characterization of Y_2O_3, CeO_2 and Y_2O_3+CeO_2 Doped FGM Tetragonal ZrO_2 Ceramics by Spark Plasma Sintering

S.G. Huang[1,a], O. Van der Biest[1,b], J. Vleugels[1,c], K. Vanmeensel[1,d] and L. Li[2,e]

[1]Department of Metallurgy and Materials Engineering (MTM), Katholieke Universiteit Leuven, Kasteelpark Arenberg 44, B-3001 Heverlee, Belgium

[2]School of Materials Science and Engineering, Shanghai University, 149 Yanchang Road, Shanghai 200072, China

[a]shuigen.huang@mtm.kuleuven.be, [b]omer.vanderbiest@mtm.kuleuven.be, [c]jozef.vleugels@mtm.kuleuven.be, [d]kim.vanmeensel@mtm.kuleuven.be and [e]liling@sh163.net

Keywords: ZrO_2-CeO_2-Y_2O_3; Reduction; SPS; Graded materials; Microstructure

Abstract. In this study, 2 mol% Y_2O_3 (2Y), 1 mol% Y_2O_3+6 mol% CeO_2 (1Y6Ce) and 12 mol% CeO_2 (12Ce)-doped tetragonal ZrO_2 ceramics were made by spark plasma sintering (SPS) for 2 min at 1450°C under a pressure of 60 MPa. The influence of stabilizers on microstructure, phase and mechanical properties of the ZrO_2 ceramics was investigated. After sintering, the 2Y and 1Y6Ce were intact, containing full tetragonal ZrO_2 phase on the polished cross-sectioned surface, whereas the 12Ce exhibited macro-cracks, corresponding to a large amount of monoclinic ZrO_2 phase. Graded microstructure and mechanical properties were observed in the 1Y6Ce, showing a gradually decreased fracture toughness from sample edge towards centre, together with the slight decreased hardness. The 2Y had a uniform microstructure and mechanical properties. The formation of the graded structure and toughness profile was explained in terms of the gradual CeO_2 reduction to Ce_2O_3 in the Y_2O_3+CeO_2 doped ZrO_2 ceramics.

Introduction

2-3 mol% Y_2O_3-doped tetragonal ZrO_2 polycrystalline (Y-TZP) represents the most widely recognized ZrO_2 structural ceramics, with high hardness and strength [1]. Recently, works were carried out by doping CeO_2 into ZrO_2-Y_2O_3 to improve the hydrothermal stability and mechanical properties [2,3]. The sintering of ZrO_2 ceramics is usually conducted in air or oxidizing atmosphere, resulting in a long sintering cycle to reach full densification and with an extensive ZrO_2 grains growth after reaching the closed porosity.

SPS is capable of sintering ceramics to high density at a relatively short cycle [4]. A reducing or inert atmosphere can be established during SPS if a graphite die is utilized for sintering materials. On sintering CeO_2-containing ZrO_2 ceramics, the reduction of CeO_2 to Ce_2O_3 must be taken into account since CeO_2 can be easily reduced into Ce_2O_3 under the SPS condition at elevated temperature. Recently, present authors analyzed the redox behavior of CeO_2 and CeO_2-ZrO_2 solid solutions from a thermodynamic point of view [5, 6]. It was thought that Ce^{+4} in CeO_2 or ZrO_2-CeO_2 solid solutions can be gradually reduced to Ce^{+3} at increasing temperature under vacuum of 10^{-1} to 10^{-2} Pa, inert atmospheres (Ar, He), or low oxygen partial pressure. As to the mechanical properties, it was reported that the fracture toughness on the surface layer of full dense post-annealed Ce-TZP can be greatly improved by an annealing process in Ar (5% N_2, 1.2 ppm O_2) at 1450°C [7]. The bending strength of 12Ce-TZP, increased from 240 to 545 MPa was detected by Heussner *et al.* [8], after annealing in N_2 at 1400°C for 2 h. The improved properties were explained in terms of the partial CeO_2 reduction and segregation of Ce, resulting in a TZP matrix with enhanced transformability. On the other hand, it was considered that sufficient CeO_2 reduction effectively increased the hardness, but decreased the toughness, explained by the low transformability of the reduced Ce-TZP phase [9]. In this study, we aims to investigate the SPS processed ZrO$_2$ ceramics. Particularly, the possible effect of CeO$_2$ reduction on the formation of a graded microstructure and mechanical properties was examined.

Experimental

The ZrO$_2$ powders were stabilized by 1Y6Ce, and 2Y, with 2 wt% Al$_2$O$_3$ as sintering additive. In addition, 12Ce (Daiichi CEZ-12) grade was prepared. The method to make Y$_2$O$_3$- and Y$_2$O$_3$+CeO$_2$ coated ZrO$_2$ powder is described elsewhere in detail [3, 10]. SPS (Type HP D 25/1, FCT Systeme, Rauenstein, Germany) was carried out in vacuum for 2 min at 1450°C under a pressure of 60 MPa. The samples were sintered in a graphite die/punch set-up. The heating and cooling rates were 200 °C/min. The samples were 20 mm in diameter and around 4 mm in height after sintering. The disc-shaped samples were cross-sectioned, ground and polished to 1 μm finish. The microstructure of thermally etched surfaces (1350°C for 20 min in air) was examined by scanning electron microscopy (SEM, XL30-FEG, FEI, the Netherlands). The average grain size was determined from SEM micrographs using *Image-pro Plus* software. X-ray diffraction (XRD, 3003-TT, Seifert, Ahrensburg, Germany) was conducted for phase identification on the polished surfaces of sintered ceramics. The Vickers hardness, HV$_{10}$, was measured (Model FV-700, Future-Tech Corp., Tokyo, Japan) with an indentation load of 98.1 N, and the fracture toughness, K$_{IC}$, was calculated from the length of the radial cracks of these indentations according to the Anstis formula [11], using an elastic modulus of 210 GPa for t-ZrO$_2$.

Results and discussion

Microstructure and phase analysis

On the polished surfaces, the 1Y6Ce and 2Y are intact, while the 12Ce exhibits macro-cracks. Up to 99% of theoretical density was obtained in the samples. Graded layers were observed on the surface of 1Y6Ce, varied from salmon pink or red on edge to grey or yellow at centre along the axial and radius directions. The CeO$_2$ reduction was suggested in this sample in terms of the changed color. A graded color layer was also observed in the 12Ce grade, while the 2Y grade had a homogeneous grey contrast. Fig. 1 shows the microstructure of the 1Y6Ce, 2Y and 12Ce in the surface and centre areas. The bright-grey and dark-grey contrasts are the ZrO$_2$ and Al$_2$O$_3$ grains, respectively. The 1Y6Ce had an average ZrO$_2$ grain size of 0.15 μm at edge and 0.31 μm at centre, accompanying with much homogeneous size distribution in the edge area, in range 0.03-0.35 μm compared to the centre in range 0.08-0.84 μm. As to the 2Y, an average size of 0.35 and 0.36 μm was obtained at edge and centre, while a similar grain size distribution were found in the 12Ce at different positions.

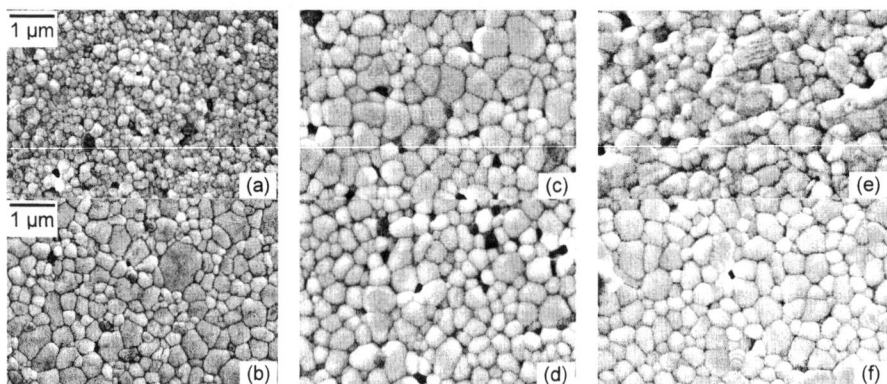

Fig. 1. SEM of the SPS processed ceramics in edge and centre areas, 1Y6Ce (a, b), 2Y (c, d) and 12Ce (e, f).

According to the microstructures, only the 1Y6Ce had a graded grain size distribution. During SPS, the temperature distribution inside of sample and graphite die is very complicated, which may be influenced by the geometry of sintering set-up and thermo-electrical behavior of investigated materials. Vanmeensel *et al.* [12] investigated the temperature distribution of an electrical insulating ZrO_2 during SPS by a mixed experimental-theoretical approach. Simulations indicated a radial temperature gradient of about 25°C inside of a full dense 3Y-TZP at 1500°C, while the gradient was less than 5°C along the central axis, in case of a Ø 40 mm graphite die with 8 mm die wall thickness was used. In this study, the temperature gradient along the central axial direction was calculated to be less than 2°C inside of a 4 mm thickness ZrO_2 ceramic. It is assumed that the temperature distribution should be similar in the 2Y, 1Y6Ce and 12Ce. On the other hand, the content and ionic radius of stabilizer in ZrO_2 solution plays an important role on grain growth beside the impact of temperature. During sintering, the edge had much lower oxygen partial pressure, leading to a continuous CeO_2 reduction to Ce_2O_3 in t-ZrO_2 matrix. Ce_2O_3 is unstable in ZrO_2 lattice because of an approximately 40% mismatch in ionic radii of Ce^{+3} and Zr^{+4}, resulting in a decrease in CeO_2 solubility in t-ZrO_2 and a high elastic lattice strain. There also existed CeO_2 reduction beneath the edge area, but not as fast and complete as that of the edge area. However, to the 12Ce, it is thought most of the CeO_2 has been converted to Ce_2O_3 according to the marco-cracks and changed color. Thus, it is considered that the graded structure of the 1Y6Ce is from the effect of CeO_2 reduction, not from the temperature gradient.

Only a trace amount of m-ZrO_2 was observed on the polished surface of the 1Y6Ce and 2Y, while the 12Ce had strong peak intensity of m-ZrO_2. Clearly, nearly all the m-ZrO_2 has been stabilized and retained as t-ZrO_2 in the sintered 2Y and 1Y6Ce. During cooling, the t-ZrO_2 grains can be transformed to m-ZrO_2 if the stabilizer content is below the critical value with respect to grain size. When sintering CeO_2-containing ZrO_2 under reducing atmosphere, the formation of Ce_2O_3 is associated with a decrease in CeO_2 solubility in t-ZrO_2 phase and a high elastic lattice strain. Therefore, the t-ZrO_2 phase is prone to be transformed in the 1Y6Ce and 12Ce due to the decreased stabilizer content and lattice strain in t-ZrO_2 phase. It has been reported that ZrO_2 with less than 1.4 mol% Y_2O_3 was full m-ZrO_2 phase [13]. In the 1Y6Ce, the CeO_2 reduction is not completed, otherwise the 1 mol% Y_2O_3 stabilized ZrO_2 phase can not be fully stabilized. Moreover, the much smaller ZrO_2 grain size in the 1Y6Ce compared to the conventionally sintered ZrO_2 also contributed to the stabilization of t-ZrO_2 phase [3].

Mechanical properties

The change of hardness and toughness of the 1Y6Ce and 2Y is plotted in Fig. 2, varied with distance along the central axial. The toughness of the 1Y6Ce, decreased dramatically from the edge of 10.82 MPa.m$^{1/2}$, towards centre of 9.65 MPa.m$^{1/2}$, together with a slight decrease in hardness, from 11.51 to 11.21 GPa. A quite homogeneous toughness and hardness was obtained in the 2Y.

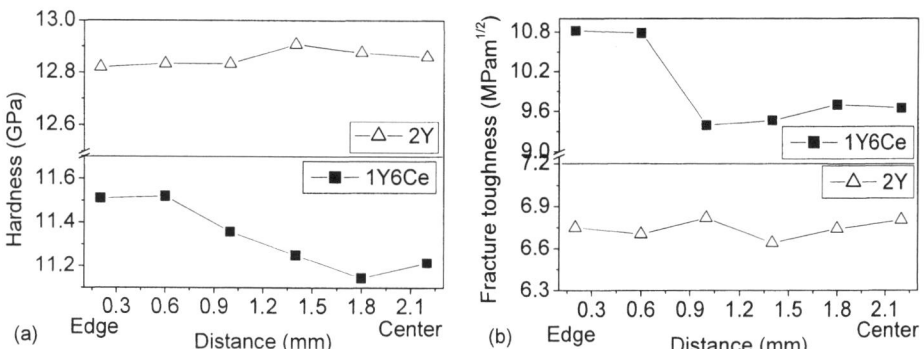

Fig. 2 Hardness (a) and fracture toughness (b) profiles of the 1Y6Ce and 2Y along the central axial direction.

As transformation induced toughening ceramics, the larger the ZrO_2 grains in TZP exhibits a higher transformability and therefore the better toughness [1]. Meanwhile, the lower the stabilizer content, the higher the toughness can be obtained if not the formation of m-ZrO_2 phase. In this study, the 1Y6Ce has higher toughness at edge and lower at centre even though with a much finer ZrO_2 grain size at edge. During SPS, a graded oxygen activity was built up in reducing environment. The gradient CeO_2 reduction resulted in gradient t-ZrO_2 phase transformability, therefore higher toughness at edge position. The toughening of CeO_2-containing TZP by reduction-induced phase transformation is consistent with the observation of Vleugels *et al.* [7]. The presence of small gradient in hardness can be from the different grain size.

Summary

The m-ZrO_2 can be fully stabilized by 1 mol% Y_2O_3+6 mol% CeO_2 or 2 mol% Y_2O_3 by spark plasma sintering for 2 min at 1450°C. Graded microstructures and mechanical properties were achieved in the Y_2O_3+CeO_2 containing ZrO_2 due to the partial CeO_2 reduction to Ce_2O_3. A higher toughness of 10.82 $MPa.m^{1/2}$ was obtained at edge of 1Y6Ce compared to 9.65 $MPa.m^{1/2}$ at centre, together with hardness value of 11.51 and 11.21 GPa. A homogeneous microstructure and mechanical properties were observed in the 2Y. The enhanced phase transformation from the t to m-ZrO_2 contributed to the improved properties of Y_2O_3+CeO_2 doped tetragonal ZrO_2 ceramics.

Acknowledgements

This work is financially supported by the GROWTH program of the Commission of the European Communities under project contract No. G5RD-CT2002-00732, the research fund of K.U.Leuven in the framework of the Flanders-China bilateral projects BIL 04/13 and BIL 04/14, as well as the Shanghai Leading Academic Discipline Project, project number T0101e.

References

[1] R.H.J. Hannink, P.M. Kelly and B.C. Muddle, J. Am. Ceram. Soc., 83[3] (2000), p. 461.

[2] Jyung-Dong Lin and Jenq-Gong Duh, Mater. Chem. Phys., 78[1] (2003), p. 246.

[3] S.G. Huang, J. Vleugels, L. Li, O. Van der Biest and P.L. Wang, J. Eur. Ceram. Soc., 25[13] (2005), p. 3109.

[4] M. Omori, Mater. Sci. Eng. A, 287[2] (2000), p. 183.

[5] S.G Huang, L. Li, J. Vleugels, P.L. Wang and O. Van der Biest, J. Eur. Ceram. Soc., 23[1] (2003), p. 99.

[6] S.G. Huang, L. Li, O. Van der Biest and J. Vleugels, Solid State Sci., 7[5] (2005), p. 539.

[7] J. Vleugels, C. Zhao and O. Van Der Biest, Scr. Mater., 50[5] (2004), p. 679.

[8] K.-H. Heussner and N. Claussen, J. Am. Ceram. Soc., 72[6] (1989), p. 1044.

[9] M. Matsuzawa, M. Abe, S. Horibe and J. Sakai, Acta Mater., 52[6] (2004), p. 1675.

[10] J. Vleugels, Z.X. Yuan and O. Van Der Biest, J. Eur. Ceram. Soc., 22[6] (2002), p. 873.

[11] G.R. Anstis, P. Chantikul, B.R. Lawn and D.B. Marshall, J. Am. Ceram. Soc., 64[9], (1981), p. 533.

[12] K. Vanmeensel, A. Laptev, J. Hennicke, J. Vleugels and O. Van der Biest, Acta Mater., 53[16], (2005), p. 4379.

[13] F.F.Lange, J. Mater. Sci., 17[1] (1982), p. 225.

Key Engineering Materials Vol. 333 (2007) pp 235-238
online at http://www.scientific.net
© 2007 Trans Tech Publications, Switzerland

Recent Advances in Material Characterization using the Impulse Excitation Technique (IET)

A.K.Swarnakar[a], S. Giménez[b], S. Salehi[c], J. Vleugels[d], O. Van der Biest[e]

Department of Metallurgy and Materials Engineering, Katholieke Universiteit Leuven,
Kasteelpark Arenberg 44, B-3001 Heverlee, Belgium

[a]Akhileshkumar.swarnakar@mtm.kuleuven.be, [b]sedigheh.Salehi@mtm.kuleuven.be,
[c]sgimenez@ceit.es,[d]jozef.vleugels@mtm.kuleuven.ac.be, [e]omer.vanderbiest@mtm.kuleuven.ac.be

Keywords: Impulse Excitation Technique, E-Modulus, Internal Friction, Hard metal, Zirconia, composite, TiB$_2$ and P/M green compact

Abstract. The Impulse Excitation Technique (IET) is a non-destructive technique for evaluation of the elastic and damping properties of materials. This technique is based on the mechanical excitation of a solid body by means of a light impact. For isotropic, homogeneous materials of simple geometry (prismatic or cylindrical bars), the resonant frequency of the free vibration provides information about the elastic properties of the materials. Moreover, the amplitude decay of the free vibration is related to the damping or internal friction of the material. At present, IET is a well-established non-destructive technique for the calculation of elastic moduli and internal friction in monolithic, isotropic materials. Standard procedures are described in ASTM E 1876-99 and DIN ENV 843-2. IET can also be performed at high temperature (HT-IET) using a dedicated experimental setup in a furnace and constitutes a valuable tool in the field of mechanical spectroscopy.

In the present work, the most recent advances in high temperature characterization using IET at K.U. Leuven are presented: the deformation behaviour of WC-Co hard metals, softening phenomena in TiB$_2$, relaxation mechanisms in ZrO$_2$ composites and "in-situ" monitoring of the damage evolution in uniaxially pressed metallic green compacts during delubrication.

Introduction

Solid objects have a characteristic set of mechanical resonant frequencies (f_r), which are related to the mass, dimensions and elastic properties of the material. The amplitude decay of the free vibration can be related to the internal friction i.e. dissipation of the vibration energy in the specimen or damping. This paper shows the recent developments of IET for the characterisation of the elastic properties and internal friction for various materials at high temperature: deformation mechanisms in WC-Co hardmetals, damage evolution of powder metallurgy (P/M) green compacts during delubrication, relaxation mechanisms in ZrO$_2$ and softening phenomena in TiB$_2$.

Experimental Procedure

The IET method relies on inducing a transit vibration in a test sample, sensing the vibration, further extracting the relevant resonance frequencies fr and the corresponding damping values (Q^{-1}). In the case of isotropic samples of rectangular shape, the elastic modulus of the materials can be calculated from the equation shown below (ASTM E 1876-97) [1].

$$E = 0.9465 \times \left[\frac{m \times fr^2}{b} \right] \times \left(\frac{L}{t} \right)^3 , \tag{1}$$

where   E = Elastic modulus, fr = Resonance frequency
               m = Mass of the sample,   L, b and t = Length, width and thickness of the sample

The damping or internal friction (Q^{-1}) can be calculated from:

$$Q^{-1} = \frac{K}{\pi.fr},$$ (2)

with the exponential decay parameter K.

IET can be applied at high temperature with a defined atmosphere. The equipment used for the measurements is from IMCE N.V. (Diepenbeek, Belgium). It can be used up to 1750°C under nitrogen and argon atmospheres or under vacuum. The vibration signals are collected by microphone or laser vibrometer according to the test conditions.

Results and discussion

The application of IET for the high temperature characterisation is presented below for four different material systems.

Hardmetals. Fig. 1 shows the damping spectrum for WC, Co and WC-Co (12 wt% Co) tested under inert atmosphere (Ar) up to 1150°C. A very small variation of the stiffness and damping is observed for pure WC (Fig. 1a). Therefore, the origin of the damping peak observed in WC-Co around 700°C (Fig. 1b) is mainly related with the cobalt phase. It has been reported [2] that this damping peak is due to the pinning and unpinning of dislocations and occurred at the brittle to ductile transition temperature range. The exponential background observed at high temperature can be associated with the creep mechanisms [2]. The damping peak of pure Co at 400°C (Fig. 1a) is supposed to be related with the hexagonal to face centred cubic transformation.

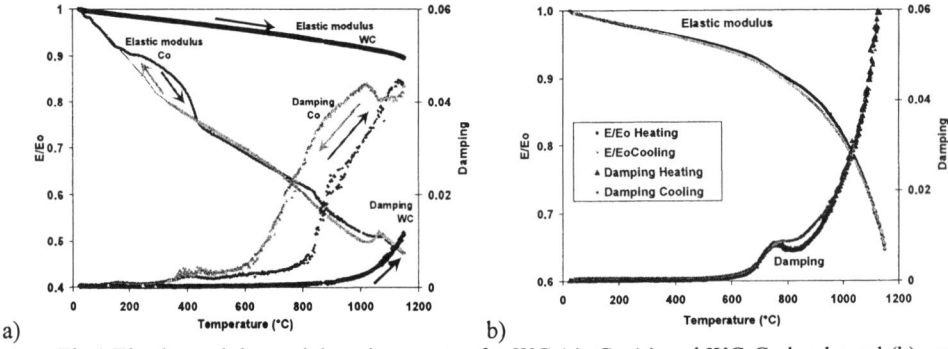

a)   b)

Fig.1 Elastic modulus and damping spectra for WC (a), Co (a) and WC-Co hardmetal (b), with temperature progress indicated by arrows in (a)

Delubrication behaviour of P/M green compacts. Fig. 2 illustrates the delubrication behaviour of P/M green compacts (Höganas Distalloy AE {Fe-4 wt%Ni steel}) at two different green densities ("+" 7.30 g/cm^3 and "-" 7.00 g/cm^3). The elastic modulus and damping were monitored during heating up to 700°C. Initially, from room temperature up to around 100°C, a strong decrease of the E modulus is coincident with a pronounced increase of damping. The signal disappears from 100°C up to around 320°C as a consequence of the very high damping caused by the softening and melting of the lubricant. The pronounced increase of E/E$_0$ (decrease of Q^{-1}) taking place from 320°C up to 450°C is associated to the increased connectivity of the powder particles as the lubricant is progressively eliminated. After regaining the signal at 320°C approximately, the differences in damping between "+" and "-" specimens progressively decrease as temperature increases. The lower elastic modulus of the "+" specimens can been correlated to the higher generation of damage occurring during delubrication. This damage is believed to be more pronounced on the denser specimens due to the lower amount of paths for the elimination of the lubricant, which results in a higher degree of gas overpressure inside the material [3].

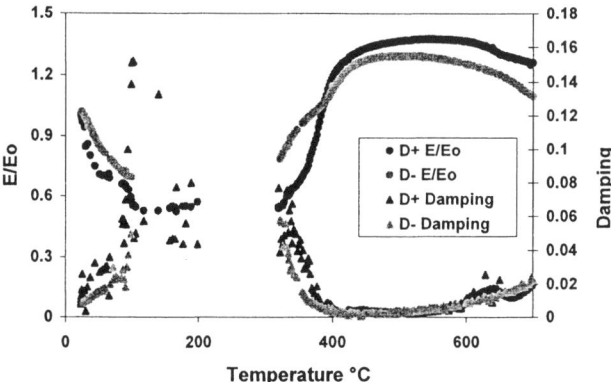

Fig.2 Delubrication of P/M green compacted at two different green densities "+" and "-"

Ceramics and composites. Experiments have been done on ZrO_2-TiN (60/40) composite and pure TiB_2 specimen. There are correlations between the damping behaviour and microstructural changes like grain boundary sliding, viscous glass formation, crack formation, crystallization, chemical reaction, etc... In general, if a damping peak observed during heating is repeated during the cooling segment, it cannot be due to irreversible phenomena like crystallization or chemical reactions. Then it should be related to grain boundary sliding or softening of amorphous pockets in a rigid ceramic skeleton.

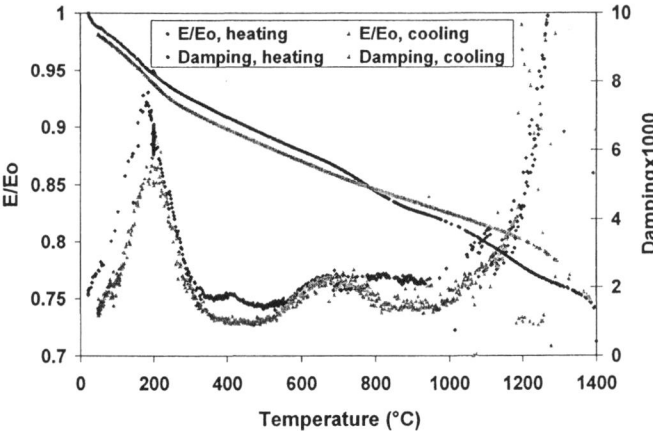

Fig. 3 Normalized elastic modulus and damping of ZrO_2-TiN (60/40) versus temperature

Fig. 3 shows the evolution of the normalized elastic modulus and damping for an yttria stabilized ZrO_2-TiN (60/40) ceramic composite. At 200°C, a visible damping peak is related to the movement of oxygen vacancies in yttria stabilized zirconia component [4, 5]. Further, during cooling at about 700°C, another distinct damping peak is observed which is due to the reverse transformation from tetragonal to monoclinic zirconia phase. It might be related to nitrogen pick-up in the t-ZrO_2 lattice which has been reported before for nitrogen atmosphere [6] or for nitride [7]. Nitrogen can partially replace oxygen in the t-ZrO_2 lattice when t-ZrO_2 reacts with nitrogen at temperatures above 1400°C [8]. At temperatures above 1000°C the damping increases exponentially due to the grain boundary sliding.

Fig. 4 shows the evolution of damping vs. temperature for TiB_2 during three thermal cycles. The tests have been conducted under inert atmosphere (Ar) up to 1200°C. Around 450°C to 600°C temperature range, the small peak (see inset) is associated with the melting of boron oxide. At around 750°C another peak exists which might be related to titanium oxide in liquid boron oxide. Yet, further investigation is required to explain this peak.

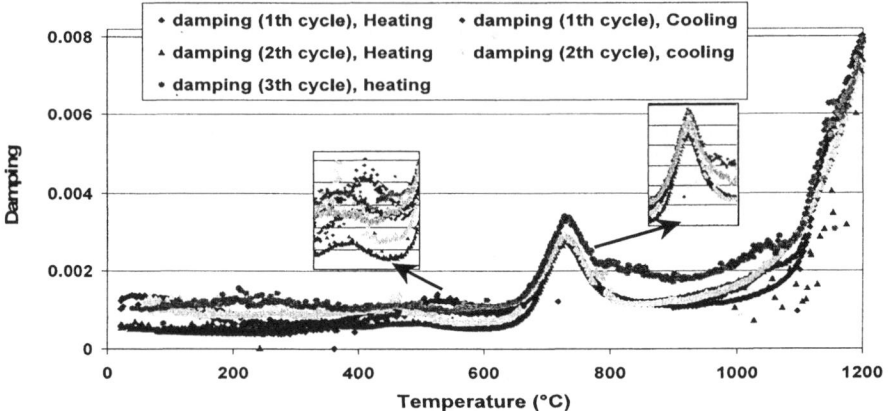

Fig. 4 Damping of TiB_2 versus temperature during three thermal cycles

Conclusion.

The high temperature IET has been successfully applied for the study of the evolution of the elastic modulus and the damping of various materials including ceramics, metals and composites. The main conclusions are:

(1) WC is stable compared with the WC-Co hardmetal and, predominately the binder phase is responsible for the deformation mechanisms, which is in good agreement with previous works [2].

(2) The damping monitoring of the P/M green compacts has evidenced the effect of green density on the damage generation occurring during delubrication.

(3) IET successfully features the relaxation mechanisms in ZrO_2-TiN (60/40) composites due to oxygen vacancies movement and in addition a tetragonal/monoclinic transformation.

(4) A damping peak observed at about 700°C in TiB_2 sample needs further investigation to be explained.

References

[1] G.Roebben, B.Bollen, A.Brebels, J.Van Humbeeck and O.Van der Biest: Rev. Sci. Instrum. 68 (1997), p. 4511.

[2] R.Schaller, J.J.Ammann and C.Bonjour: Mat. Sci. Eng. A105/106 (1988), p. 313.

[3] S. Giménez, A.Vagnon, D.Bouvard and O.Van der Biest: accepted for publication Materials Science and Engineering A.

[4] G.Roebben, B.Basu, J.Vleugels and O.van der Biest: J. Europ. Ceram. Soc. 23 (2003), p. 481.

[5] M.Weller and H.Schubert: J. Am. Ceram. Soc. 69 (1986), p. 573

[6] G.Roebben and O.Van der Biest: Key Engineering Materials 206-213 (2002), p. 621.

[7] Y. Cheng and D.P.Thompson: J. Am. Ceram. Soc. 74 [5] (1991), p. 1135

[8] J.-S. Lee, J. Fleig, J. Maier, D.Y. Kim and T.-J. Chung: J. Am. Ceram. Soc. 88 (2005), p. 3067

Key Engineering Materials Vol. 333 (2007) pp 239-242
online at http://www.scientific.net
© *2007 Trans Tech Publications, Switzerland*

Strain Mismatch in Ceramic Multilayers: Determination by Strength Measurements

Javier Pascual[1,a], Francis Chalvet[2], Tanja Lube[1,b], Goffredo de Portu[2,c]

[1]Institut für Struktur- und Funktionskeramik der Montanuniversität Leoben. Peter-Tunner-Straße 5, 8700 Leoben. Austria

[2]Istituto di Scienza e Technologia dei Materiali Ceramici-CNR. Via Granarolo 64, 48018 Faenza. Italy

[a]isfk@unileoben.ac.at, [b]tanja.lube@mu-leoben.at, [c]deportu@irtec1.istec.cnr.it,

Keywords: strength, residual stresses, asymmetrical laminates.

Abstract. In asymmetrical 3-layer laminates with constant overall and inner layer thickness, the residual compressive stresses in the two outer layers are not longer the same: compression is higher in the thinner layer. Therefore, it can be expected that the strength also depends on the outer layer thickness. Experimental evidence of this behavior was found by measuring the bending strength of asymmetrical tri-layers. A value of the thermal expansion mismatch was determined by fitting the theoretical expression to the experimental data.

Introduction

Usual techniques to experimentally determine the residual stresses include indentation techniques [1], X-ray diffraction [2], neutron diffraction [3] and piezo-spectroscopic analyses of photo-stimulated fluorescence or Raman bands [4]. An analytical assessment is also possible [5] but requires the knowledge of the strain mismatch occurring during cooling upon the sintering temperature, or what is the same, the temperature at which the residual stresses can not be relaxed at high temperature (sometimes called "frozen stress temperature").

A possibility to estimate the strain mismatch is the analysis of the strength of asymmetrical laminates. For 3-layer laminates (see Fig. 1) with residual compression in the outer layers, the residual stresses have to be redistributed in such a way that the residual compression is higher in the thinner outer layer and lower in the thicker outer layer to maintain mechanical equilibrium. It can be expected that the strength of such specimens -with a constant overall thickness- is a function of the thickness of the outer layer t_1' (its compressive stress): the thinner t_1' is, the higher the strength will be.

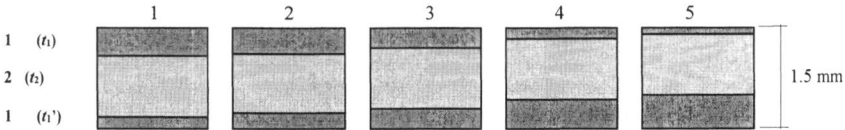

Figure 1: Cross-section sketch of different asymmetrical 3-layer with the same inner layer and overall thickness. t_1' in placed under tension during bending test.

Residual stresses in asymmetrical laminates. The biaxial compressive stress, included in a plane parallel to the interfaces, in a 3-layer system, consisting of two materials 1 and 2 with different coefficients of thermal expansion, $\alpha_1 < \alpha_2$, can be estimated using [6]

$$\sigma_1' = \frac{-E_1 E_2 t_2 \varepsilon_m}{(1-\upsilon)((t_1+t_1')E_1 + t_2 E_2)} \quad \text{and,} \quad \sigma_2' = -\sigma_1' \frac{(t_1+t_1')}{t_2} \tag{1}$$

where E_1 and E_2 are the elastic moduli of layer 1 and 2 respectively, υ is Poisson's ratio and t_1 and t_2 are the thicknesses of layers of material 1 and 2, as depicted in Fig. 2. ε_m is the thermal strain mismatch which is defined by the integral between the frozen stress temperature T_{sf} and the room temperature ($T_0 = 20°C$):

$$\varepsilon_m = \int_{T_o}^{T_{sf}} (\alpha_1 - \alpha_2)\, dT .$$

(2)

Eq. 1 results from the force balance in a symmetric laminate assuming Hookean behavior and a plane stress state. It is valid as a first approximation, since it assumes a constant stress state within each layer.

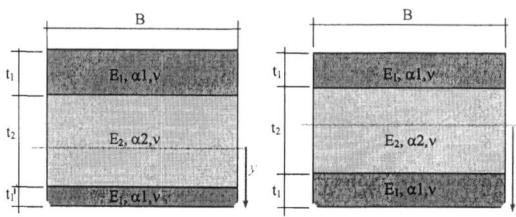

Figure 2. Cross sections of (a) an asymmetrical and (b) a symmetrical tri-layer. The dashed line indicates the position of the neutral axis.

For asymmetrical laminates, the residual stress field is modified due to a bending moment M that appears to maintain mechanical equilibrium. For the asymmetrical case Eq. 1 has to be slightly modified. The bending moment M has been calculated in the literature [7]

$$M = -\sigma_1' \cdot f\left(t_1', \text{geometry, materials}\right).$$

(3)

where σ_1' has already been defined and the function f depends on the materials elastic properties and the architecture of the non-symetrical multilayer. The total residual stress in layer 1' at any depth y measured from the neutral axis can be expressed by

$$\sigma_1(y) = \sigma_1' + \frac{M\,y}{I}$$

(4)

where I is the moment of inertia. Only if the laminate fails due to surface flaws during the bending test it could be assumed that the strength in bending is given by

$$\sigma_{tri} = \sigma_0 + \sigma_1(\bar{y}).$$

(5)

where σ_0 can be interpreted as the bending strength of specimens made from pure material 1 (i.e. residual stress-free) and $\bar{y}$ is the value of y at the tensile fiber.

Values of σ_{tri} can be measured for specimens with different t_1' and plotted as strength versus t_1'. Eq. (5) can be fitted to this data. As a result, an estimation for ε_m and σ_0 can be obtained.

Experimental procedure

Tests were carried out on ceramic laminates produced by tape casting. The architecture of the as received trilayers was as follows. 8 tapes distributed in a symmetrical way: 2A/4AZ/2A, where A is 100 vol.% Al_2O_3, and AZ means 60 vol. % Al_2O_3 + 40 vol.% tetragonal stabilized $Zr(3Y)O_2$. Material properties of the layers are summarized in Table 1. To determine the strength of the different asymmetrical tri-layers, 4-point bending tests were carried out with a Zwick Z010 device. The tests were performed on 1.5 × 2 × 16 mm chamfered specimens (13/3 and 13 mm span lengths). The chosen test speed was 1.5 mm/min. Temperature and relative humidity were recorded during tests as 23°C and 36% respectively. Fracture surfaces were analyzed by stereomicroscopy and SEM in order to reveal the fracture origins.

Table 1. Relevant material properties of the layers.

layer	grade	$\alpha\,[°C^{-1}]$	$E\,[GPa]$	ν
A	Alcoa A16 (A)	$9\,10^{-6}$	391 [8]	0.24
AZ (40A60Z)	A + 60% TZ3Y-S Tosoh (AZ)	$10\,10^{-6}$	340	0.24

Results and discussion

A plot of the strength depending on the outer layer thickness t_1' for asymmetrical tri-layers made from two different laminated plates is shown in Fig. 3a (plates are referred to as #11 and #12) Microscopy investigations on the tensile surfaces revealed that with the exception of two specimens, all specimens of batch #11 failed due to large grains of alumina located at the tensile surface. The two specimens for which this was not the case (presumably because the material that originally contained the large grains was ground off completely) are indicated in Fig. 3a in open small squares. The typical fracture surface for specimens from the plate #11 is shown in Fig. 3. Large grains about 50 μm size were found in that sample. Fracture origins for specimens of batch #12 were also surface defects although no abnormal large grains were found in this material.

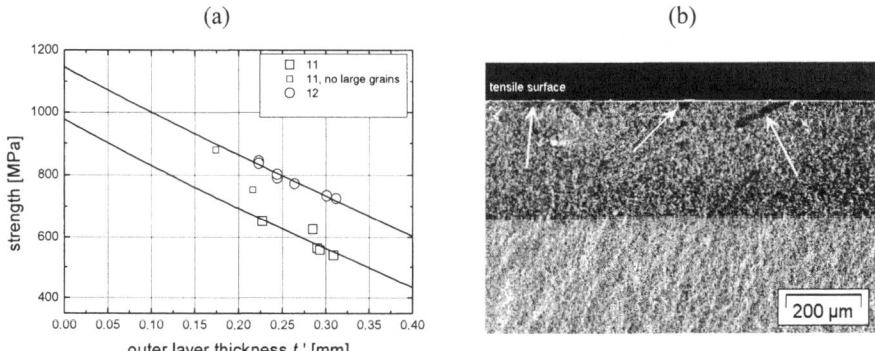

Figure 3. (a) Plot of strength as a function of the outer layer thickness t_1' for batches #11 and #12. Specimens of #11 that did not fail due to large grains and were excluded from the fitting are marked with small square symbols. The lines correspond to a simultaneous fit of Eq. 5 on both data sets with common ε_m (b) Fracture surface of a specimen showing abnormal large alumina grains

The strength values plotted in Fig. 3a were fitted to the Eq. 5 by means of Origin 7.5 where ε_m and σ_0 were the parameters to fit. The specimens from #11 that did not fail from large grains were excluded from the fit. The results for the fitting are presented in Table 2. Since the value for ε_m is supposed to be the same for both data sets, a simultaneous fit is performed on both data-sets where the value for ε_m is forced to be the same for both #11 and #12 while σ_0 is allowed to be different for both sets.

The results for σ_0 can be interpreted as the strength of two microstructurally different aluminas. The value of $\sigma_0 = 417$ MPa for #12 is a reasonable value for the strength of alumina. A similar value was determined on bend bars made from pure A-alumina [8]. The value for #11 is rather low, but it can be explained by the existence of the large surface defects. With the fracture toughness $K_{IC} = 3.8$ MPa√m, measured by SEVNB, the size of a semicircular surface flaw that leads to failure at this stress can be estimated to be approx. $c = 47$ μm. This size corresponds satisfactory with the observations on fracture surfaces, see Fig. 3b.

The results for ε_m leads to a $T_{sf} = 1160°C$ that is slightly lower than the value $T_{sf} = 1200°C$ which has been reported [9] for similar materials. This temperature is smaller than the sintering temperature as the AZ-layers present a high plasticity at high temperature that relaxes the stresses [10]. From Eq. 1 and Eq. 2 the crucial influence of this temperature on the magnitude of the residual stresses is obvious.

For symmetrical tri-layers, the thickness of the external A layers is around $t_1 = t_1' = 265$ μm and $t_2 = 970$ μm for the inner AZ layer. According to Eq. 1, with ε_m determined here and assuming the material properties presented in Table 1, a compressive stress of approx. -360 MPa is expected in the external A-layers while a tensile stress of 197 MPa arises in the inner AZ layer.

Table 2. Results for σ and ε_m from the fitting procedure.

Specimen	fracture origins	σ_0 [MPa]	ε_m []	T_{sf} [°C]
11	surface large grain	247	0.00114 ± 0.00015	1160
12	machining defects	417		

Summary

The strain mismatch that is responsible for the build-up of residual stresses in laminates was determined by mechanical testing of asymmetrical 3-layer laminates. Specimens were made from two different production batches. An obvious difference in strength was measured on these two plates. The importance of microstructure on mechanical properties was clearly observable. The presented results indicate that the strength of multilayers in the investigated alumina/alumina-zirconia system is determined by the strength of the residual-stress free alumina plus the contribution of the compressive residual stress. The determined value for the stress-free temperature is $T_{sf} = 1160°C$ (the strain mismatch $\varepsilon_m = 0.00114 ± 0.00015$).

Acknowledgments

This work was supported in part by the European Community's Human Potential Programme under contract HPRN-CT-2002-00203, [SICMAC]
Javier Pascual acknowledges the financial support provided through the European Community's Human Potential Programme under contract HPRN-CT-2002-00203, [SICMAC]

References

[1]   A. Tarlazzi, E. Roncari, P. Pinasco, S. Guicciardi, C. Melandri, G. de Portu: Wear 24 (2000), p. 29 .

[2]   T. Adachi, T. Sekino, T. Nakayama, T. Kusunose, K. Niihara: Materials Letters 57 (2003), p. 3057.

[3]   A. Stacy, H. J. MacGillivary, G. A. Webster, P. J. Webster, K. R. A. Ziebeck: J. Strain Anal. 20 (1985), p. 93.

[4]   G. De Portu, L. Micele, Y. Sekiguchi, G. Pezzotti: Acta Mater 53 (2005), p. 1511.

[5]   X. C. Zhang, B. S. Xu, H. D. Wang, W. Y.X: Thin Sol Films 488 (2005), p. 274.

[6]   S. Ho, C. Hillman, F. F. Lange, Z. Suo: J Am Ceram Soc 78 (1995), p. 2353.

[7]   A. V. Virkar, J. L. Huang, R. A. Cutler: J Am Ceram Soc 70 (1987), p. 164.

[8]   J. Pascual, F. Chalvet, T. Lube, G. de Portu: Mat Sci For 492-493 (2005), p. 581.

[9]   D. J. Green, P. Cai, G. L. Messing: J Eur Ceram Soc 19 (1999), p. 2511.

[10] M. Jimenez-Melendo, C. Clauss, A. Dominguez-Rodriguez, G. de Portu, E. Roncari, P. Pinasco: Acta Mater 46 (1998), p. 3995.

Key Engineering Materials Vol. 333 (2007) pp 243-246
online at http://www.scientific.net
© *2007 Trans Tech Publications, Switzerland*

Residual Stress Assessment in the Al$_2$O$_3$\Mullite based Laminated System

L. Micele[1,3,a], M. Brach[1,b], F. Chalvet[1,c], G. de Portu[1,3,d] and G. Pezzotti[2,3,e]

[1]Institute of Science and Technology for Ceramics, CNR-ISTEC, Via Granarolo 64, I-48018, Faenza, Italy

[2] Ceramic Physics Laboratory, Kyoto Institute of Technology, Sakyo-ku, Matsugasaki, 606-8585, Kyoto, Japan

[3]Research Institute for Nanoscience, RIN, Sakyo-ku, Matsugasaki, 606-8585, Kyoto, Japan

[a] micele@istec.cnr.it, [b] Mylene.BRACH@hermes.com, [c] fchalvet@wanadoo.fr, [d] deportu@istec.cnr.it, [e] pezzotti@chem.kit.ac.jp

Keywords: Al$_2$O$_3$\mullite composite, laminated structure, piezo-spectroscopy, Raman quantitative analysis

Abstract. To improve mechanical properties of mullite, a mullite-Al$_2$O$_3$\mullite laminate composite was prepared. Lamination generates residual stresses within the structure, measured by piezo-spectroscopy. A preliminary and complete piezo-spectroscopic characterization of the Al$_2$O$_3$\mullite system was carried out. A method to determine the concentration of Al$_2$O$_3$ in the composite by Raman spectrum was proposed and used to assess the composition of the laminated structure along the cross section. The experimental results evidenced a gradual change of composition and residual stress state between the two layer.

Introduction

Mullite (3Al$_2$O$_3$·2SiO$_2$) is used as a high-temperature structural ceramics, because of its excellent physical properties (high melting point, high resistance to creep, low thermal expansion, high temperature mechanical stability and its high thermal shock resistance). In addition, mullite can be obtained from low cost natural materials. However its massive use is hindered by its relatively poor mechanical properties. Laminated structures have been studied and designed in order to improve mechanical properties of ceramic materials [1,2]. To improve the properties of a material through a layered architecture, it is important to stimulate an appropriate residual stress distribution in the structure. To do this the different layers used must have different Coefficient of Thermal Expansion (CTE). An excess of alumina in mullite increases the CTE of this composite with respect of pure mullite.

In this contribution multilayered composites were prepared by tape casting starting from kaolin and an excess of Al$_2$O$_3$ as raw materials. The macroscopic residual stress arisen by difference in layers CTE was assessed using the fluorescence piezo-spectroscopy (PS) based on the fluorescence of Cr^{3+} impurities in Al$_2$O$_3$. PS was already applied to laminated structures by previous authors [3,4], who demonstrated the importance of knowledge of the composition to perform reliable measurements. For that reason, a method for a quantitative analysis of the progress of the reaction between silica and Al$_2$O$_3$ to form mullite by Raman spectroscopy was for the first time proposed and utilized.

Experimental procedure

The study entailed the use of kaolin powder (Kaolin SSP, ECC International, Imerys, Paris, France) and high purity (99.7%) alumina powder (Alcoa A16-SG, Alcoa Aluminium Co., N.Y., USA) with an average particle size of 0.46 μm and 0.30 μm respectively. Using experiences of previous authors

[5-8], two types of slurries have been prepared: a slurry of pure kaolin (hereinafter named "MS") and a slurry of kaolin and alumina (35/65 wt%, "MA"). Tape casting was performed with a laboratory bench with a stationary double-blade system on a carrier band of siliconated mylar. After drying, green kaolin and kaolin-alumina laminae were punched and the different sheets stacked by thermocompression at 80°C for 30 minutes at 30 MPa pressure. The laminated composites were prepared by thermocompression with the stacking sequence MS/(MA/MS)$_6$. The coefficient of thermal expansion (CTE) of the two composites were 5.6 and 7.6x10^{-6}K^{-1} for MS and MA respectively. Debonding was carried out at 600°C (heating rate 3°C/h) and sintering 1h at 1550°C (heating and cooling rate 30°C/h). As a consequence of heat treatment, alumina and silica present in the kaolin powder react to give mullite.

As reference materials for Raman and PS characterization a series of 15 mullite-alumina composites were prepared. The volumetric percentage of Al$_2$O$_3$ in the reference materials was 0, 10, 16.7, 20, 30, 34.8, 40, 50, 54.5, 60, 70, 76.2, 80, 90 and 100%. To avoid reaction and change in composition during sintering, the same Al$_2$O$_3$ already used for tape casting and pure commercial mullite (Baikalox SASM) were selected. Powders were mixed, cold isostatic pressed at 250 MPa and sintered 1h at 1550°C. After sintering, specimens were cut in bars of approximately 3x4x20mm.

Spectroscopic measurements and PS calibration of standards were carried out according to well-assessed procedure previously reported [4, 5] using a spectrometer apparatus T 64000 Jovin-Yvon.

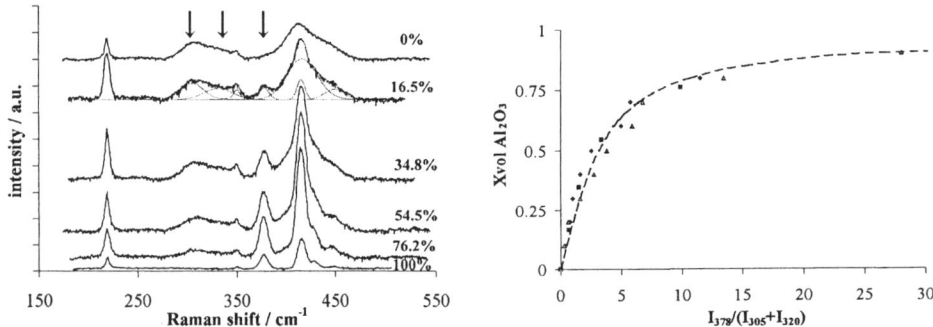

Fig. 1 (left). Raman spectra of the reference mullite-Al$_2$O$_3$ composites. For the composite with 16.5 vol% of Al$_2$O$_3$ the fitting is shown. The arrows indicate the three bands used in Eq. (3)

Fig. 2 (right). The volumetric fraction of Al$_2$O$_3$ vs the intensity ratio of selected Raman bands.

Results and discussion

Raman and PS analysis of standards. Fig. 1 shows the Raman spectra in the region between 150 and 550 cm^{-1} of some reference mullite-alumina composites. After a suitable fitting, a relation between the intensity ratio of some bands (305 and 320 cm^{-1} for mullite and 377 cm^{-1} for alumina) and the volumetric percentage of Al$_2$O$_3$ was found to be (Fig. 2):

$$X^{vol}_{Al_2O_3} = \frac{2}{\pi}\arctan\left(\frac{I_{377}}{3(I_{305} + I_{320})}\right) \tag{1}$$

where I$_{xxx}$ is the intensity of the bands whose Raman shift in cm^{-1} is given by the subscript. The error related to the determination of composition with Eq. 1 was estimated to be ±3-4% in a wide range of composition. However, owing to the high number of bands in the considered spectral window and to the weakness of spectra, the error associated to the calculation significantly increase when the quantity of one of the components is lower than 10%, making the composition assessment merely qualitative.

The PS characterization of standard materials proceeded with the study of the correlations between the composition and the unstressed peak position (Fig. 3) and the PS coefficient (Fig. 4). In both cases a simple monotonic trend was found: experimental data regarding unstressed peak position were fitted by a straight line:

$$v_0 = (-3.1 \pm 0.2)X_{Al_2O_3}^{vol} + v_0^{mul} \tag{2}$$

where v_0^{mul} is the unstressed peak position of the Al_2O_3 traces in the commercial mullite (14415.7 ± 0.1 cm^{-1}). PS coefficients Π, instead, had an increase for small quantities of Al_2O_3 present in the mullite matrix. For higher quantity of Al_2O_3 the increasing is fainter. This trend is described by Eq. 3:

$$\Pi_{composite} = 0.68\ln(X_{Al_2O_3}^{vol} + 0.085) - 2.61 \tag{3}$$

It should be pointed out that Eq. 1, 2 and 3 are empirical equations whose parameter were optimized using the least square method.

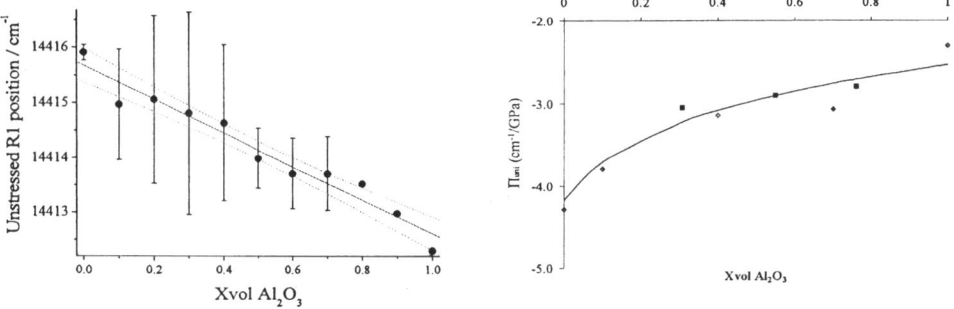

Fig. 3 (left). Unstressed R1 position as function of the Al_2O_3 content in the reference composites
Fig. 4 (right). Uniaxial PS coefficient vs Al_2O_3 content in the reference composites

Microstructure and residual stress analysis. After sintering, the micrograph of the cross section showed a new interlayer (Fig. 5). The plot of Fig. 6 illustrates the change in composition across the layers. Within the MS layer the quantity of Al_2O_3 was negligible. Alumina concentration gradually increased in the interlayer toward to MA layer, where it reached the approximate value of 30 vol%, as expected. The concentration gradient may be explained with the large quantity of SiO_2 in kaolin, that completely reacted with the Al_2O_3 present in the MS layer to form mullite + SiO_2. The excess of SiO_2 probably diffused in the MA layer partially reacting with the excess of Al_2O_3 and creating the gradient.

The profile of residual stress measured by piezo-spectroscopy is plotted in Fig. 7. To obtain the appropriate PS parameters by Eq. 2 and 3 the Al_2O_3 volumetric fraction was considered constant within each layers with a value of 0, 10 and 30 vol% for MS, interlayer and MA respectively. As expected by CTE mismatch, the MS layer was under compressive stress and MA layer was under tensile stress. The interlayer's stress state was intermediate respect to MS and MA, of weak tension. The highest value of tension was reached close to MS, the layer under compression. On the opposite side, nearby the MA layer, the stress of interlayer is almost completely released. This trend doesn't follow what previously observed [4,5] in alumina\zirconia laminated structures. The reason, still under investigation, may reside in the simplification used about the constant average concentration of the interlayer. Moreover, also some spectroscopic effects due to the weakness of the fluorescence spectrum in the regions with low content of Al_2O_3 might effect the measured stress value.

Observing simultaneously composition and stress data, the formation of functionally graded structure can be envisaged. In fact there is no abrupt change in composition between layers, and also the stress mismatch between MS and MA layers is moderated by interlayer.

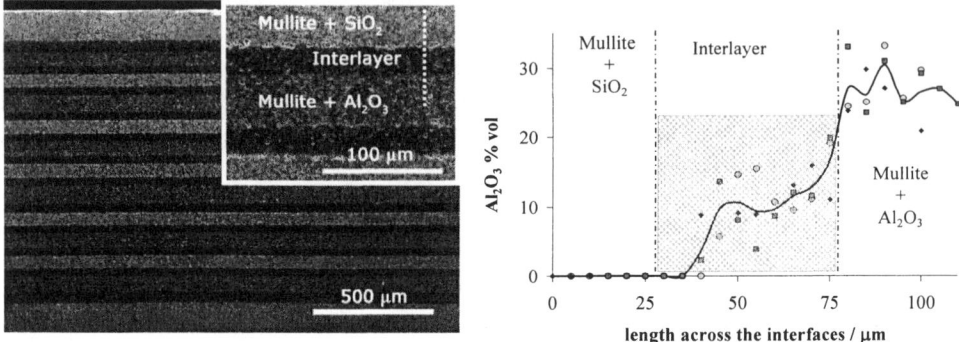

Fig. 5. SEM micrograph of the laminate's cross section. The dotted line in the insert refers to the composition profile plotted in Fig. 6.

Fig. 6. Layers' composition along the dotted line of the insert of Fig. 5.

Conclusions

The Al_2O_3\mullite system was deeply analysed because of its potential use as high performance refractory material. Laminates structures were prepared by tape casting starting from kaolin and Al_2O_3 as raw materials. The Al_2O_3\mullite system was characterized from a PS point of view. Also a procedure for Raman quantitative analysis was for the first time suggested. The analysis of composition and residual stress distribution in the laminated structure evidenced the creation of a functionally graded material.

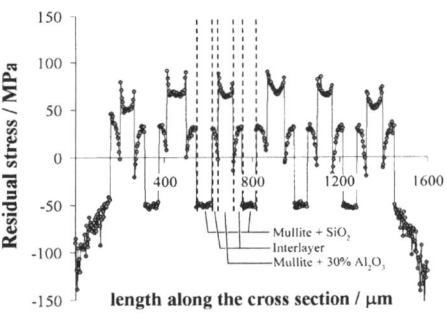

Fig. 7. Residual stress profile along the laminate's cross section.

References

[1] P.Z. Cai, D.J. Green and G.L. Messing: J. Am. Ceram. Soc. Vol. 80 (1997), p, 1929

[2] F. Toschi, C.Melandri, P.Pinasco, E.Roncari: J. Am. Ceram. Soc. Vol. 86 (2003), p. 1547

[3] V. Sergo, D.M. Lipkin, G. de Portu and D.R. Clarke: J. Am. Ceram. Soc Vol 80 (1997), p. 1633

[4] G. de Portu, L. Micele, Y. Sekiguchi and G. Pezzotti: Acta Mat. Vol. 53 (2005), p. 1511

[5] F. Chalvet, G. De Portu, SICMAC Research Training Network Newsletter, 1, Nov. 2003

[6] F. Chalvet, G. De Portu, SICMAC Research Training Network Newsletter, 2, Feb. 2004

[7] M. Brach, G. De Portu, SICMAC Research Training Network Newsletter, 5, Feb 2005

[8] M. Brach, G. De Portu, SICMAC Research Training Network Newsletter, 6, Jun. 2005

Key Engineering Materials Vol. 333 (2007) pp 247-250
online at http://www.scientific.net
© 2007 Trans Tech Publications, Switzerland

Roughness Effect on the Mechanical Properties of Ceramic Materials Measured from Nanoindentation Tests

A. Rico, M. A. Garrido, E. Otero, J. Rodríguez

Departamento de Ciencia e Ingeniería de Materiales. Universidad Rey Juan Carlos. C/ Tulipán s/n
28933 Móstoles – Madrid.

alvaro.rico@urjc.es

Keywords: Nanoindentation, Roughness effect, ISE, Ceramic materials.

Abstract. An experimental study was performed to evaluate the roughness effect on the determination of hardness and Young's modulus of ceramic materials from nanoindentation tests. Several specimens polished at various stages were tested at different peak load values. Local roughness measurements have been done by means of atomic force microscopy. Results indicate that roughness and size effects are joined. Proportional Specimen Resistant Modified model (PSRM) was applied to avoid the scale effect, isolating the roughness influence. Indentations where the average roughness-elastic displacement ratio is lower than a critical value are needed to get consistent results.

Introduction

Sensing indentation techniques have allowed determining mechanical properties like Hardness and Young's Modulus from small samples. The method developed by Oliver and Pharr [1] estimate these properties from the unload branch of the load-displacement curve. The resulting hardness and effective Young's modulus could be calculated from:

$$H = \frac{P_{max}}{A_c} \quad \text{and} \quad E* = \frac{\sqrt{\pi}}{2} \frac{1}{\beta} \frac{S}{\sqrt{A_c}}, \tag{1}$$

where S is the slope of the load-displacement curve at the initial unloading, A_c is the projected contact area (usually evaluated from the contact depth, h_p, and an indenter shape function, which for an ideal Berkovich indenter is $A_c = 24.5h_p^2$), β is a correction factor depending on the tip geometry (1.034 for a Berkovich indenter) and E* the reduced modulus defined as:

$$\frac{1}{E*} = \frac{1-\upsilon^2}{E} + \frac{1-\upsilon_i^2}{E_i}, \tag{2}$$

where E, υ, E_i and υ_i are Young's modulus and Poisson ratio of specimen and indenter, respectively. The Oliver and Parr methodology is well known and can be consulted elsewhere [1].
Several problems arise when these equations are applied to ceramic materials. A strong dependency on the indentation size and on the surface roughness is observed [2]. Hardness results are lower as the indentation load increases. This phenomenon is known as Indentation Size Effect (ISE) and several theoretical models have tried to explain it in metal materials, based on dislocations mechanism in the near of the indentation [3]. These models have also been successfully applied to ceramics [4, 5]. Nevertheless, in the alumina polycrystalline samples like used in this work the plot H^2 vs.$1/h_t$, (where h_t is the total penetration of the tip in the sample) is not a straight line as predicted by this model. On the other hand, one of the most common approaches is the Proportional Specimen Resistant Modified (PSRM) model, in which, the load, P, and the slope of load-displacement curve at the initial unloading is written in terms of the contact depth, h_p, as following:

$$P = a_0 + a_1 h_p + a_2 h_p^2,$$ (3)

$$S = A + B h_p.$$ (4)

The ISE is associated with the coefficients a_0 and a_1, while a_2, is related to the size independent hardness. Although its origin is purely empirical, PSRM model can be partially justified in terms of an energy balance analysis [6]. The quadratic term is interpreted as a measure of the energy used to induce permanent deformation in the material. The linear term is considered a contribution of that part of the work done used to increase the surface area of the sample during the test (formation and propagation of microcracks, migration of the grain boundaries, deformation of porous, etc). The independent term is included just to deal with experimental errors.

Surface roughness is another aspect with considerable relevance in ceramics nanoindentation, because the projected area evaluated by Oliver and Pharr method does not consider the superficial asperities. The influence is more significant for very stiff and hard materials, where the indentation depth is rather small and the plastic deformation limited.

In this work, an experimental study of the surface rougness effect on the nanoindentation data of alumina samples is presented. Roughness characteristics were determined at two different scales by using a scanning profilometer and atomic force microscopy. The PSRM model was applied to elude the ISE, isolating the roughness effect.

Experimental techniques

Commercial alumina samples was chosen to study the roughness effect: alumina with a relative density of 96,6 %, hardness of 19.8 GPa and Young´s modulus of 325 GPa [7]. Metallographic techniques were applied to induce different polish grades in alumina samples. Abrasive powders of 45 μm, 3 μm, 1 μm y 0.25 μm size were used in the final polish grade. Roughness measures were carried out with a *Starret Sigma V$_B$ 400* scanning profilometer. Due to the small dimensions of the actual contact, a more local roughness measurement was done by means of a Digital Instruments Nanoscope IIIa atomic force microscope. Nanoindentation tests were done at different maximum loads (5 mN, 20 mN, 50 mN, 250 mN and 400 mN) in a MTS XP Nanoindenter. In every test, a Berkovich tip with a final radio of 100 nm was used, and a calibration of the contact area was performed using fused silica as reference material. The machine compliance was also corrected to minimize experimental errors.

Results and discussion

Table 1 summarizes the two groups of roughness data and the figure 1 includes the bearing areas, a concept which can be qualitative related to the real contact area.

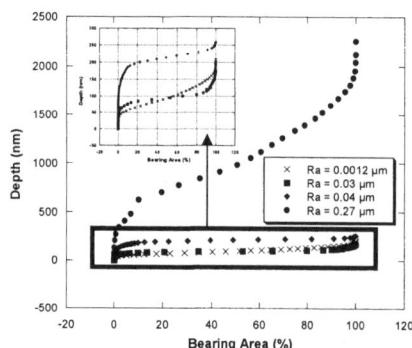

Figure 1 Depth in sample vs. Bearing area.

Table 1. Roughness data obtained from a scanning profilometer and Atomic Force Microscope.

Abrasive size (μm)	Scanning profilometer		AFM	
	Ra (μm)	Rp (μm)	Ra (μm)	Rp (μm)
45	0,39	1,25	0,27	0,99
3	0,09	0,48	0,04	0,36
1	0,09	0,24	0,03	0,18
0,25	-	-	0,0012	0,088

An example of the load-displacement curves is included in figure 2. A summary of the hardness and Young´s modulus, determined by Oliver and Pharr methodology, is shown in figure 3. It is evident that scale and roughness effects are joined phenomena, which interestingly, can provide reliable data under very different and indeterminate conditions.

To evaluate separately the roughness effect, PSRM model was applied to avoid the scale effect. In figure 4, experimental data were fitted to Eq. (3) to obtain a_2 parameter for each roughness. Additionally, linear fit of the experimental slope - contact depth data lead to E* from Eq. (1). Table 2 summarizes the values for the fitting parameters.

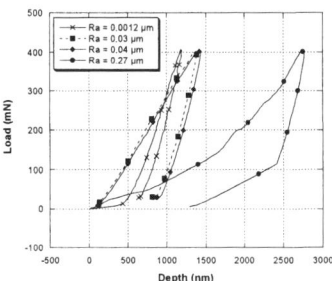

Figure 2 Typical nanoindentation curves.

By applying equations (3) and (4), the size independent hardness (H_0) and Young's modulus (E_0) can be determined. Figure 5 exhibits the variation of both properties with the average roughness parameter Ra.

Samples with Ra values of 0.27 µm, 0.04 µm and 0.03 µm, lead to hardness and Young's modulus smaller than those found in bibliography. Only very smooth samples (Ra = 0.0012 µm) exhibit a good agreement between H_0 and E_0, and references values. But, what is the meaning of the "smooth" in the last statement?. Taking into account that Oliver and Pharr method makes use only of the unloading branch of the load-displacement curve, it seems reasonable to consider that the characteristic length to compare Ra should be the elastic displacement h_e, associated with the unloading branch, that is, the total displacement minus the residual one.

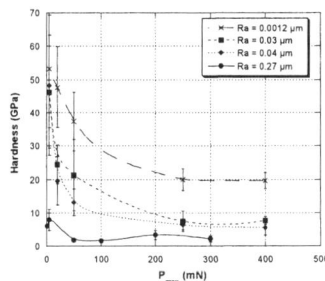

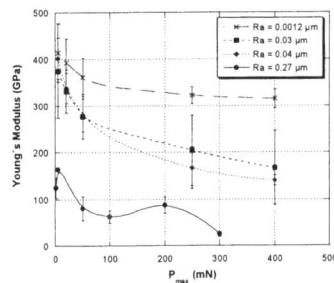

Figure 3. Hardness and Young's Modulus of the alumina samples versus the nanoindentation peak-load.

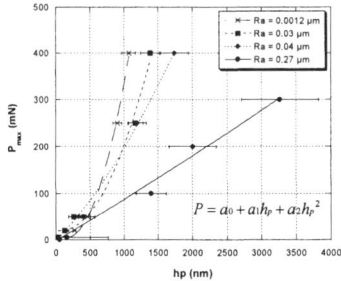

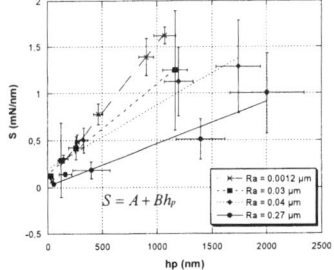

Figure 4. PSRM fits to obtain independent-load mechanical properties.

Table 2. Fit parameters from figure 4.

		Figure 4a			Figure 4b		
Rp (µm)	a_0 (mN)	a_1 (mN nm^{-1})	a_2 (mN nm^{-2})	B (mN nm^{-2})	A (mN nm^{-1})	E* (GPa)	
0,088	14,663	-0,137	0,0004563	0,0014790	0,056	264,81	
0,240	20,421	-0,028	0,0002096	0,0009498	0,015	169,97	
0,480	-2,531	0,155	0,0000443	0,0006757	0,201	124,66	
1,250	-2,331	0,088	0,0000016	0,0004491	0,016	79,70	

In figure 6, the hardness and Young's modulus results are plotted against the average roughness-elastic displacement ratio. To get consistent results, only tests where (R_a/h_e) is lower than a critical value should be considered. In the material analyzed in this work (R_a/h_e) should be lower than 3 %.

A remaining question not considered in this contribution should be afforded in subsequent studies. If the unloading branch is the only part of the experimental load-displacement curve considered to determine hardness and modulus, the role of the residual roughness seem to be the crucial aspect rather than the initial surface state. AFM measurements could significantly contribute to this issue.

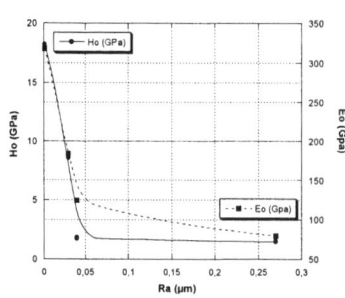

Figure 5. True mechanical properties vs. R_a.

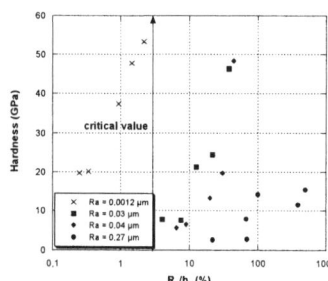 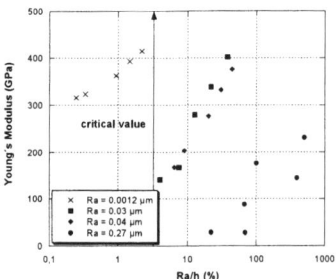

Figure 6. Mechanical properties vs. R_a/h_e ratio.

Conclusions

- Scale effects were observed for alumina samples tested at different peak load values. Proportional Specimen Resistant Modified model was applied to circumvent these effects providing size independent properties.
- Surface roughness strongly affects the calculated hardness and Young's Modulus from nanoindentation tests. Its effect is joined with that of the indentation size.
- Only tests where (R_a/h_e) is lower than a critical value can provide consistent results.

References

[1]   W.C. Oliver and G. M. Pharr: J. Mater. Res. Vol. 7, (1992) pp. 1564 – 1583.
[2]   Jianghong Gong, Hezhuo Miao, Zhijian Peng, Longhao Qi: Mater. Sci. Eng. A, Vol. A354 (2003) 140 – 145.
[3]   William D. Nix and Huajian Gao: J. Mech. and Phys. of Sol. Vol. 46 (1998) 411 – 425.
[4]   Ya. M. Soifer, A. Verdyan, L. Rapoport: Material Letters. Vol 56 (2002) 127 – 130.
[5]   Gang Feng and William D. Nix: Scripta Mater. Vol 45 (2004) 599 – 603.
[6]   Jianghong Gong, Ying Li: Journal of Materials Science, Vol. 35 (2000), pp. 209 – 213.
[7]   William E. Lee, D. Phil and W. Mark Rainforth: *Ceramic microstructures. Property control by processing*. Published by Chapman and Hall, Bondary Row, London UK (1994).

Key Engineering Materials Vol. 333 (2007) pp 251-254
online at http://www.scientific.net
© 2007 Trans Tech Publications, Switzerland

Geometry Effect on the Thermal Shock Response of Al$_2$O$_3$/ZrO$_2$ Multilayered Ceramics

R. Bermejo, P. Supancic and T. Lube

Institut für Struktur- und Funktionskeramik, Montanuniversität, Leoben, A-8700 Leoben, Austria

raul.bermejo@mu-leoben.at

Keywords: Alumina, Zirconia, Thermal Shock, Multilayer, Geometry Effect, Water Quench Test.

Abstract. In this work, the geometry effect on the thermal shock behaviour of a nine layered Al$_2$O$_3$-5%tZrO$_2$/Al$_2$O$_3$-30%mZrO$_2$ ceramic fabricated by slip casting has been studied. A finite element model has been used to estimate the magnitude and location of the maximum thermal stresses in the layered material as well as the influence of the variation of this layered architectural design in the thermal shock crack initiation and extend throughout the specimens of study. Experimental tests on various samples have been carried out to validate the model. The residual stress distribution profile in the laminate, due to the elastic mismatch of the different layers along with the zirconia phase transformation on the Al$_2$O$_3$-30%mZrO$_2$ layers, conditions the thermal shock response of the material. It is demonstrated how the variation of the outer most layer thickness in the laminates modifies the stress state in the surface, affecting the thermal shock crack initiation.

Introduction

Thermal shock resistance in ceramic materials can be achieved by either making it more difficult for defects within a body to start growing or by limiting the extent to which they can grow once cracks initiate. Several processing routes have been investigated to obtain tough ceramics such as doping, fiber and/or particle reinforcement, functional grading and layered architectural designs [1]. In particular, ceramic composites with a layered structure such as alumina/zirconia have been reported to exhibit an increased apparent fracture toughness and energy absorption as well as non-catastrophic failure behavior. The thermal shock response of these layered materials influenced by the presence of compressive layers was studied on a previous work [2], showing a higher resistance to the penetration of a crack under thermal shock conditions compared to the reference monoliths. In this work, the geometry effect on alumina-zirconia layered composites under severe thermal shock conditions has been studied. In doing so, a 3D finite element model has been implemented to account for the effect of the outer most layer thickness on the maximum stress location and distribution within the laminate. Additionally, experimental water quench tests have been accomplished in the different laminates to be compared with the model and study the role of the outer layer thickness in the thermal shock crack initiation and extend throughout the multilayered structure during abrupt temperature changes.

Experimental

Sample Architecture

Specimens of study were fabricated using the slip casting technique [3] by mixing the appropriate powders of Alumina (Condea, d$_{50}$ = 0.29 µm, HPA05 USA) and both Y$_2$O$_3$-free and Y$_2$O$_3$-stabilized zirconia (TZ-0 & TZ-3YS, d$_{50}$ = 0.60 µm d$_{50}$ = 0.37 µm, Tosoh, Japan) with distilled water as described elsewhere [4]. This laminate is composed of five alternated layers of alumina with 5% tetragonal zirconia (referred to as ATZ) of 530 µm thickness, fused together with four layers of alumina with 30% monoclinic zirconia (named as AMZ) of 100 µm thickness.

Three different multilayer geometries with various outer layer thicknesses were chosen for this investigation. A geometry with outer layers of 750 µm has been previously studied [2]. The other two cases consist of identical specimens with outer layers of 530 μm and 300 μm, respectively; as shown schematically in Figure 1.

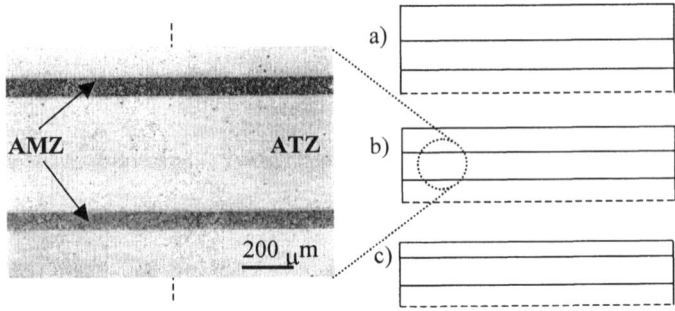

Figure 1. Scheme of the half of the different multilayer geometries with outer layers of a) 750 μm, b) 530 μm, and c) 300 μm.

Thermal Shock Experiments and 3D-FE simulation
The water quench test has been employed in the laminates to visualize the thermal shock cracks developed during quenching at a temperature difference, ΔT, of 300 °C. When samples cooled to room temperature they were checked for cracks by impregnating the surfaces with a red penetrating dye. In order to estimate the magnitude and distribution of the maximum stress during thermal shock tests and thus predict the crack initiation, a 3D finite element analysis has been implemented using the commercial code ANSYS 8.1 [5]. Both physical and thermal properties have been taken from previous material characterization [2]. Thermal shock simulations have been carried out for the different geometries using a heat transfer coefficient, h_f, of 70000 W/m²K and a ΔT of 300 °C as for the experiments.

Results and Discussion

The results obtained with the simulation of the thermal shock test showed a different value for the maximum stress developed in the specimens of various geometries due to the re-distribution of the thermal stresses, mainly in the outer layer. In addition, the maximum stress values on the first steps of the thermal shock varied from z (longitudinal) to x (transversal) component depending on the geometry chosen. As it can be inferred from Figures 2-4, the specimens with thicker outer layers exhibit maximum stress values associated with the longitudinal stress direction (x-component). However, the maximum stress shifts to the z-component as the thickness of the outer layer decreases. The thermal shock experiments accomplished at a ΔT of 300 °C showed a different crack pattern for the various architectures, as seen in the referred figures, being transversal for the specimens with thicker outer layers and longitudinal for the architectures with thinner ones. This phenomenon is associated with the location and component of the maximum stress during the first steps of the thermal shock tests. In this regard, thermal shock predictions using a 3D-FE model and water quench experiments are in good agreement in terms of describing the location where cracks initiated in the laminates depending on the maximum stress component. For specimens with thick outer layers, thermal shock transversal cracks initiate at the long edge of the material surface (Figure 2). This can be explained due to the higher z-stresses developed at that site. Very remarkable is that these transversal cracks extend up to the first interlayer and arrest because of the high compressive stresses inherent to the thin AMZ layers, while the rest of the structure remains intact (as seen in Figure 2). In the case of laminates with constant ATZ thickness, maximum x and z-stresses are of the same magnitude giving rise to the formation of both transversal and longitudinal cracks (Figure 3). However, for samples designed with thinner outer layers, cracks are mainly longitudinal due to the higher x-stress developed in the small edge of the specimen surface as well as in the central region (Figure 4). In addition, some *spalling* is also observed at the outer most AMZ layer of some bars (Figure 3). This might be caused by the extent of the edge cracks, formed during sintering, due to the thermal shock stresses normal to the layer plane.

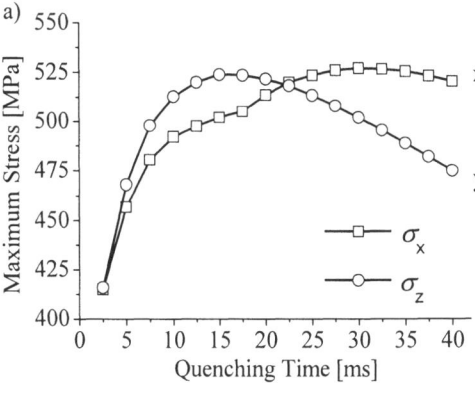

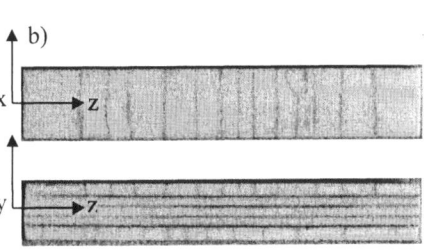

Figure 2. a) Curves of the maximum x and z stresses developed in the laminate with an outer layer of 750 μm, and b) top and front view of the transversal cracks developed in the laminate for a ΔT of 300°C.

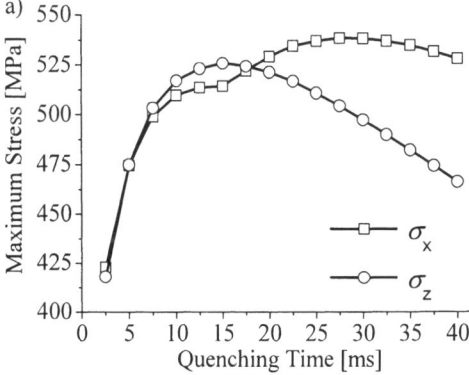

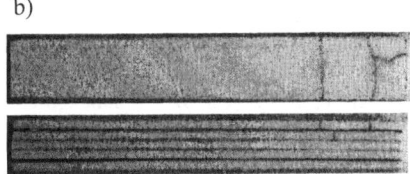

Figure 3. a) Curves of the maximum x and z stresses developed in the laminate with an outer layer of 530 μm during the thermal shock, and b) top and front view of the cracks developed at both edges in the laminate for a ΔT of 300°C.

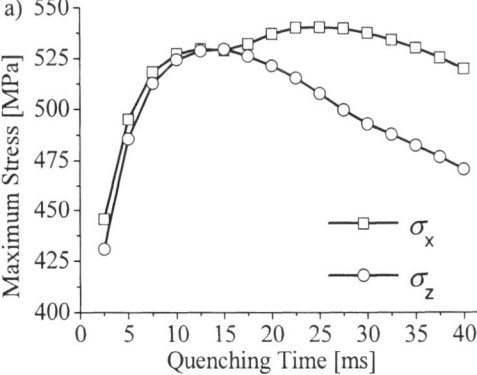

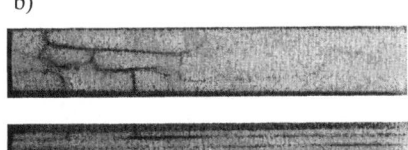

Figure 4. a) Curves of the maximum x and z stresses developed in the laminate with an outer layer of 300 μm during the thermal shock, and b) top and front view of the preference longitudinal cracks for a ΔT of 300°C.

Conclusions

Thermal shock predictions using a 3D-FE model and water quench experiments were in good agreement in terms of describing the magnitude and distribution of the thermal shock stresses and thus the location where thermal shock cracks initiated in the laminates of study. It is demonstrated that the thickness of the outer most layer of an Al_2O_3-ZrO_2 laminate influences the crack initiation under thermal shock conditions. This is due to the change in the maximum stress component developed during the quenching test. For the case of laminates with thick outer layers an effective crack arrest of the thermal shock cracks is observed owed to the high compressive residual stresses in the thin layers. This phenomenon is of an extremely importance since it points out that the multilayered structure may withstand severe thermal shock conditions by limiting the thermal shock cracks to the extent of the outer most ATZ layer thickness, while the rest of the structure remains intact after the thermal loading.

References

[1] Lange, F.F., "Powder Processing Science and Technology for Increasing Reliability", J. Am. Ceram. Soc., 72, 3-15 (1989)

[2] Bermejo, R., Llanes, L., Anglada, M., Supancic, P. and Lube, T., "Thermal Shock Behaviour of a Al_2O_3/ZrO_2 Multilayered Ceramic with Residual Stresses due to Phase Transformations", Key. Eng. Mat., 290, 191-8 (2005)

[3] Tiller, T.M. and Tsai, C., "Theory of Filtration of Ceramics: I, Slip Casting", J. Am. Ceram. Soc., 69, 882-7 (1986)

[4] Bermejo, R., Torres, Y., Sánchez-Herencia, A.J., Baudín, C., Anglada, M. and Llanes, L., "Fracture Behavior of an Al_2O_3-5%tZrO_2/Al_2O_3-30%mZrO_2 multilayer system fabricated by slip casting", Fatigue Fract. Engng. Mater. Struct., 29, 71-8 (2006)

[5] ANSYS, Inc. Theory Reference. Release 8.1., Ed. by P. Kohnke, Canonsburg, Pennsylvania, USA

Key Engineering Materials Vol. 333 (2007) pp 255-258
online at http://www.scientific.net
© *2007 Trans Tech Publications, Switzerland*

Influence of The Cone Crack Geometry on the Strength Degradation

Luca Ceseracciu [a], Marc Anglada[b] and Emilio Jiménez-Piqué[c]

Dpt. of Materials Science and Metallurgical Engineering
Technical University of Catalonia, 647 Diagonal av., 08028 Barcelona, Spain

[a]luca.ceseracciu@upc.edu, [b]marc.j.anglada@upc.edu, [c]emilio.jimenez@upc.edu

Keywords: ceramic laminates, numerical analysis, cone crack

Abstract. The presence of surface compressive residual stress in a laminated material enhance the resistance of the component by reducing the stress intensity factors acting on the cracks -either natural or artificial- existing in the surface.

Fissures in the form of cone crack are often generated by blunt contact in service, that can affect the functionality as well as the strength of the material.

In this work, a two-steps analysis of the effect of residual stresses on the geometry of cone crack and how this change in geometry influences the far-field strength of the material was performed by means of a Finite Elements model and of experimental observations.

In the first part, an automatic incremental model was formulated, which allowed to establish the crack shapes that were used in the second part for simple four-points test models.

It was observed that residual stresses change considerably the crack shape, with important implications in the design of contact-damage tolerance, and that this reflects on corresponding changes in the strength.

Introduction

Among the ways explored for improving toughness and/or reliability of ceramic materials, laminated structures are especially important and promising for the interesting enhancements, in terms either of toughness[1], flaw tolerance[2], work of fracture[3], they offer respect to their monolithic counterparts.

This improved resistance provided, for example, by multilayers with residual stresses[4], can be appreciated in common strength tests, such as flexion tests, as well as in more specific, application-related tests, such as spherical indentation, which well represents the blunt contacts, very common in service, e.g. in tribological or biomedical applications[5].

An analysis combining both aspects, the local damage and its effect on the far-field stress degradation, would produce interesting considerations on the effective performances of a ceramic laminated structure in actual applications.

In this work, the influence of residual stresses on the formation of cone crack under Hertzian indentation is studied by means of Finite Elements analysis and verified by some aimed experimental observations, and the influence of the obtained geometries on the fracture strength of the material qualitatively evaluated with a simple comparative FE model. This latter was designed as to consider only the contribution to strength deriving from the different geometries, and not the intrinsic contribution from the laminated structure.

Cone crack in brittle materials

Crack geometries. When a brittle material is indented with a blunt body, the first damage visible on the surface is a ring crack, originating from some favorable flaw by the tensile stress near the contact area[6]. Such crack, then, extends outward following, on a first approximation, the σ_3 direction[7]. However, this latter approximation does not produce results in good agreement with experimental observations. More realistic models consider the evolving stress field as the crack propagates. Numerical simulations with incremental crack growth is, therefore, very suitable to obtain accurate results[8,9].

In unstressed materials the angle of cone crack is almost constant and depends mainly on the Poisson coefficient[9]. If residual stresses are present, their contribution can be expressed, on a first approximation, as from the case of inclined crack on a semi-infinite plate, so that the stress intensity factors at the crack tip can be written as:

$$K_I \approx \chi \frac{P}{c^{3/2}} + Y\sigma_r \sin^2 \alpha \sqrt{c} \qquad (1)$$

$$K_{II} \approx K_{II}^i + Y\sigma_r \sin\alpha \cos\alpha \sqrt{c} \qquad (2)$$

The difference in K_I will produce a change in the crack length: a shortening if the residual stresses are compressive, a lengthening otherwise. The difference in K_{II} will produce a change in the propagation angle[10]. It can be seen that the contribution from residual stresses increases as the crack grows, while the one from the indentation stress field, intuitively, decreases. Therefore, a change in propagation angle can be expected.

An axisymmetrical model was programmed in the commercial software ABAQUS 6.5.1, with the built-in scripting language Python. Flat punch indentations were represented with the vertical displacement of the contact area nodes; load corresponding to the chosen penetration depth was calculated from the reaction forces at the contact area nodes. A short vertical crack was placed near the contact radius, at whose tip the stress intensity factors were calculated, and the propagation angle with the Maximum Energy Release Rate criterion. Quadratic elements with the "quarter-edge-nodes" method were employed to represent the tip singularity[11]. After the calculation, the crack geometry was updated of a small increment, calculated by convergence tests, very short in the first steps, then progressively increased as the crack reached a low-gradient stress field.

Simulations were run until the condition $K_I > K_{Ic}$ was fulfilled. Equilibrium cracks for different values of residual stresses were calculated: two compressive values from actual laminated systems, one high-compression and one low-tension arbitrary, but not unrealistic, values.

Strength degradation. The strength of the systems studied was evaluated from simple 2D models of four-points flexion tests. The longitudinal geometry was taken from that of actual samples (thickness $t=2$ mm, spans of 10 and 20 mm), although homogeneous material was employed, so that the differences in the computed strength could be attributed only to the cone crack geometries, that were taken directly from the ones calculated in the first part of the procedure.

The output of the simulations was the elastic energy G at the crack tip, calculated from the displacement field. Considering a general formulation, G can be expressed as:

$$G = AP^2 \qquad (3)$$

Where the term A considers all the geometrical factors, in the cases studied the differences derive only from the crack geometry. The A factor was calculated from polynomial fitting of the FE data, and the critical load Pc for the crack propagation from the condition $G=G_c$, with G_c the fracture energy. It should be noted that the 2D model does not provide realistic values of load, since it does not consider the momentum of inertia of the beam. Therefore, the values reported are normalized to the unstressed material, as a comparison between the different crack geometries.

Experimental procedures

The multilayered materials studied in this work, showed in fig. 1, were produced by *tape casting* technique at ISTEC, CNR (Faenza, Italy) starting from high purity powders of Al_2O_3 and ZrO_2. 13-layers geometries were stacked by alternated alumina and alumina-zirconia 60-40%wt. plies. Monolithic alumina samples were produced from the same starting powder under similar processing conditions. Details of the processing are reported elsewhere[12]. Residual stresses on the surface were measured by indentation techniques and μ-Raman spectroscopy of approx. –200 MPa.

Indentations were performed with a 1.25 mm radius WC-Co spherical indentator. The indentation load for tests on both materials was chosen of 800 N, approx. 20% higher than the critical load for the apparition of surface ring crack, in order to guarantee well developed cone cracks.

After the indentations, samples were ground and polished in the direction normal to the indented surface for a transversal observation of the cracks generated.

Results

The cone crack geometries calculated in the first part of the FE procedure are presented in fig. X. The changes both in crack length and angle are apparent. Some interesting considerations can be done:

1) In presence of relatively high compressive residual stresses, the crack deflection can be as high as to propagate almost parallel to the layers interface. In that case, the risk of catastrophic failure for the crack reaching the underlying tensile layer would be reduced.

2) Tensile stresses, even low, produce a lengthening of cone crack relatively higher than the compressive counterpart. This suggests that tensile stresses should be avoided, as they worsen significantly the contact behavior of the material.

G-curves corresponding to Eq. 3 are presented in Fig. X+1, and the principal data in Table Y.

It can be seen that the different geometries in materials with residual stresses produce great changes in the theoretical flexion strength, up to twice as high, for large compressive stresses. If the same amount of stresses present are tensile, albeit low ones, the weakening of the material is relatively higher than the respective improvement.

In the model employed, the effects of the laminated structure, such as composite elastic modulus, surface stresses, R-curve, are not considered. It is supposed, therefore, that the global result could be in some measure different, although this difference is not quantifiable, because it depends on the specific structure, not only on the stresses present.

σ_r	200	0	-200	-400	-600
crack length [μm]	453	280	229	196	172
Rel. load	0.54	1	1.35	1.66	1.96

Table Y. FE calculations of the crack length and relative equivalent load as a function of the residual stress

Conclusions

The influence of uniform biaxial residual stresses in multilayered ceramics on the geometry of cone crack by Hertzian indentation was studied by means of an automated incremental Finite Elements model. Both length and propagation angle of the crack resulted strongly influenced by the presence of such stresses, with important implications on the damage tolerance of the composite, especially if the stresses are compressive, in which case the crack will tend to reduce its angle, so that the probability of reaching the second, tensile layer, is reduced.

The more important consequence of the different evolution of crack geometry is the change in the fracture energy G under flexion test. Improvement up to two times the load necessary for fracture in unstressed materials can be reached with realistic values of residual stresses. Vice versa, an albeit low tensile residual stress would produce a significant diminution of fracture load, that suggests that tensile stresses should be carefully avoided in applications where contact loading is expected.

Acknowledgements

Work supported in part by the European Community's Human Potential Programme under contract HPRN-CT-2002-00203, [SICMAC] and by the Spanish Ministry of Science and Culture through grant MAT2005-01168.

References

[1] J.S. Moya: Adv. Mater. 7.2 (1995), p. 185

[2] A.J.Sánchez-Herencia, C.Pascual, J.He and F.F.Lange: J. Am. Ceram. Soc. 82.6 (1999), p.1512

[3] J.W. Clegg, K. Kendall, N.M.Alford, T.W. Button and J.D. Birchall: Nature 347 (1990), p.455

[4] H.M. Chan: Annu. Rev. Mater. Sci. 27 (1997), p.249

[5] Y.-G. Jung, I.M. Peterson, D.K. Kim and B.R. Lawn: J. Dent. Res. 79.2 (2000), p.722

[6] P.D. Warren, D.A. Hills and D.N. Dal: Tribol. Int. 28.6 (1995), p.357

[7] B.R. Lawn: J. Am. Ceram. Soc. 81.8 (1998), p. 1977

[8] C. Kocer: Fin. Elem. Anal. Des. 39 (2002), p. 639

[9] C. Kocer and R.E. Collins: J. Am. Ceram. Soc. 81.7 (1998), p. 1736

[10]T. Fett and D. Munz: Int. J. Fract. 115 (2002), p. 69

[11]ABAQUS 6.5.1 Analysis User's Manual, section 7.10.2

[12]E. Jiménez-Piqué, L. Ceseracciu, F. Chalvet, M. Anglada and G. de Portu: J. Eur. Ceram. Soc. 25.15 (2005), p. 3393

Key Engineering Materials Vol. 333 (2007) pp 259-262
online at http://www.scientific.net
© 2007 Trans Tech Publications, Switzerland

Residual Stresses in Laminar Functionally Graded Ceramic Materials

Pavol Hvizdoš[a], Kristoffer Norberg

Department of Materials Science and Metallurgical Engineering/ETSEIB, Universitat Politecnica de Catalunya, Diagonal 647, 08028-Barcelona, Spain

[a]pavol.hvizdos@upc.es

Keywords: layered ceramics, functionally graded materials, residual stresses.

Abstract. A simple analytical model of residual stresses far from edges in symmetrical planar functionally graded material (FGM) is presented. The model is based on elastic plate theory and neglects the influence of edges and free surfaces. The results are compared to analytical model of laminar ceramics and to finite element model of FGM. The influence of various geometrical and material parameters on the internal stress state is discussed.

Introduction

The descriptions of the stress states inside functionally graded materials usually use some convenient model based on small deformation, small strain plate/beam theories of continuum mechanics. In a structure of uniform plates of dissimilar materials bonded together without cracks or macroscopic defects along their interfaces state of biaxial loading exists away from the free edges. Material undergo normal stresses only in the in-plane directions and is free of shear stresses or out-of-plane stresses [1]. Such plate theory approach was used in [2, 3] for analytical strain analysis of symmetrical multilayered laminates. Consider a system consisting of ($2n$-1) alternating layers of 2 types: a (n layers of thickness t_a) and b (n-1 layers of thickness t_b), as depicted in Fig.1.

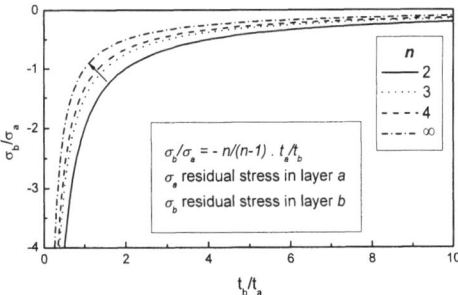

Fig.1. Schematic representation of laminar multilayered material.

Fig. 2. The ratio σ_b/σ_a is determined principally by ratio of layers thicknesses. The role of number of layers is illustrated.

The two materials are characterized by their respective material properties: Young's modulus E, Poisson ratio v, and thermal expansion coefficient α. The thermal expansion mismatch strain for two materials bonded together cooling from the processing (joining) temperature T_2 down to the ambient temperature T_1 is

$$\Delta\varepsilon = \varepsilon_a - \varepsilon_b = \int_{T1}^{T2}(\alpha_a - \alpha_b)dT \,, \tag{1}$$

where the subscripts a and b denote the respective materials, and ε is the thermal strain. The thermal stresses arising from the strain mismatch are [2]:

$$\sigma_a = \Delta\varepsilon E_a' \big/ \big[1 + (E_a' n t_a)/(E_b'(n-1)t_b)\big], \qquad \sigma_b = -\Delta\varepsilon E_b' \big/ \big[1 + (E_b'(n-1)t_b)/(E_a' n t_a)\big], \qquad (2)$$

where $E' = E/(1-v)$ is the biaxial modulus, as the strains are biaxial, parallel with the layers. The ratio σ_b/σ_a depends only on layers' thickness ratio t_b/t_a, and, to a degree, on the number of layers, as it is illustrated by Fig. 2. This ratio is always negative, since the signs of the residual stresses are always opposite. The magnitudes of the stresses then are determined by moduli of elasticity and by $\Delta\varepsilon$.

Functionally Graded Material

To model the stress state inside a FGM without necessity to apply computational methods, some simplifications have to be made, but most of them are analogous to those made usually also in finite element analyses. As before, let us consider a symmetrical layer configuration consisting of homogeneous layers of materials a and b with gradient layers in between (Fig. 3). Here, too, no bending takes place and momenta cancel each other. Assuming that the gradient profile is linear, we apply the linear rule of mixture for all basic material characteristics. Further, we assume that the

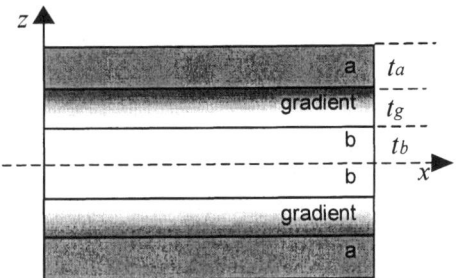

Fig. 3. Schematic of symmetrical planar functionally graded material.

suffered thermal strain is a function of z only [4]. Then the effective mechanical properties within the gradient layer change linearly from the ones of the layer a to those of the layer b in following way:

$$E'(z) = E_a' + \frac{E_b' - E_a'}{t_g} z, \quad \alpha(z) = \alpha_a + \frac{\alpha_b - \alpha_a}{t_g} z, \quad \varepsilon(z) = \varepsilon_a + \frac{\varepsilon_b - \varepsilon_a}{t_g} z \quad \text{for } z \in (0, t_g). \qquad (3)$$

Assuming the linear transition from α_a to α_b with respect to composition, for the thermal strain mismatch the Equation (1) remains valid. Again, only elastic deformation is considered:

$$\sigma_a = \varepsilon_a E'_a, \qquad \sigma(z) = \varepsilon(z)\, E'(z), \qquad \sigma_b = \varepsilon_b E'_b. \qquad (4)$$

In absence of external load the forces are in equilibrium:

$$2\int_{ta}\sigma_a dz + 2\int_{tg}\sigma(z)dz + 2\int_{tb}\sigma_b dz = 0. \qquad (5)$$

By setting $\sigma(z)$ according to (3) and (4), the relation between the respective thermal strains can be found from Eq. (5) as:

$$\varepsilon_a = -\varepsilon_b \frac{E_b' t_b + E_b' t_g/3 + E_a' t_g/6}{E_a' t_a + E_a' t_g/3 + E_b' t_g/6}. \qquad (6)$$

Then using Eqs. (1) and (6) the residual stresses are:

$$\sigma_a = \Delta\varepsilon E_a' \frac{E_b' t_b + (E_a' + 2E_b')(t_g/6)}{E_a' t_a + E_b' t_b + (E_a' + E_b')(t_g/2)}, \qquad \sigma_b = -\Delta\varepsilon E_b' \frac{E_a' t_a + (2E_a' + E_b')(t_g/6)}{E_a' t_a + E_b' t_b + (E_a' + E_b')(t_g/2)}, \qquad (7)$$

and the ratio between them:

$$\frac{\sigma_b}{\sigma_a} = -\frac{t_a + t_g/3 + (E_b'/E_a')(t_g/6)}{t_b + t_g/3 + (E_a'/E_b')(t_g/6)}. \qquad (8)$$

The stress profile within the gradient layer, $\sigma(z)$, then can be calculated using Eq. (4).

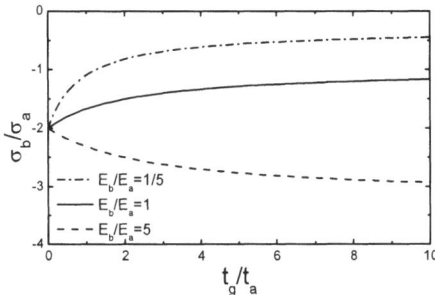

Fig 4. Example illustrating the influence of the thickness of the gradient layer on the stress ratio for various elasticity ratios. $t_a=2t_b$.

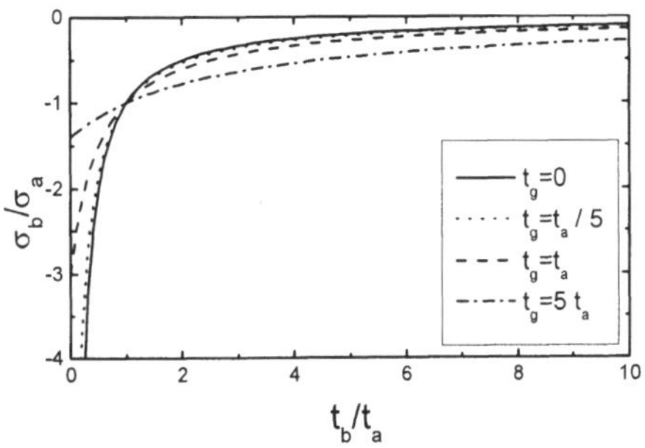

Fig. 5. The ratio of the residual stresses as a function of thickness ratio for various t_g. $E_a=E_b$.

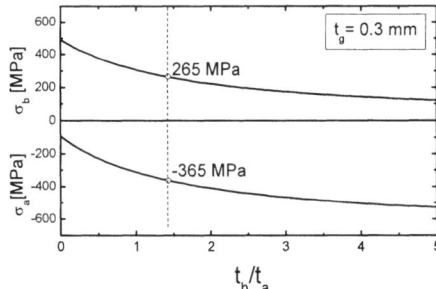

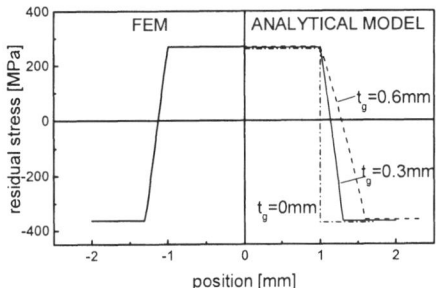

Fig.6. Residual stresses developed in alumina / zirconia composite FGM vs. the thickness ratio. Compressive stresses (σ_a) in exterior layers ($Al_2O_3+10\%ZrO_2$), tensile stresses (σ_b) in the centre ($Al_2O_3+30\%ZrO_2$). Comparison with the FEM values for particular geometry [5].

Fig. 7. The residual stresses profile in an alumina / zirconia FGM calculated by FEM [5] compared with results obtained by the present model. For illustration, profiles for different t_g are plotted.

For $t_g=0$ the formulas (6), (7), and (8) reduce to forms identical to those for simple laminate. Generally however, the ratio σ_b/σ_a now depends also on the t_g, and on the Young's moduli, as shown in Fig. 4. There the plot is made for $t_a=2t_b$; for different thickness ratios the graph shifts up or down, for $t_g \to 0$ the lines converge. Another representation of role of the gradient interlayer can be seen in Fig. 5, which is analogous to Fig. 2. Naturally, this layer makes the change of the stress state more gradual, and with increasing t_g the difference between σ_b and σ_a decreases – the line becomes flatter.

As an example of actual residual stresses Fig. 6 shows their magnitudes in an alumina/zirconia FGM ($Al_2O_3+10\%ZrO_2/ Al_2O_3+30\%ZrO_2/ Al_2O_3+10\%ZrO_2$, [5]) as functions of the thickness ratio. The results for a particular geometry (t_a=0.7mm, t_g=0.3mm, t_b=1mm) calculated using finite element model (FEM) [5] are included. The comparison shows that for similar geometry the bulk residual stresses found using the present model are in perfect agreement with the FEM. Fig. 7 shows the stress profiles across the layers of the bulk residual stresses calculated using both models, and also illustrates the influence of the gradient layer's thickness.

Conclusion

The presented model conveniently describes the residual stresses arisen due to thermal expansion mismatch in the bulk of planar functionally gradient material in absence of bending. It corresponds well to laminar and to finite element numerical models. It allows quick design considerations when particular residual stresses in an FGM are required.

Acknowledgement

The work was supported by the Spanish Ministry of Science and Culture through the "Ramón y Cajal Programme". The experiments were performed in the framework of the EU Training Network "SICMAC" (contract HPRN-CT-2002-00203).

References

[1] S. Suresh and A. Mortensen: *Fundamentals of Functionaly Graded Materials*, (ASM International and the Institute of Materials, Cambridge 1995).

[2] A.V. Virkar, J.L. Huang and R.A. Cutler: J. Amer. Ceram. Society, 70 (1987), p. 164.

[3] T. Chartier, D. Merle and J.L. Besson: J. Europ. Ceram. Society, 15 (1995), p. 101.

[4] A.E. Giannakopoulos, S. Suresh, M. Finot and M. Olsson: Acta Metall. Mater., 43 (1995), p. 1335.

[5] P. Hvizdoš, D. Jonsson, M. Anglada, G. Anné and Omer Van Der Biest: Mechanical Properties and Thermal Shock Behaviour of an Alumina/Zirconia FGM prepared by EPD. J. Europ. Ceram. Society, (2006), in press, available online June 2006.

Key Engineering Materials Vol. 333 (2007) pp 263-268
online at http://www.scientific.net
© 2007 Trans Tech Publications, Switzerland

Modelling of Crack Growth Near the Metallic-Ceramic Interface During Thermal Cycling of Air Plasma Sprayed Thermal Barrier Coatings

A. Casu[1, a], J.-L. Marqués[1,b], R. Vaßen[1,c] and D. Stöver[1,d]

[1]Institut für Werkstoffe und Verfahrenstechnik (IWV1), Forschungszentrum Jülich GmbH, Jülich/Germany

[a]a.casu@fz-juelich.de, [b]j.l.marques@fz-juelich.de, [c]r.vassen@fz-juelich.de, [d]d.stoever@fz-juelich.de

Keywords: thermal barrier coating, sub-critical crack growth, lifetime, metallic-ceramic interface

Abstract. The lifetime under thermal cycling of a system consisting of an air plasma sprayed thermal barrier coating (TBC) deposited on a metallic bondcoat (BC) is determined by the sub-critical growth of micro-cracks near the interface between both coatings. This growth mainly occurs during the cooling down phase, as shown by the acoustic emission monitoring during the thermal cycling. The factors controlling the stress level leading to the crack growth are the local curvature of the metallic-ceramic interface, the growth of an oxide scale (TGO) at such interface and the sintering of the TBC, the two last processes occurring during the high temperature cycle phase. Implementing all these factors, a model based on Finite Element Method (FEM) calculations is presented where growing cracks are incorporated by assigning soft properties to the FEM cells occupied by the cracks. Determining the growth direction for the maximum energy release rate at every cooling down step, the current crack extension during the cycling is tracked until it reaches a characteristic length corresponding to the TBC failure. The influence by the metallic-ceramic interface roughness and by the temperature gradient across the TBC is discussed.

Introduction

For a higher efficiency in a gas turbine, the increasing service temperature requires an efficient thermal insulation of the metallic components. This is achieved through the deposition of a ceramic TBC on the metallic component, usually with an additional metallic interlayer BC between TBC and component. This BC has the double function of enhancing the adhesion of the deposited ceramic topcoat and of building an alumina oxide scale (TGO) which protects the underlying metallic component against oxidation. The whole system consisting of metallic component, bondcoat and thermal barrier coating is denoted TBC system. The most extended technique for the BC deposition is vacuum plasma spraying (VPS), and for the TBC air plasma spraying (APS); the most used ceramic material for the TBC is yttria partially stabilized zirconia (YSZ). Due to the different thermal expansion coefficients of the metallic and ceramic coatings, high stresses develop near to the BC-TBC interface during strong temperature changes. This stress level becomes further modified when the TGO grows between BC and TBC. Upon thermal cycling and for heating temperatures below 1300°C, the growth of the micro-cracks already present in the sprayed TBC eventually leads to the failure of the ceramic coating near the BC interface [1]. A reliable lifetime estimation for the sprayed TBC system is still missing because of the complicated interplay of BC-TBC profile roughness, TGO growth and TBC sintering. Incorporating all these factors and calculating the thermal stress with the FEM software ANSYS (Ansys Inc., Canonburgh, Pittsburgh, PA, USA), a model is presented which fully tracks the growth of one crack in a plasma sprayed TBC under thermal cycling. The model details can be found in [2]; here the main results regarding the influence of BC/TGO-TBC roughness and sintering gradient across the TBC will be discussed.

Model definition

The surface of the plasma sprayed metallic BC, on which the ceramic TBC is subsequently deposited, is rough as result of the disordered flattening, stacking and fragmentation of molten particles. The local curvature of the BC-TBC interface controls the strength and orientation of the thermal stress near such interface, which leads during thermal cycling to the growth of micro-cracks already existing in the sprayed TBC. Notwithstanding its rough character, the BC-TBC profile is not completely random. For plasma sprayed surfaces the Fourier transformation of the height profile, represented double logarithmically, does not consist of one single straight line as it would be the case for a pure random signal. It characteristically displays an inflection point at a length scale λ_0 separating the range of long length scales from that corresponding to the very short length scales, the latter describing the random fluctuations in the height profile [3]. This inflection point is thus the smallest non-random length scale characteristic to the BC-TBC profile.

For modelling the local growth of the TBC micro-cracks near the BC-TBC interface, the curved BC-TBC profile will be considered as periodic of height given by $y(x) = A_0 \cos(2\pi x / \lambda_0)$, with wavelength λ_0 since this, as the shortest characteristic length, yields the strongest curvature and stress level in the TBC. Within the range of this wavelength, a single representative micro-crack is considered under the assumption that similar cracks are present along the whole BC-TBC profile but without interfering with each other. The failure of the TBC system is assumed when the initial micro-crack has grown to a length equal to λ_0, thus connecting to the neighbour cracks. The amplitude of the periodic profile is obtained from the average roughness $R_a = \dfrac{1}{L}\int_0^L |h(x)| dx$,

measuring the local height deviation $h(x)$ (with respect to the line of averaged height) of the BC surface along a sampling length L. For the introduced cosine function as profile height, the amplitude and the roughness are related through $A_0 = \pi R_a / 2$. For the BC considered in this paper, deposited by VPS of MCrAlY (M=Ni,Co) powder with size distribution d_{10}=19.8μm, d_{50}=30.9μm and d_{90}=47.0μm, the inflection in the Fourier spectrum is located at λ_0=50±10μm and has a roughness R_a=5.9±0.4μm, corresponding to an amplitude $A_0 \approx 10$μm.

Table 1. Thermo-mechanical parameters for the TBC system materials at room temperature [4].

	metallic substrate, IN378	metallic BC, MCrAlY	TGO, Al$_2$O$_3$	ceramic TBC, YSZ
thermal expansion coeff. α	15.8x10^{-6}K^{-1}	17.5x10^{-6}K^{-1}	8.0x10^{-6}K^{-1}	10.7x10^{-6}K^{-1}
elastic modulus E	191GPa	140GPa	360GPa	25GPa
Poisson number ν	0.3	0.3	0.22	0.22

The TBC system to be investigated consists of a flat metallic substrate 3mm thick made of IN738, a 150μm thick VPS metallic BC with the above mentioned wavelength and amplitude at the interface to the TBC, and a 300μm APS ceramic TBC made of YSZ. The FEM mesh modelled is 2 dimensional (plane strain state) and contains half wavelength of the BC-TBC profile (one hill and one valley) with a cell size of 0.5x0.5μm for the 60μm of the TBC region located nearest to the BC. Since the splat structure in the sprayed TBC allows for a relative high creep rate at high temperature, only the thermal stress occurring during the cooling phase of the thermal cycling will be considered [4,1]. Upon cooling down to room temperature the behaviour is assumed elastic for all the materials, whose material properties (considered homogeneous) are listed in Table 1. Additionally, during the thermal cycling, a progressive TGO growth as well as a local sintering in the TBC is implemented in the FEM modelling according to the following equations

$$d_{TGO} = A_{TGO} e^{-E_{TGO}/k_B T} \sqrt{t} \quad \text{and} \quad E_{TBC}(t)/E_{TBC}(t=0) = 1 + A_{sint} e^{-E_{sint}/k_B T} \sqrt{t}, \tag{1}$$

with A_{TGO}=3.715x10^{-3}m/s$^{1/2}$ and A_{sint}=6.225x10^{8}s$^{-1/2}$, and activation energies E_{TGO}=1.435eV and E_{sint}=3.10eV [4]; T is the temperature during the stationary hot temperature phase, at the BC/TGO-TBC interface for TGO growth or locally within the TBC for the TBC sintering. The elastic modulus for YSZ in as-sprayed state is taken as $E_{TBC}(t$=0)=25x10^{9}Pa [5].

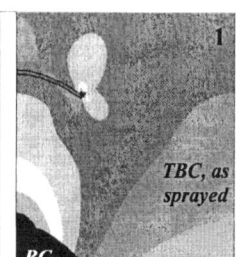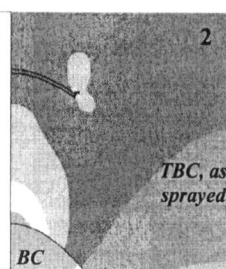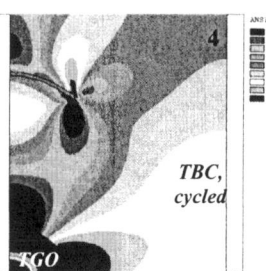

Figure 1. Vertical thermal stress σ_{yy} distribution (–200 to +200MPa) upon cooling down for a TBC crack of length 15μm located 17μm above the BC/TGO-TBC hill. As-sprayed situation (1-2) and after 5μm TGO growth (3-4). Graphs 1 & 3, cooling down from a homogeneous temperature distribution at T=1000°C. Graphs 2 & 4, cooling down from a stationary temperature gradient between T_{surf}=1200°C at the TBC external surface and T_{TGO}=1000°C at the BC/TGO-TBC interface.

The implementation of a crack in the TBC is carried out by giving very soft properties (elastic modulus of 100Pa, vanishing thermal expansion coefficient) to the mesh cells occupied by the crack. Two cases are considered within the TBC: a crack centered above the BC/TGO-TBC profile hill, and above the valley. The vertical separation between an initial TBC micro-crack and this profile is taken as the thickness of one sprayed lamella, approximately 17μm, which corresponds to 18 spraying cycles for an average TBC thickness of 300μm. For a given temperature distribution across the TBC system, the combined influence of TBC sintering and TGO growth determines the stress level acting on the crack tip and thus the amount of energy released by the crack growth. As reference, a TBC system will be taken with a temperature of 1000°C at the BC-TBC interface during a stationary thermal cycling's heating phase of 5min duration and subsequently cooled down and maintained at room temperature during 2min. In the as-sprayed state (no TGO grown on the BC) and for a homogeneous temperature distribution through the whole system, the stress level upon cooling is higher than in the case of a gradient temperature with a topcoat surface temperature of T_{surf}=1200°C (Fig. 1, Graphs 1 & 2). The reason is that in the gradient case, all layers located below the TBC suffer a lower temperature change at cooling and produce a lower thermal stress than in the homogeneous temperature case. Since all these layers have a much higher elastic modulus than the TBC, they mainly determine the stress level even in the TBC itself. Nevertheless this situation is inversed after a cycling time long enough: the sintering of the TBC makes the thermal stress from the BC/TGO-TBC reach deeper into the TBC for the temperature gradient case, since upper layers are relatively strong sintered. Thus in the temperature gradient situation, the stress acting on the crack gradually increases and eventually exceeds that for the homogeneous temperature distribution. It is worth noting that with an increasing TGO growth, the negative (compressive) vertical stress above the valley gradually changes into positive (tensile) stress, which leads to crack opening of horizontal cracks located centered above the valley.

Crack growth modelling and results

The implementation of crack growth in the FEM model is carried out as follows: starting with an initial TBC micro-crack of length a_0 placed 17μm above the BC-TBC profile, at every crack growth step the elastic energy U is calculated by means of a FEM simulation for the current crack state and

9 further extensions: crack extended one cell ahead; one cell up/down; one cell up/down, one cell ahead; one cell up/down, two cells ahead; one cell up/down, three cells ahead. The stress distribution at the cell located before the current crack tip is read and the direction β for the maximum opening stress on the crack tip is determined. Then the energy release rate for each of the 9 cases is calculated, defined as the relative energy change between the current crack state of length a and each crack extension $a+\delta a$, $G_I(\sigma,a) = \dfrac{U(\sigma,a)-U(\sigma,a+\delta a)}{\delta a}$, and the maximum value and its corresponding crack extension direction is selected, always fulfilling to be compatible with the β angle previously calculated. The sub-critical growth is determined through

$$\frac{\Delta a}{\Delta t} = v_0 \left(\frac{G_I(\sigma,a)}{G_{I,crit}} \right)^{m/2} \quad \text{with} \quad G_{I,crit} = \frac{1-v_{TBC}^2}{E_{TBC}} K_{I,crit}^2 , \qquad (2)$$

and v_0 the crack velocity at critical conditions, equal to 7.6×10^{-5} m/s for YSZ, and $m=18$ [4]. $K_{I,crit}$ is the fracture toughness for mode I, equal to 1MPa m$^{1/2}$ for plasma sprayed YSZ [5], which yields a critical energy release rate $G_{I,crit}$=38.1Pa m. This $G_{I,crit}$, corresponding to the effective bonding which withstands the crack opening, is assumed to remain unaffected by the TBC sintering. The time step Δt in (2) is adjusted such that the crack increase Δa becomes equal to the full crack extension δa or, if Δa is lower than $\delta a/2$, the current Δa is accumulated and the time step is increased by an amount corresponding to a TGO thickness growth by one FEM mesh cell (0.5µm). The growth will be continued until the crack covers a whole profile wavelength λ_0 or the crack tip enters into a region of pure compressive stress and becomes stopped.

Table 2. TGO thickness and location at failure for different BC/TGO-TBC interfaces.

A_0=10µm,λ_0=50µm	heating phase: isothermal T=1000°C		gradient T_{surf}=1200°C, T_{TGO}=1000°C	
a_0, y_0=17µm	hill crack	valley crack	hill crack	valley crack
10µm	25.46µm, in TBC	28.72µm, in TGO	3.02µm, in TBC	3.48µm, in TGO
15µm	16.61µm, in TBC	35.04µm, in TGO	2.51µm, in TBC	3.48µm, in TGO
20µm	8.18µm, in TBC	no failure	1.98µm, in TBC	3.48µm, in TGO
A_0=10µm,λ_0=70µm	heating phase: isothermal T=1000°C		gradient T_{surf}=1200°C, T_{TGO}=1000°C	
a_0, y_0=17µm	hill crack	valley crack	hill crack	valley crack
10µm	24.38µm, in TBC	15.96µm, in TGO	3.63µm, in TBC	3.48µm, in TGO
15µm	29.07µm, in TBC	19.80µm, in TGO	3.40µm, in TBC	3.48µm, in TGO
20µm	12.87µm, in TBC	22.04µm, in TGO	2.48µm, in TBC	3.48µm, in TGO
A_0=14µm,λ_0=50µm	heating phase: isothermal T=1000°C		gradient T_{surf}=1200°C, T_{TGO}=1000°C	
a_0, y_0=17µm	hill crack	valley crack	hill crack	valley crack
10µm	20.80µm, in TBC	19.53µm, in TGO	2.48µm, in TBC	4.98µm, in TGO
15µm	7.18µm, in TBC	13.02µm, in TGO	2.13µm, in TBC	3.98µm, in TGO
20µm	8.69µm, in TBC	>35µm, in TBC	2.11µm, in TBC	2.98µm, in TGO
A_0=10µm,λ_0=50µm	heating phase: isothermal T=1000°C		gradient T_{surf}=1200°C, T_{TGO}=1000°C	
a_0, y_0=8µm	hill crack	valley crack	hill crack	valley crack
5µm	>35µm, in TBC	14.54µm, in TGO	no failure	no failure
10µm	20.00µm, in TBC	15.38µm, in TGO	2.48µm, in TBC	6.48µm, in TGO
15µm	7.77µm, in TBC	14.48µm, in TGO	2.41µm, in TBC	4.98µm, in TGO
20µm	6.55µm, in TBC	10.96µm, in TGO	2.67µm, in TBC	4.57µm, in TGO

The resulting TGO thickness at failure is listed in Table 2 for different BC-TBC profile parameters and initial micro-crack lengths. As expected from the discussion in the previous section, the TBC sintering under a strong temperature gradient during the heating phase is able to over-compensate the lower thermal stress upon cooling created by a reduced temperature jump in the metallic layers. This leads to a high reduction in the time and TGO thickness required for failure. A possibility to reduce the unphysical rapid crack growth under temperature gradient is to consider a slight increase in $G_{I,crit}$ due to sintering. Additionally, the TBC sintering parameters in (1) were measured for a free standing TBC. Sintering in a TBC system under the stress created by the underlying metallic BC may impede the sintering rate at the nearest lamellae, where crack growth actually occurs. In general, a decrease in the profile wavelength or an increase in its amplitude leads a faster crack growth, due to the enlargement of the local profile curvature. Hill cracks grow more rapid than valley cracks; for the former, the failure takes place inside the TBC, for the latter the crack penetrates into the TGO. Nevertheless, considering that also small cracks can be present *within* the first lamella (and not only *above* it), also the situation for a small initial crack $a_0 = 5\mu m$ has been considered for a crack separation to the BC equal to half the lamella thickness. In that case and under a homogeneous temperature distribution at heating, growth is accomplished in a faster way for the valley cracks and failure occurs in the TGO.

References

[1] R. Vaßen, F. Traeger and D. Stöver: J. Thermal Spray Technol. **13** (2004), p. 396

[2] A. Casu, J.-L. Marqués, R. Vaßen and D. Stöver: *Proceedings of the 30th International Conference on Advanced Ceramics and Composites (Cocoa Beach 2006)*, to appear in July 2006.

[3] S. Giesen: "Characterization of plasma sprayed coatings by means of Fourier analysis and stochastic equations," diploma thesis, Fachhochschule Aachen/Jülich (2005), in German.

[4] F. Traeger, M. Ahrens, R. Vaßen and D. Stöver: Mater. Sci. Eng. A **358** (2003), p. 255

[5] S.R. Choi, D. Zhu and R. Miller: Int. J.Appl. Ceram. Technol. **1** (2004), p. 330

Key Engineering Materials Vol. 333 (2007) pp 269-272
online at http://www.scientific.net
© 2007 Trans Tech Publications, Switzerland

Microstructural Evolution of Thermal Barrier Coatings during Isothermal Oxidation

A. Salazar, J. Gómez-García, P. Poza and V. Utrilla

Departamento de Ciencia e Ingeniería de Materiales. ESCET.
Universidad Rey Juan Carlos.
C/ Tulipán s/n, 28933 Móstoles (Madrid). Spain

alicia.salazar@urjc.es

Keywords: thermal barrier coatings, thermal spray, oxidation, heat treatment, scanning electron microscopy, transmission electron microscopy, X ray diffraction.

Abstract. This paper analyses the degradation of a ceramic top coating $70\%ZrO_2 - 30\%CaO$ deposited onto a stainless steel AISI 304 by thermal spray, using Ni-6%Al-5%Mo as overlay coating. These thermal barrier coatings were heat treated for 48, 120 and 288 h at 800 °C to evaluate the degradation of these materials by isothermal oxidation. The microstructure evolution during oxidation was analysed by environmental scanning electron microscopy, transmission electron microscopy and X ray microanalysis. A thermally grown oxide layer was observed between the overlay coating and the ceramic top coating after oxidation. This layer was formed by a mixed Al, Ni and Mo oxides.

Introduction

Thermal barrier coatings (TBC) are widely used in turbines for propulsion and power generation where materials to withstand increasing operating temperatures, mechanical loads and chemical degradation are required [1-2]. TBC comprises at least two layers: a ceramic top coating which provides thermal insulation and a metallic overlay coating. The mismatch between the thermal expansion coefficient (CTE) of the ceramic coating [3] and the metallic substrate generates high interfacial stresses during coating manufacture and in-service thermal cycling that leads ceramic failure by spalling. To solve this problems of CTE mismatch and to protect the metallic component from chemical attack (the ceramic coating is permeable to oxygen and other species at high temperature) an intermediate metallic "bond coat" has been used. To improve the oxidation and corrosion resistance of the protected component, this overlay coating should develop a surface oxide layer thermodynamically stable, slow growing and adherent which inhibits continued ingress of active oxygen and other species. TBC durability is mainly controlled by this thermally grown oxide layer (TGO) and, consequently, detailed analysis of TGO development is critical to understand the mechanical and chemical degradation of TBCs during service [4].

The aim of this paper is to describe the TGO development on a ZrO_2-CaO ceramic top coating deposited onto a stainless steel by thermal spray, using NiAlMo as overlay coating. These coatings are extensively used in coal-fired power stations where the maximum operating temperature is 800°C.

Experimental Techniques

Ni-6wt.%Al-5wt.%Mo metallic overlay coating (Ultrabond 51000) and 70wt.%ZrO_2-30wt.%CaO ceramic top coating (MetaCeram 28085) were deposited onto an AISI 304 stainless steel substrate by oxyfuel gas powder spraying, using a Castolin DS 8000 gun. Further details about specimen's preparation have been published elsewhere [5]. Some of the coated materials were heat treated in air at 800 °C for 48, 120 and 288 h.

Metallographic samples from the as-sprayed and oxidized coatings were prepared in the longitudinal section, parallel to the spraying direction. The microstructure was analysed in a Philips

XL30 environmental scanning electron microscope (ESEM) equipped with energy dispersive X ray microanalysis (EDX).

The porosity of the ceramic top coat and the thickness of the different layers comprising the TBC have been analysed using an image analysis software Image Pro-Plus 4.5.

X ray diffraction (XRD) analysis was performed in a Philips PW3040/00 X'Pert MPD/MRD diffractometer using Cu-K$_\alpha$. Patterns were obtained from the top surface of the as-sprayed and oxidized coatings. In addition, oxidized samples were mechanically grinded to obtain XRD traces at 50, 100 and 150 μm deep from the surface to analyse the structural changes during oxidation near the ZrO$_2$-CaO/NiAlMo interface.

Transmission electron microscopy (TEM) samples in cross sections, from the as-sprayed and heat-treated coatings, were prepared following the guidelines proposed by Nagata *et al.* [6]. These specimens were examined using a Philips TECNAI 20 TEM employing a combination of bright field (BF) and centered dark field (CDF) imaging, as well as selected area (SADP) and nano-beam electron diffraction (NBDP) patterns analysis. EDX microanalysis was also carried on in the TEM.

Results and discussion

Microstructure of the as-sprayed materials

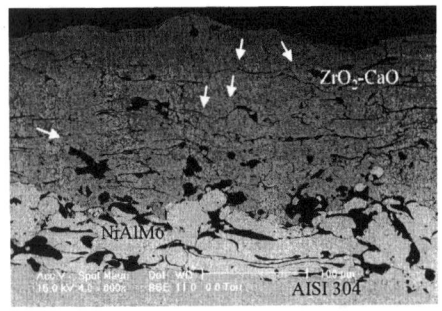

Fig. 1. Back scattered electron image of a typical cross section of the as-sprayed material. The arrows point to the pores and cracks of the ceramic layer.

The coating comprised a 60 μm width metallic bond coating and a 160 μm width ceramic top coating. The microstructure of both coatings was inhomogeneous, containing thin splats associated with the deposition of individual molten droplets with porosity between splats and un-molten particles ≈15 μm in diameter. Porosity in the ceramic top coating was analysed quantitatively and it was 16.8 ± 2.1%.

The bond coating microstructure is mainly formed by Ni grains which contain Al and Mo. TEM analysis identified these grains as γ-Ni (Fm3m space group, a=0.352 nm). γ'-Ni$_3$Al (ordered L1$_2$ structure coherent with γ) phase grew as small precipitates (~10 nm) inside the γ-Ni grains (Fig. 2a). Dark areas rich in Al and O are also observed around the Ni grains in Fig. 1. This suggests the formation of Al$_2$O$_3$ during spraying. TEM analysis disclosed polycrystalline zones (~15 μm) formed by γ-Al$_2$O$_3$ grains (Fd3m space group, a=0.791 nm) of ~300 nm in size (Fig. 2b). γ-Al$_2$O$_3$ is a metastable phase [3] which formed by Al oxidation during oxyfuel gas powder spraying.

TEM investigations showed CaZrO$_3$ (Pnma space group, a=0.55912 nm, b=0.80171 nm and c=0.57616 nm) as the main constituent of the ceramic coating, formed by grains typically ~300-500 nm in size. XRD traces also evidenced a minority presence of cubic ZrO$_2$ (Fm3m space group, a=0.52 nm). These results were similar to those presented in Fig. 4 for the surface of the oxidized samples.

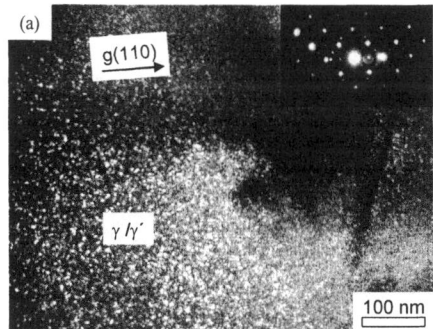
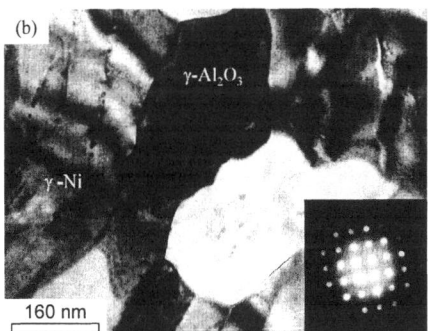

Fig. 2. TEM images of the NiAlMo bond coat: (a) CDF image showing γ'-Ni₃Al grown within γ-Ni grain. The SADP shows the L1₂ reflections characteristic of γ' and the image was formed with one of these reflections. (b) BF image and the corresponding SADP (B=<100>) of γ-Al₂O₃ close to a γ-Ni grain.

Microstructure of the oxidized materials

During thermal oxidation, a TGO layer was formed between NiAlMo bond coat and ceramic top coating (Fig. 3). The thickness of this TGO increased from 12 μm, for 48 hours treatment, to 22 μm, for 288 hours treatment.

Fig. 4 shows XRD traces from a TBC heat treated for 120 hours obtained at different depths from surface. The TBC's surface showed peaks associated with the orthorhombic $CaZrO_3$ and cubic ZrO_2, observed as well in the as-sprayed samples. The XRD trace performed at 150 μm deep from surface exhibited peaks associated to the oxides comprising the TGO: NiO (Fm3m space group, a=0.417 nm), MoO_2 (P21/c space group, a= 0.561 nm, b= 0.486 nm, c=0.563 nm and β=120.9°) and $NiAl_2O_4$ spinels (Fd3m space group, a= 0.805 nm). The spinel (311) peak is superimposed to the (111) NiO peak and an asymmetric peak centred at 37° is observed. The same happens at 45°, where the spinel (400) peak is superimposed to the $CaZrO_3$ (202) peak.

TEM analysis of the TGO showed $NiAl_2O_4$ spinels ~ 2μm in size located close to the bond coat (Fig. 5a). NiO was also observed close to the ceramic interface and they grew as small grains ~250 nm in size (Fig. 5b).

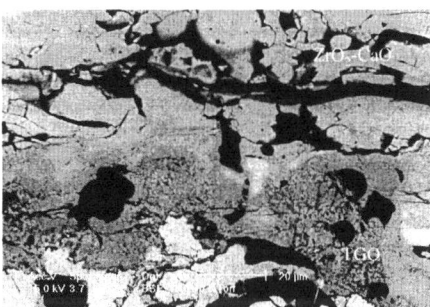

Fig. 3. Back scattered electron image of the TBC oxidized at 800°C for 288 h. The TGO is clearly observed.

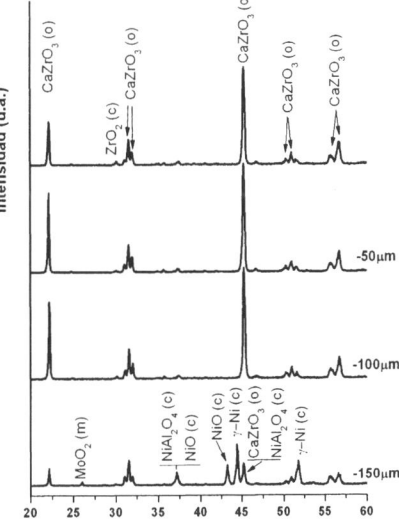

Fig.4. XRD traces of the TBC treated at 800°C for 120 hours obtained at different depths from surface.

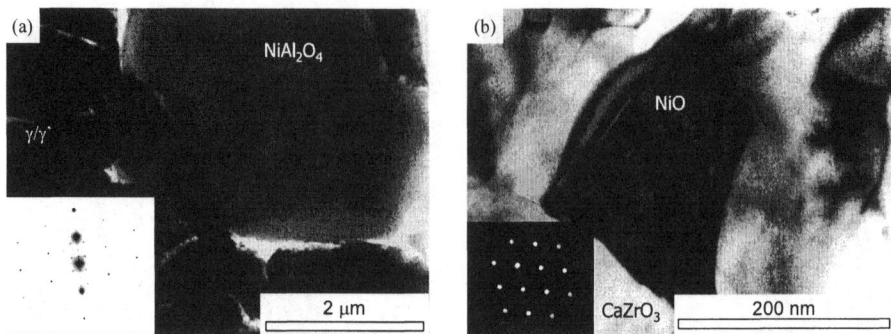

Fig.5. BF images of the TGO associated to the oxidized TBC at 800°C for 48 h: (a) NiAl₂O₄ spinel grain formed close to the bond coat. SADP with B=<102> is included, (b) NiO grains formed close to the ceramic coating. NBDP with B=<0$\overline{1}$1> is included.

The low Al content (6%) in the bond coating was insufficient to form a continuous and uniform Al_2O_3 layer. This is detrimental for the oxidation behaviour of this system, as the oxygen diffuses to the metallic substrate. The oxidation behaviour could be improved increasing the Al content [7] in the overlay coating. Recent works have corroborated the improvement in the oxidation behaviour of NiCoCrAlY coatings processed by plasma spraying when the substrate is previously aluminized [8].

Conclusions

The bond coat of the TBC was mainly formed by γ-Ni. γ'-Ni₃Al grew inside γ grains as small precipitates ~10 nm in size. Most of the Al was oxidized during thermal spraying leading the formation of γ-Al_2O_3 in polycrystalline areas (~ 20 μm) comprising grains of 300 nm in size.

During isothermal oxidation, a TGO was formed. This layer includes a mixture of Al, Ni and Mo oxides. In no case a continuous and uniform Al_2O_3 layer was observed due to the low content of Al in the bond coat. These coatings do not guarantee the protection of metallic substrates at 800°C for long exposure times in oxidizing atmospheres.

Acknowledgements

The authors would like to thank *Instituto de Cerámica y Vidrio (CSIC)* for supply the materials, and the Spanish government MCYT through grant MAT 2001-1123-C03-03 for financial support.

References

[1] R. A. Miller, J. Thermal Spray Technol., Vol. 6, N° 1, 35-42 (1997).

[2] S. Bose, J. DeMasi-Marcin, J. Thermal Spray Technol., Vol. 6, N° 1, 99-104 (1997).

[3] W. E. Lee, W. M. Rainforth, in: Ceramic microstructures, edited by Chapman and Hall, London (1994).

[4] V. Sergo, D. R. Clarke, J. Am. Ceram. Soc., Vol. 81, 3237-3242 (1998).

[5] J. Gómez-García, P. Poza, V. Utrilla, Bol. Soc. Esp. Ceram. V., Vol. 45, N° 2, 70-74 (2006).

[6] F. Nagata, T. Shimotsu, H. Kakibayashi, in: Approach to Atomic Structure (High Resolution Transmission Electron Microscopy), Supplemental Issue of Hitachi Instrument News, March 1989.

[7] M. J. Stiger, M. N. Yanar, M. G. Topping, F. S. Pettit, G. H. Meier, Zeitschrift fur Metallkunde, Vol. 90, N° 12, 1069-1078 (1999).

[8] P. Poza, P. S. Grant, Surf. Coat. Tech., accepted.

Key Engineering Materials Vol. 333 (2007) pp 273-276
online at http://www.scientific.net
© 2007 Trans Tech Publications, Switzerland

Development of Thin Ceramic Coatings for the Protection against Temperature and Stress Induced Rumpling of the Metal Surface of Turbine Blades

Bernd Baufeld[a] and Omer van der Biest[b]

Katholieke Universiteit Leuven, MTM, Kasteelpark Arenberg 44, B-3001 Leuven, Belgium

[a]Bernd.Baufeld@mtm.kuleuven.be, [b]Omer.Vanderbiest@mtm.kuleuven.be

Keywords: ceramic coating; electrophoretic deposition; impulse excitation technique; Young's modulus; damping

Abstract. In order to obtain a protection against temperature and stress induced detrimental rumpling of the metal surface of turbine blades, thin ceramic coatings are suggested. As a cheap and fast method for the fabrication of a ceramic zirconia coating, electrophoretic deposition on a Ni based superalloy is described. Crack free, 0.15 mm thick coatings with homogenous morphology were obtained. The Young's modulus and the damping property of the ceramic coating, derived from the impulse excitation technique, are investigated as a function of the temperature up to 1000°C.

Introduction

In service turbine blades as used in gas turbines for aviation and stationary power production frequently experience extensive surface rumpling due to the severe cyclic thermal and mechanical loading. This surface rumpling decreases the aerodynamic efficiency of the turbine and in addition may lead to early failure of the component [1, 2]. It was observed that thin ceramic thermal barrier coatings prepared by electron beam physical vapor deposition prevent this surface rumpling [2]. However, this coating process is expensive and time consuming.

Electrophoretic deposition (EPD), on the other side, is known to be a cheaper and faster deposition method [3, 4]. However, due to crack development during drying and sintering in the case of thicker coatings, only relatively thin ceramic coatings can be obtained. While this is a drawback for most applications, it is of no disadvantage for the rumpling prevention. Any additional mass on the turbine blades reduces the efficiency of the turbine. Hence, as long as the protective function is given, as thin as possible coatings are preferential.

EPD, comparable to the galvanic deposition for metal atoms, is a method to deposit ceramic powder, which is homogenously suspended in a liquid, via a constant electric field on a substrate acting as an electrode. This is possible, since the ceramic powder acquires an electric surface charge in the liquid. After coagulation of the powder particles on the substrate, the substrate is removed from the suspension, dried and sintered in order to obtain a stable ceramic coating.

In the following, the EPD of a zirconia coating is illustrated and the structure of this coating is described. Furthermore, the determination of one important material property, the Young's modulus, applying the impulse excitation technique (IET) is explained.

Experimental

Specimen preparation. The substrate material was polycrystalline Ni based superalloy IN625 and specimens were prepared in the form of cylindrical rods, with a length of 65 mm and a diameter of 6 mm.

Partially stabilized zirconia powder has been used (5 mol% Y_2O_3 stabilized grade Melox 5Y XZO 99.8%) with the addition of 0.75 wt% cobalt oxide nanopowder as a sintering aid (Aldrich Cobalt (II, III) oxide 99.8%). This powder was mixed and ball milled with zirconia balls in methyl-ethyl keton (MEK) with a multidirectional mixer (Turbula type) for one day. For each EPD session fresh suspensions were prepared from this powder/MEK mixture by adding the respective components in order to obtain a suspension consisting of 80 vol% MEK, 20 vol% n-butylamine, 1 wt% nitro-cellulose, and 63 g/l powder. The suspensions were first mechanically stirred for 10 minutes, ultrasonically stirred for 15 minutes, and finally again mechanically stirred for further 10 minutes.

The EPD cell consisted of the substrate as one electrode, which was placed horizontally in the

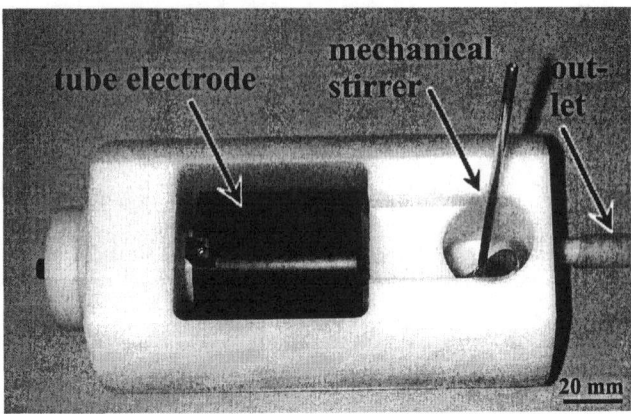

Fig. 1 Top view of the EPD cell for cylindrical specimens, positioned horizontally in the center of the cylindrical tube electrode consisting of conductive plastic. The specimen is not visible, since it is located in the center of the tube.

center of a cylindrical conductive plastic tube with a diameter of 56 mm and a length of 55 mm, which acted as the second electrode (Fig. 1). This set-up was installed in a non-conductive container with a fluid volume of 400 ml. During EPD, the suspension was subjected to further mechanical stirring and for a specified time to an electric field with the strength of 17 V/mm. After EPD, the container was drained and the specimens dried in air.

The green specimens were sintered in a conventional resistance furnace in hydrogen atmosphere at 1200°C for 6 h.

Density measurements of sintered and spalled ceramic coatings were performed in ethanol, based on the Archimedes principle (Sartorius Balance).

Impulse excitation technique. IET is a standard test method for the assessment of the dynamic Young's modulus of isotropic homogeneous materials [5]. Fundamentally, the mechanically excited resonance frequency of a specimen is measured. With the known specimen geometries, the dynamic Young's modulus can be calculated, which is proportional to the square of the resonance frequency. In the case of coated cylindrical specimens it has been shown by Schrooten et al. [6] that the Young's modulus of the coating $E_{coating}$ can be derived from the average modulus of the whole system E_{total}, provided coating thickness t, radius of the substrate R, and substrate stiffness $E_{substrate}$ are known:

$$E_{coating} = \frac{E_{total}(R+t)^4 - E_{substrate}R^4}{(R+t)^4 - R^4}$$ (1)

The high temperature IET tests were performed in a graphite furnace HTVP-1750C in Ar atmosphere. All tests were analyzed by the RFDA software (both IMCE, Diepenbeek, Belgium). More details about this set up can be found in [7].

Results

Electrophoretic deposition. EPD was performed for times between 30 and 300 s. The yield increased with deposition time (Fig. 3). For the longer deposition times, however, the green coating exhibited multiple cracks after drying. Yet, deposition times below 1 minute resulted in crack free coatings. Subsequent sintering of such crack free coatings did not create cracks.

The cross-section (Fig. 2) proofs the good coating quality for the case of a 50 s long EPD. The surface is smooth and no cracks are visible. It has a high porosity, but the pores are evenly distributed, and not many agglomerates exist. The density of the coating after sintering was determined to be 2.4 g/mm^3, which corresponds to a porosity of about 60 %. The coverage of the substrate surface is good and the adherence of the coating is sufficient, allowing mechanical cutting of the specimen. The sintered coating thickness after 50 s EPD was 0.15 mm.

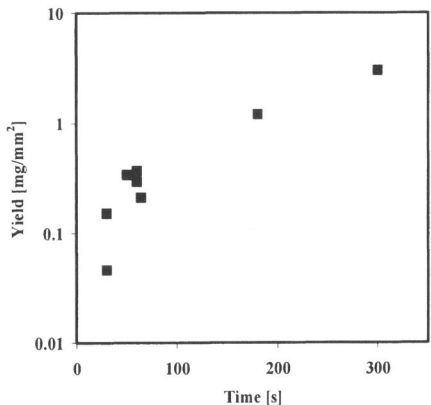

Fig. 3 Green EPD yield, i. e. the weight change between uncoated and green coated specimen, in dependence on the deposition time.

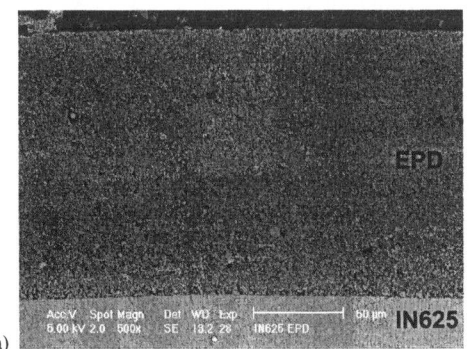
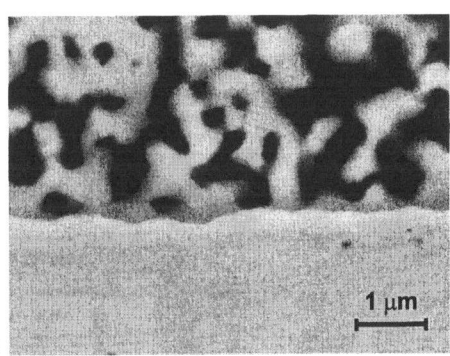

Fig. 2   Cross section of EPD coating (deposition time 50 s) after sintering, giving an overview (a) and detail of the interface between coating and substrate.

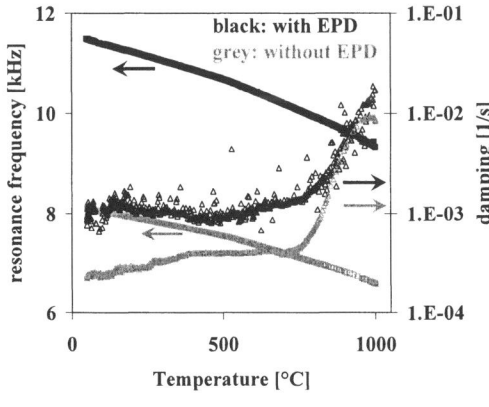
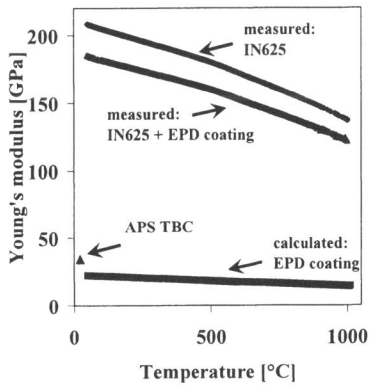

Fig. 4   Temperature dependence of the resonance frequency and damping for a specimen with a 0.15 mm thick zirconia EPD coating and of an uncoated IN625 specimen (a), and the temperature dependence of the resultant Young's modulus of the coated system, the substrate and the coating, the latter calculated using Eq. 1. For comparison, the Young's modulus of an air plasma sprayed zirconia coating at room temperature is given [8].

Impulse excitation technique.   Fig. 4a shows the temperature dependence of the resonance frequency and the damping of a coated and uncoated specimen as measured by IET. From these measurements the Young's modulus of the coated and the uncoated system was derived (Fig. 4b). Both Young's moduli decrease significantly with increasing temperature. The Young's modulus of the ceramic coating, also shown in Fig. 4b, was calculated using Eq. 1. It is much smaller (22 GPa at room temperature), and decreases only weakly with increasing temperature (18 GPa at 1000°C). While the Young's modulus of the EPD coating is much smaller than the one of dense zirconia, it is in the same order as reported for air plasma sprayed zirconia coatings [8].

At temperatures below 850°C, the damping of the coated specimen is significantly higher than for the uncoated specimen, supposedly due to energy dissipating processes within the porous ceramic coating. An increased damping is a beneficial property, since this may reduce vibration or noise in a turbine. For example, air plasma sprayed zirconia coatings designed for this task were studied by Yu et. al. [9].

Conclusions

The fabrication of zirconia coatings on IN625 Ni-based superalloy was successful. For coatings thinner than 0.15 mm no cracks were observed and a homogenous morphology was achieved. Due to the high porosity of about 60 %, the ceramic coating proved to have a relatively small Young's modulus of 22 GPa at room temperature, with almost no decrease with increasing temperature. A beneficial increase of damping for the coated system compared to the substrate was observed.

Further work has to be done to proof the adherence of the coating under demanding thermal and mechanical loading and to show its effectiveness to suppress surface rumpling under these conditions.

Acknowledgments.   B. Baufeld acknowledges the individual Marie-Curie fellowship of the European Commission Nr. MEIF-CT-2005-010277.

References

[1] R. C. Pennefather, D. H. Boone, Surf. Coat. Tech. 76-77 (1995), p. 47.

[2] M. Bartsch, B. Baufeld, edited by P. D. Portella, H. Sehitoglu, K. Hatanaka, Fifth International Conference on Low Cycle Fatigue Berlin, 2003, p. 183.

[3] O. Van der Biest, L. Vanderperre, An. Rev. Mat. Sci. 29 (1999), p. 327.

[4] A. R. Boccaccini, I. Zhitomirsky, Cur. Op. Sol. St. Mat. Sci. 6 (2002), p. 251.

[5] ASTM, E 1876-99 (1999), p. 1075.

[6] J. Schrooten, G. Roebben, J. A. Helsen, Scr. Mat. 41 (1999), p. 1047.

[7] G. Roebben, B. Bollen, A. Brebels, J. Van Humbeeck, O. Van der Biest, Rev. Sci. Instrum. 68 (1997), p. 4511.

[8] J. S. Wallace, J. Ilavsky, J. Therm. Spr. Tech. 7 (1998), p. 521.

[9] L. Yu, Y. Ma, C. Zhou, H. Xu, Mat. Sci. Eng. A 408 (2005), p. 42.

Key Engineering Materials Vol. 333 (2007) pp 277-280
online at http://www.scientific.net
© 2007 Trans Tech Publications, Switzerland

Nucleation of Shear Bands on EB-PVD Thermal Barriers Coatings under Hertzian Indentations

Y. Gaillard[1, a], E. Jimenez-Piqué[1,b], M. Bartsch[2,c] and M. Anglada[1,d]

[1]Department of Materials Science and Metallurgy (CMEM), Universitat Politécnica de Catalunya, Avda. Diagonal 647 (ETSEIB), 08028 Barcelona, Spain.

[2]German Aerospace Center, Linder Höhe, 51147, Köln, Germany.

[a]yves.gaillard@upc.edu, [b]emilio.jimenez@upc.edu, [c]marion.bartsch@dlr.de, [d]marc.j.anglada@upc.edu

Keywords. Shear band, EB-PVD, thermal barriers coatings, Hertzian contact.

Abstract. In this paper the formation of shear bands in columnar EB-PVD thermal barriers coatings is studied. In particular, critical parameters of nucleation of shear bands, such as contact pressure and initiation of cracks in the columns, are extracted from the experimental results. The pertinence of these parameters is discussed respecting to the stress field induced in the material during the indentation.

Introduction

Yttria stabilised zirconia (YSZ) thermal barriers coatings (TBC) deposited by electron-beam physical vapour deposition (EB-PVD) exhibit a columnar microstructure together with high porosity [1]. Particularly these coatings show very stiff columns but also very poor adhesion between these columns due to the inter-columnar porosity. Under local (point) loading, such as indentation, impact test or foreign object damage (FOD), these materials show different deformation mechanisms. The first one which is activated is the sinking-in between the columns [2] which transmits the stress directly to the substrate. This mechanism appears is very important in the case of hard coatings deposited on soft substrate, as the substrate is susceptible to deform plastically before the coating [3]. The second mechanism is buckling of columns to accommodate the plastic deformation due to the local loading inside the ceramic layer [4], this deformation mechanism produces fracture shear band, formed by an alignment of buckled columns [5-6]. The third and last mechanism is the classical plastic deformation of the YSZ inside the columns. This type of deformation has been only reported at high temperature [6] or under very high local stresses such as indentations performed with sharp indenter [2, 7]. This paper is devoted to the study of the shear band formation in yttria stabilised zirconia EB-PVD TBC. The nucleation of this particular defect in the ceramic layer appears to be fundamental as this nucleation is characteristic of the failure of the TBCs [8]. In fact, when this shear band propagates and reaches the interface, it provokes the delamination of the ceramic layer, which exposes the substrate to a potential aggressive environment. Spherical indentations, at room temperature, were performed using indenters with large radius of curvature in order to avoid the formation in the ceramic top coat of any type of plastic zone.

Experimental

YSZ EB-PVD coatings containing 7-8%wt Y has been studied. These tri-layered materials are composed of an intermetallic bond coat (BC) (exhibiting a thickness of 60 microns) deposited on a substrate, here a super-alloy INCONEL 625, and a ceramic top coat (with a 200 microns thickness) deposited on the bond coat. The columnar microstructure of this ceramic layer has been described elsewhere [7, 9]. The porosity of this type of coating has two origins: a) inter-columnar porosity consisting of a space between columns; b) intra-columnar nano-sized porosity. The presence of dendrites localised at the column boundaries contribute also to the increase of porosity. The surface of the samples was polished in order to eliminate the as-received roughness of several micrometers. Hertzian indentations were performed using a REVETEST scratch tester (CSM Switzerland) and a FRANCK indenter. The combined used of these two equipments allows to perform indentations in a large range of loads, between 10 to 2000 N using spherical indenters with radius of curvature of 0.5, 0.75, and 1.25 mm. For each residual imprint, plane and cross-sectional observations were performed to characterise the microstructure of deformation.

Results and discussion

Table 1: Apparition load range of the shear band and corresponding indentation stress.

Curvature radius of the indenter	Apparition load range	Corresponding indentation stress
0.5 mm	49 – 98 N	2 - 2.3 GPa
0.75 mm	98 - 147 N	2 - 2.2 GPa
1.25 mm	196 - 294 N	1.9 - 2.1 GPa

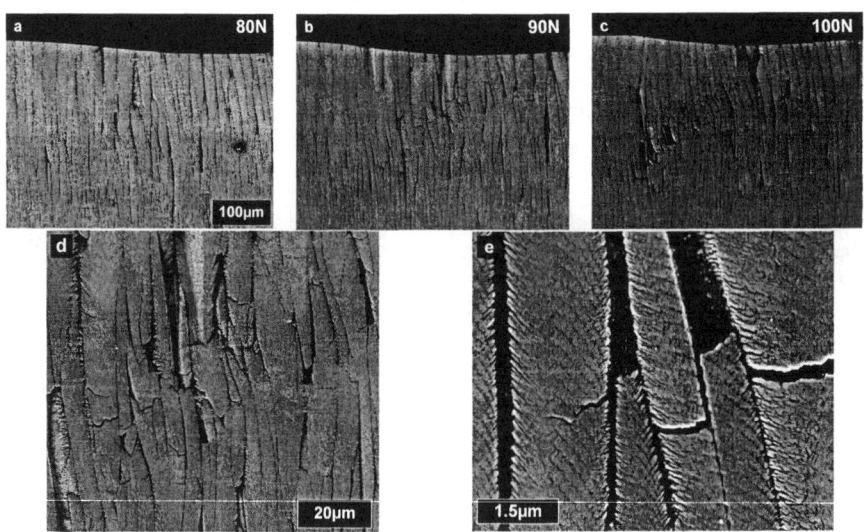

Fig. 1: (a, b, c) Cross-section observation by SEM of the deformation microstructure produced by the 0.5mm indenter at 80, 90 and 100N. (d) Magnification of the cracks zone observed in (b), and (e) crack path in the columns at the end of a shear band obtained at 120N.

Due to the poor adhesion between columns, the application of a local loading in the ceramic top coat always results in the formation of a residual imprint, even for the lowest load. For these low loads indentations, the ceramic layer is deformed elastically, and the deformation is produced at the BC, and no trace of shear band or plastic zone in the ceramic columns is observed. This behaviour is similar to the observed in the case of hard film deposited on soft substrate [3]. Upon further loading, a critical load is reached when a shear band is nucleated in the top coat. This critical load depends of the radius of the indenter. The load range apparition for each indenter has been determined by cross-sectional observations, and the results are summarized in table 1. The indentation stress has been calculated from the residual radius of the imprint [10]. It appears that, although the critical load values increase with the radius of curvature, the corresponding stress is nearly constant and around 2 GPa. Therefore, contact stress appears to be a parameter controlling the nucleation of shear band under indentation.

By performing indentations in the load range of the shear band apparition for the 0.5mm indenter, it has been observed that the deformation of the top coat remains elastic until 80N (see Fig. 1a). For 90N load, the formation of a crack network under the contact area is clearly observed (see Fig. 1b). Finally, for a load of 100N the apparition of a shear band is clearly observed (Fig. 1c). It is important to note that for a 90N load, isolated segments of columns are broken and inclined but the shear band is not yet well developed. In the magnification of this region shown on Fig. 1d, these inclined segments are clearly observed, and also the origin of the fracture, localised at the columns boundaries. On Fig. 1e, the extremity of a shear band obtained under an indentation performed at 120N is shown. It is clear that the cracks are initiated at the tip end of the dendrites. This remark appears to be crucial as it allows understanding the processes of nucleation and propagation of the shear bands. The dendrites play the role of pre-existing cracks that are susceptible to open during the indentation. Respecting to the dendrites geometry (see Fig. 2), the opening of these cracks can occur in mode I, mode II or in mixed mode. Depending on the mode of deformation, the component of the stress field responsible of crack propagation on the dendrites is different. Following the scheme presented on Fig. 2, it is seen that the component of the stress field which act on the crack opening on mode I is $\sigma_{x'x'}$, while $\tau_{x'z'}$ acts on the mode II. As the ceramic behaves elastically before the nucleation of the shear bands and assuming that this thick layer behaves as a bulk material*, it is possible to determine the stress field induced by the indentation in the ceramic layer before the nucleation [10, 11]. In particular, $\sigma_{x'x'}$ and $\tau_{x'z'}$ can be calculated. The contour plots of these two components are represented on Fig. 2 for a dendrite having an angle of 45°, a common angle observed in our coatings. Comparing the shape and the magnitude of these stresses components with the experimental results two important remarks can be made: first the values of the maximum stresses are proportional to the mean contact pressure ($1.3P_m$ and $0.46P_m$ respectively for $\sigma_{x'x'}$ and $\tau_{x'z'}$) which is coherent with the fact that the shear band is nucleated at a given indentation stress. Secondly, the experimental observations have shown that the initial cracks are formed in depth under the contact zone and not directly at the surface of the material indicating a clear effect of the $\tau_{x'z'}$ component which is maximal in depth along the indentation axis. However, the propagation of the shear band in depth appears to be closely related to the particular shape the $\sigma_{x'x'}$ component.

---

* This assumption is supported by the very high thickness of the ceramic layer but does not take into account the columnar structure

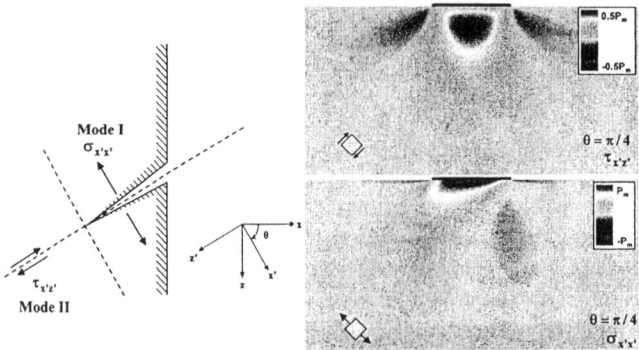

Fig. 2: 2D scheme of the typical dendrites observed on the EB-PVD coatings in the columns boundaries and contour plot of the $\sigma_{x'x'}$ and $\tau_{x'z'}$ components of the stress field represented in plane perpendicular to the indented surface.

Summary

The nucleation of shear band in YSZ columnar thermal barrier coatings has been studied. It has been shown that this nucleation is strongly dependant of the contact pressure in accordance with the stress induced in the material before the nucleation. Furthermore the fundamental role of the dendrite as initial defect has been also demonstrated. The initiation of cracks at the tip end of the dendrites appears to be critical to initiate trans-columnar cracks but also in the propagation of the shear band inside the ceramic layer.

Acknowledgments Work supported by the European Community's Human Potential Programme under contract HPRN-CT-2002-00203, [SICMAC]. Y.G. acknowledges the financial support provided through the European Community's Human Potential Programme under contract HPRN-CT-2002-00203, [SICMAC].

References

[1] A.G. Evans, D.R. Mumm, J.W. Hutchinson, G.H. Meier, F.S. Pettit, *Progress in material science*, 46 (2001), p. 505.
[2] R. G. Wellman, A. Dyer and J. R. Nicholls *Surface and Coatings Technology*, 176 (2004), p. 253.
[3] S. Bhowmick, Z-H. Xie, M. Hoffman,V. Jayaram, S.K. Biswas, *Journal of material research*, 19 (2004), p. 2616.
[4] R.G. Wellman, M.J. Deakin, J.R. Nicholls, *Wear*, 258 (2005), p. 349.
[5] X. Chen, R. Wang, N. Yao, A.G. Evans, J.W. Hutchinson, R.W. Bruce, *Material science and Engineering A*, 352 (2003), p. 221.
[6] M. Watanabe, C. Mercer, C.G. Levi, A.G. Evans, *Acta Materialia*, 52 (2004), 1479.
[7] Y. Gaillard, E. Jiménez-Piqué, and M. Anglada, *Phil. Mag.*, (2006), in press
[8] A.G. Evans., N.A. Fleck, S. Faulhaber, N. Vermaak, M. Maloney, R. Darolia, *Wear*, 260 (2006), p. 886.
[9] U. Schulz, M. Schmucker, *Material science and Engineering A*, 276 (2000), p.1.
[10] B.R. Lawn, *Journal of the American ceramic society*, 81 (1998), p. 1977.
[11] M.T. Huber, Ann. Phys. **43** (1904), p.153.

Key Engineering Materials Vol. 333 (2007) pp 281-284
online at http://www.scientific.net
© 2007 Trans Tech Publications, Switzerland

Impression Creep in TBC and Advanced Ceramics Materials

Františka Dorčáková[a], Vít Jan[b], Lucia Hegedűsová[c], Ján Dusza[d]

Institute of Materials Research of the Slovak Academy of Sciences
Watsonova 47, 043 53 Košice, Slovakia

[a]fdorcakova@imr.saske.sk, [b]jan@fme.vutbr.cz, [c]lhegedusova@imr.saske.sk,
[d]jdusza@imr.saske.sk

Keywords: Indentation creep, activation energy, TBC topcoat, Y-ZrO$_2$ layer.

Abstract. The indentation creep of free-standing Y-ZrO$_2$ layer and 20Sc-60Si-20Mg-80O-20N oxynitride glass has been investigated. Creep experiment has been performed with flat cylindrical indenter (hot pressed SiC) in the temperature range from 860 °C to 1300 °C at the loads from 20 to 100 MPa. The strain-time relationship was registered and the creep exponent and activation energy of creep have been calculated. The microstructure changes have been observed and documented. Viscosity as a function of temperature and the glass transition temperature (T_g) were determined in oxynitride glass and compared with values from compressive creep.

Introduction

For the investigation of time dependent mechanical properties of ceramics materials and glasses (e.g. creep, viscous flow) the four-point bending test respectively compression/tensile test are generally applied. Although the phenomenological interpretation of these tests is relatively simple, the preparation of relevant samples is difficult and expensive, especially in the case of advanced structural ceramics. The indentation methods have also been applied successfully for the investigation of the creep process and for the determination of the creep parameters and for viscosity measurements on glasses [1-3]. Indentation creep tests are technically simpler than the four-point bending or tensile test, but the correct physical interpretation of the deformation process taking place during impression tests is very intricate, because the stress field under the indenter is multiaxial and inhomogeneous.

The aim of the current work is a short description of our indentation creep test equipment and demonstration of some results obtained from its application to free-standing Y-ZrO2 layer and 20Sc-60Si-20Mg-80O-20N oxynitride glass. In the case of oxynitride glass, comparing the results obtained by different methods (indentation resp. compressive creep) it will be shown that indentation measurements can also be applied to the determination of the viscosity of glasses.

Theoretical background

The indentation creep test performed with flat ended cylindrical indenter was originally proposed by Yu and Li in the 70s for investigation of creep properties of materials [4]. During indentation creep measurements due to a constant load, F, an indenter is pressed into the surface of the sample. The indentation depth, h, of the indenter is registered as a function of the elapsed time, t, so the velocity of the indentation, dh/dt, can be calculated. Applying indentation test, an important question is that how the equivalent tensile stress, σ, and equivalent strain rate, $\dot{\varepsilon}$, can be given from other parameters characterizing the indentation measurements. So far, for cylindrical indenter these equivalent parameters have been calculated by the following formulas [2,5]:

$$\sigma = c_1\, p_{ind} \quad \text{and} \quad \dot{\varepsilon} = c_2\, \dot{\varepsilon}_{ind} \tag{1}$$

where c_1 and c_2 are constants, p_{ind} is the pressure under the indenter defined as:

$$p_{ind} = \frac{F}{A_p} = \frac{4F}{\pi d^2} \tag{2}$$

where A_p is the projected contact surface and d is the diameter of the cylinder and $\dot{\varepsilon}_{ind}$ is given as

$$\dot{\varepsilon}_{ind} = \frac{dh/dt}{d} \qquad (3)$$

For the analytical determination of $\dot{\varepsilon}_{ind}$ the knowledge of the h-t relationship is required.

The viscosity of glass can be expressed from tensile stress, σ, and strain rate, $\dot{\varepsilon}$ [6]:

$$\eta = \frac{\sigma}{2(1+\upsilon)\dot{\varepsilon}} \qquad (4)$$

where υ is the Poisson's ratio. The value of $\upsilon = 0.3$, which was reported for similar glasses in the literature [6], was used in Eq. (4). The glass transition temperature, T_g, was determined as the average temperature corresponding to the range of viscosities 10^{12} Pa.s and $10^{12.6}$ Pa.s from the temperature dependence of the corresponding viscosities. The activation energies were calculated from the linear fits of the Arrhenius dependence of the glass viscosity on inverse temperature.

Experimental materials and methods

Free-standing Y-ZrO$_2$ layer were prepared at the IMR DLR Köln, Germany. Normally, this ceramic material (partially stabilized zirconia with 7 wt.% yttria) has been applied on the metal substrate surface as a thermal barrier coating and the ceramic layer is usually about 200 μm thick. For the creep behavior investigation a thicker sample consist of several consecutive layers of Y-ZrO$_2$ were prepared by the EB-PVD method as a free standing material without metal substrate [7]. The overall thickness of these samples was about 1 mm.

Oxynitride glass was prepared at the Institut für Keramik im Maschinenbau Universität of Karlsruhe, Germany, from the mixture of powders of SiO$_2$, MgO, α-Si$_3$N$_4$ and Sc$_2$O$_3$ [6]. The amount of silicon nitride powder was calculated in such a way that the resulting nitrogen content was 20 e/o. Specimens with the thickness 2 mm were cut and polished for indentation creep, and specimens with the size of approximately 2.5 mm x 2.5 mm x 7.5 mm were used for compressive creep experiments.

Creep experiments were performed in the modified bending creep furnace (Model HTTF 2, SFL Ltd./Instron, Inc., UK) in air. Flat cylindrical indenter with diameter of 2 mm was used. Monolithic SiC was used for the preparation of indenter with the creep resistance much higher than tested materials. Indentation depth was measured as a difference between two LVDT elements and recorded as a function of time by a PC. Y-ZrO$_2$ samples were tested in the temperature range from 1100°C to 1300°C at the loads 30, 50 and 70 MPa, oxynitride glass was investigated in the temperature range from 850°C up to 910°C under stresses of 20 MPa. The viscosity of Sc-Si-Mg-O-N glass has been determined from the compressive creep experiment too. Compressive creep has been performed in the temperature range from 840°C up to 910°C under stresses of 10 MPa.

Results and discussion

Free-standing Y-ZrO$_2$ layers after creep experiments are shown at the Fig. 1a) and b). Three different structural changes were possible to identify comparing the undeformed and deformed specimens [7]. *First*, the lower part of each layer consisting of fine crystals with high porosity was compressed and the crystals sintered. The *second* mechanism was the bending of the lower ends of the bigger crystals above this newly sintered layer. *Third*, the long stiff crystals did resist to the load very well, however, when they were not exactly parallel with the applied force, the crystals slip on each other and started to lay down resulting in further overall deformation of the specimen.

The indentation depth – time curves of free-standing Y-ZrO$_2$ are illustrated in Fig. 2a). Figure shows series of indentation curves at different temperatures under constant load. Indentation curves contain two evident stages characterizing the primary and secondary creep. The maximum increase of the indentation rate is about one order of magnitude for the temperature interval from 1100°C to 1300°C. On the other hand, the indentation rate of oxynitride glass increases more than two orders of magnitude at the narrow temperature interval from 870°C to 910°C (Fig. 2b)). Stage of primary creep isn't evident for curves of glass with the high indentation rate.

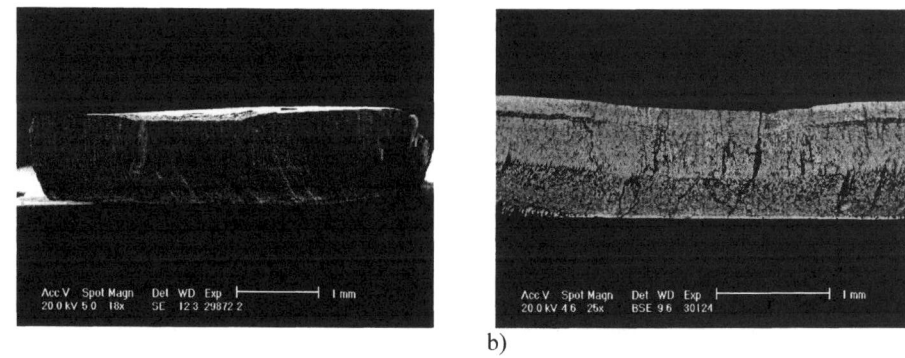

a)   b)

Fig. 1: a) Microstructure of free-standing Y-ZrO$_2$ layers, b) polished section of deformed specimen, cracks and densifying of the base sub-layers is clearly visible

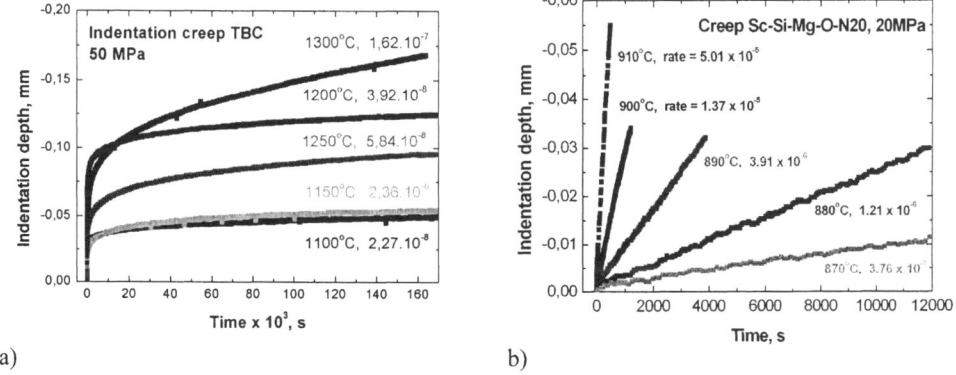

a)   b)

Fig. 2: Series of the indentation creep curves of the studied materials

Based on the experimental results the activation energies and stress exponents have been calculated. For the case of free-standing Y-ZrO$_2$ layers (Fig. 3a) both the activation energies and stress exponents are lower comparing to the published values measured for the conventional zirconia measured using classical creep tests, ($n \sim$ 1-2, Q $\sim$ 460 kJ/mol for ordinary sintered 8YSZ) [8,9]. The activation energy Q $\sim$ 220kJ/mol, obtained for the investigated material, is lower than activation energies for grain boundary diffusion [8]. We assume that the difference in our n and Q values compared to the values measured using conventionally prepared materials can be explained by the different microstructure and maybe by the different testing procedure.

The comparison of indentation and compression viscosity of oxynitride glass is shown in the Fig. 3b). Viscosity and the glass transition temperature determined from the indentation creep experiments achieve approximately 1/2 order of magnitude in terms of viscosity and ~12°C in terms of T$_g$ than these determined from compression creep experiments. The activation energies obtained by the different methods are similar. It can be caused that the complete sample is loaded at the compressive creep, but only the subtle part of the sample under the indenter is loaded at the indentation creep. Unloaded volume of glassy sample puts up resistance for the viscous flow and results to higher viscosity whereas activation energy remains the same.

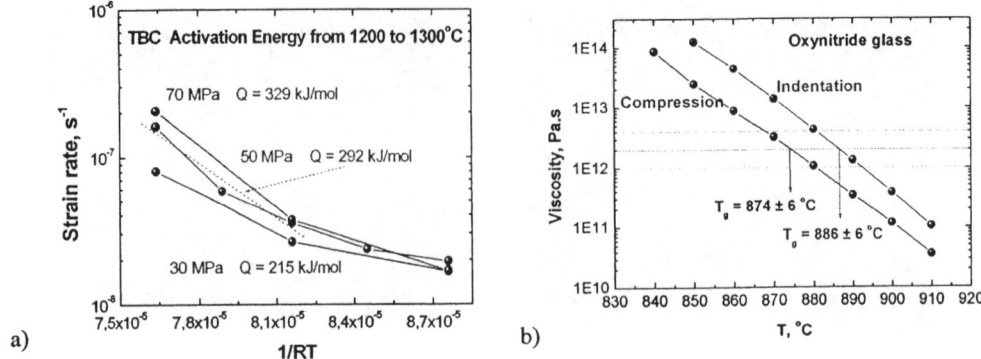

Fig. 3: a) Apparent activation energies of the free-standing Y-ZrO$_2$, b) viscosity of glass as a function of temperature in indentation and compression mode

Conclusion

It was shown that indentation creep measurements performed with cylindrical indenter can be successfully used for the determination of creep resistance of thin layers and viscosity of glasses.

Microstructure investigation and indentation creep tests of a free-standing samples of 8-YSZ EB-PVD deposited layers as TBC topcoats achieved the three mechanisms that undergo during the indentation creep test. The values of deformation exponents and activation energies are lower comparing to the published values of the conventional 8YSZ obtained using traditional creep test.

Viscosity values obtained on the glass from indentation tests are slightly higher than those determined by compressive creep method. It was also shown that the activation energy of the viscous flow can be assessed from the indentation measurements without the evaluation of the viscosity coefficient.

Acknowledgments

Work supported in part by the European Community's Human Potential Programme under contract HPRN-CT-2002-00203, [SICMAC], by NANOSMART, Centrum of Excellence of SAS, Slovak Grant Agency for Science via grant No. 2/4173/04 and by the Science and Technology Assistance Agency under the contract No. APVT – 51 – 049702.

References

[1] W.T. Han and M. Tomozava: J. Am. Ceram. Soc. Vol. 73 (1990), p. 3626.

[2] A.L. Yurkov: J. Mat. Sci. Letters Vol. 12 (1993), p. 767.

[3] G. Cseh, N.Q. Chinh, P. Tasnádi, P. Szommer and A. Juhász: J. Mat. Sci. Vol. 32 (1997), p. 1733.

[4] E.C. Yu and J.C.M. Li: J. Mat. Sci. Vol. 12 (1977), p. 2200.

[5] G. Cseh: Indentation creep for the investigation of high temperature plastic behavior, PhD thesis, Eötvös Loránd University (2000).

[6] F. Lofaj and F. Dorčáková: Metalurgija Vol. 42 (2003), p. 229.

[7] V. Jan, F. Dorčáková, J. Dusza and M. Bartsch, in: Deformation and Fracture in Structural PM Materials, edited by Ľ. Parilák and H. Danninger, ÚMV SAV Košice (2005).

[8] J. Wolfenstine: Journal of Power Sources Vol. 111 (2002), p. 183.

[9] A.H. Chokshi: Scripta Materialia Vol. 48 (2003), p. 791.

Key Engineering Materials Vol. 333 (2007) pp 285-288
online at http://www.scientific.net
© 2007 Trans Tech Publications, Switzerland

Ti – Al Intermetallic Layers Produced on Titanium Alloy by Duplex Method

P.Wieciński[1,a], H.Garbacz[1,b], M. Ossowski[1, c], T.Wierzchoń[1,d], K.J.Kurzydłowski[1,e]

[1]Warsaw University of Technology, Faculty of Materials Science and Engineering, Woloska 141, 02-507 Warsaw Poland

[a]degi4@wp.pl, [b]haga@inmat.pw.edu.pl, [c]ossomac@wp.pl [d]twierz@inmat.pw.edu.pl, [e]kjk@inmat.pw.edu.pl

Keywords: duplex treatment, Ti-Al intermetallic layer, composite layer, glow discharge assisted heat treatment, titanium alloy

Abstract. The paper presents the results concerning the microstructure of Ti – Al intermetallic layers produced on a TA6V titanium alloy by the "duplex method". This method combines vacuum evaporation coating of aluminum with glow discharge assisted heat treatment of the deposited films. It has been found that this combination of surface engineering techniques yields multi – layered films of the diffusive character. The films contain intermetallic phases from Ti – Al system which ensure a high microhardnes and good wear resistance. It is finally suggested that these properties can significantly widen application range of titanium alloy parts in aerospace.

Introduction

Titanium alloys have been used for industrial application for about 50 years. They are widely used in automobile, aircraft, chemical and food industries. Such a wide range of applications is related to the good mechanical and physical properties of titanium alloys: low density, a high relative strength, good creep resistance, and good corrosion behaviour in many environments. However, titanium alloys exhibit poor frictional wear resistance due to adhesive welds that are formed on their surface and the relatively high friction coefficient[1]. It has also been known that the mechanical properties of many titanium alloys rapidly decrease above 500°C because of their strong oxidation [2,3]. In order to improve wear resistance and to reduced absorption of oxygen, titanium and its alloys are subjected to various surface treatments.

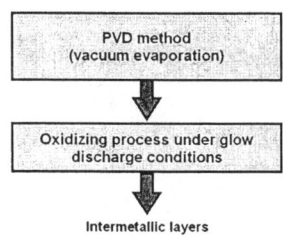

Fig. 1. Scheme of the duplex treatment

One of the surface engineering techniques which have been used for producing coatings on titanium alloys is the duplex method[4,5]. This method consists of the successive use of two (or more) surface engineering processes to produce composite surface layers. The good properties of the composite surface layers cannot be achieved by single surface methods. One of the techniques which appears to be most attractive is producing surface layers under glow discharge conditions (Fig.1). This treatment is widely used for nitriding and oxynitriding of titanium alloys[6-8].

The aim of this study was to produce Ti-Al intermetallic layers on TA6V titanium alloy by a duplex method. This applied method consisted of vacuum evaporation of aluminum and glow discharge assisted heat treatment of the deposited films. This paper presents the results of the investigation of the microstructure of these layers.

Experimental procedure

The material used was a TA6V titanium alloy (wt%: 6,4Al, 4,1V, 0,1C, 0,16Fe, 0,18O$_2$, 0,01N$_2$, 0,003H$_2$, balance Ti). The alloy was delivered in the shape of rolled bars of 12mm in diameter. The samples were cut from the bars in slices of 3 mm thickness and then polished. The aluminum coating on the titanium alloy substrate was formed by vacuum evaporation in vacuum of 10^{-4} Pa. The thickness of the aluminum coating was assessed to be equal to 5-6 µm. Samples with aluminum film were annealed under glow discharge conditions at 680°C in an argon atmosphere.

The microstructure, phase and chemical composition of the layers were examined by a optical microscope, a scanning electron microscope (SEM) with an energy dispersive spectroscopy (EDS) unit, and by an X-ray diffraction analysis. The surface topography was investigated using a scanning profilometer.

Results and discussion

Fig. 2 shows the microstructures of TA6V in the initial state (Fig. 2a) and after the duplex treatment (Fig. 2b, 2c). The microstructure of TA6V in the initial state consists of two phases (α+β). The equiaxed microstructure of titanium alloy did not change during the surface treatment. The thickness of the layers was discerned to be about 10µm.

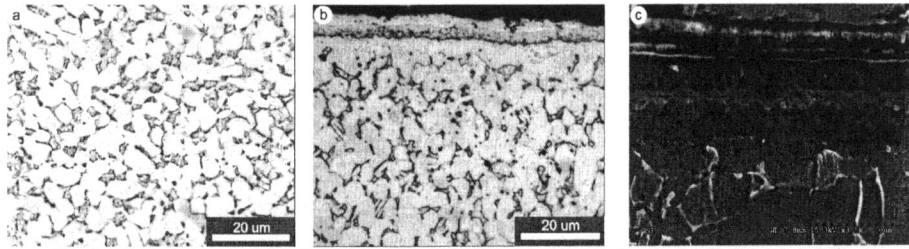

Fig. 2. Microstructure of TA6V titanium alloy in initial state (a) and after duplex heat treatment (b, c)

The structure of the layers is composed of a number of sublayers (Fig.2c). The composite, diffusion character of these layers was confirmed by SEM examination. An energy dispersive spectrometer (EDS) analysis of the chemical composition revealed the presence of aluminum, titanium, oxygen and vanadium. The chemical composition of these sublayers is given in Table 1. Fig 3 shows the microstructure of the layer, indicating the measurement point of the chemical analysis.

Table 1. Chemical composition of the layers

measurement	Chemical composition [%at.]			
	Ti	Al	O$_2$	V
1	9,53	42,53	47,49	0,44
2	30,35	68,06	-	1,59
3	34,83	63,62	-	1,55
4	50,04	47,65	-	2,31
5	72,94	23,99	-	3,06
6	85,15	12,38	-	2,47
7	85,69	11,68	-	2,63
8	87,78	9,66	-	2,56

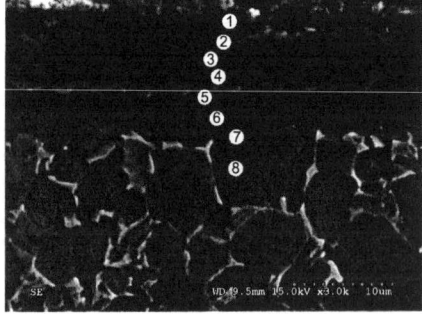

Fig. 3. Microstructure of the layer (SEM)

As we can see, in the first region of the chemical analysis, the most abundant elements are oxygen and aluminum. This suggests that the oxide present on the surface is Al_2O_3. It was only this region where oxygen was detected. Going through the layer, the content of titanium was evidence to raise and that of aluminum to decrease. The chemical content ratio of titanium and aluminum was evaluated to be equal to 1:2,25 in the region "2", 1:1 in the region "4", 3:1 in the region "6". The phases present in the layer were identified using an X-Ray phase analysis (Fig. 4). We can see that the glow discharge assisted heat treatment process applied to the TA6V alloy pre - coated with an aluminum film gives surface layers of the $Al_2O_3+TiAl_3+TiAl+Ti_3Al$ type (Fig. 5).

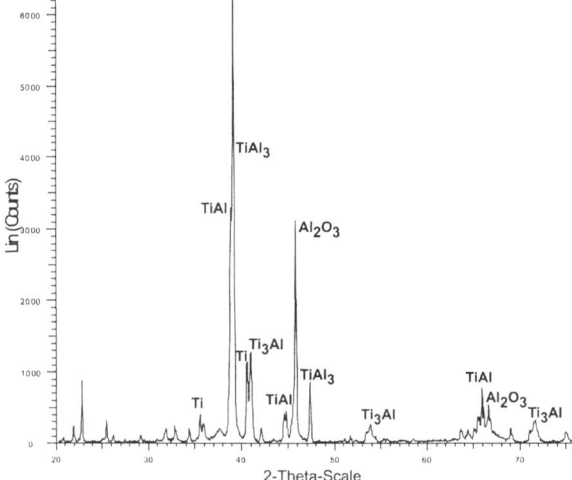

Fig. 4. X-Ray diffraction patterns of the TA6V titanium alloy coated with a composite layer

Fig. 5. Scheme of the produced layer

Fig. 6. shows the surfaces of titanium alloy substrate after polishing (Fig.6a), after the deposition of aluminum (Fig.6b) and after glow discharge assisted heat treatment (Fig.6c). As we can see the topography of the aluminum film is similar to the topography o the substrate surface (Fig. 6b). The glow discharge assisted heat treatment, on the other hand, resulted in increased development of the sample surface. This treatment yielded a thin Al_2O_3 layer, free of pores and cracks.

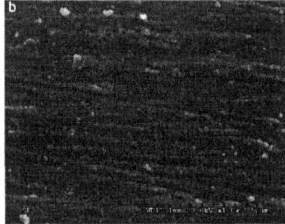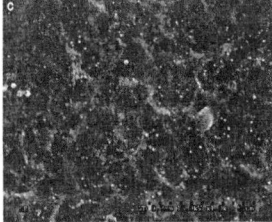

Fig. 6. The surfaces of TA6V titanium alloy after polishing (a), after aluminum deposition (b) and after glow discharge assisted heat treatment (c)

The roughness of the surface was determined using a scanning profilometer (Fig 7, Table2). It can be seen that, after the deposition of aluminum, the surface roughness is similar to that after polishing. The glow discharge assisted heat treatment, on the other hand, increases the roughness, which is due to the cathode sputtering effect that occurs when the layers are heated under glow discharge conditions.

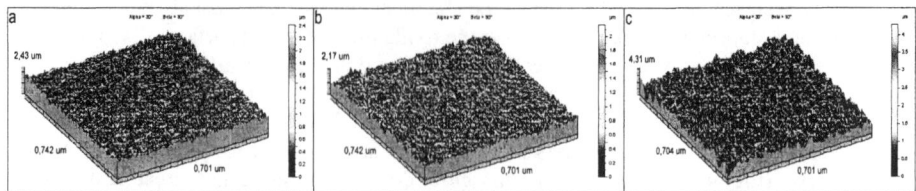

Fig. 7. Surface topography: a) after polishing; b) after aluminum deposition; b) after glow discharge heat treatment

Table 2. Roughness of surface after polishing, aluminum deposition and glow discharge heat treatment; Ra-average arithmetical roughness; Rq-mean square deviation of roughness; Rt-maximum height of roughness

	Ra [μm]	Rq [μm]	Rt [μm]
After polishing	0,111	0,15	2,43
After aluminum deposition	0,106	0,144	2,17
After glow discharge heat treatment	0,37	0,465	4,31

The effect of the layer, thus produced, on the mechanical properties of the TA6V titanium alloy was examined by measuring the surface microhardness of the samples (Fig.8).

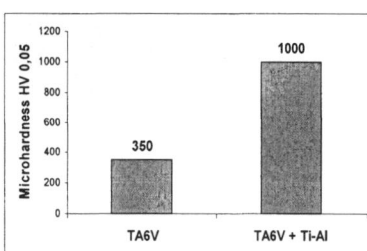

Fig. 8. Microhardness of TA6V in the initial state and when coated by composite layer

The microhardness of the material with the composite layer formed on its surface appeared to be three times as high as that of the material in the initial state. This substantial increase of the surface microhardness can be attributed to the present of Al_2O_3.

Conclusion

The following conclusions can be drawn from present study. The duplex method, in which the TA6V alloy is first pre-coated with aluminum and then subjected to the glow discharge treatment, permits producing multicomponent intermetallic Ti – Al layers with an external zone of Al_2O_3. The layers have diffusion character and their phase structure is: Ti_3Al + TiAl + $TiAl_3$ + Al_2O_3. The layers increase substantially the surface microhardness of the TA6V alloy. The improvement of the mechanical properties of the alloy achieved applying the duplex treatment may widen significantly its application range.

Acknowledgment
The study was supported from grant no. PW-004/ITE/02/2005

References
[1] A. Fleszar, T. Wierzchoń, Sun Kyu Kim, J.R. Sobiecki: Surface and Coating Technology 131 (2000), p.62
[2] I. Gurrapa, A.K. Goagia: Surf. Coat. Technol. 139 (2001), p.216
[3] C. Leyens, M. Peters, W.A. Kaysser: Surf. Coat. Technol. 94-95 (1997), p.34
[4] T. Wierzchoń: Surf. Coat. Technol. 180-181 (2004), p.461
[5] A. Czyrska-Filemonowicz, P.A. Buffat, T. Wierzchoń: Scri. Materialia 53 (2005), p.1439
[6] J. R. Sobiecki, T. Wierzchoń: Vacuum 79 (2005), p.203
[7] E. Czarnowska, T. Wierzchoń, A. Maranda-Niedbała, E. Kaczmarewicz: Journal of Material Science: Materials In Medicine 11 (2000) p.73
[8] T. Wierzchoń, A. Fleszar: Surf. Coat. Technol. 97 (1997), p.205

Key Engineering Materials Vol. 333 (2007) pp 289-292
online at http://www.scientific.net
© 2007 Trans Tech Publications, Switzerland

Monocomponent Powders and Thin Films in the Binary System SiO₂ – TiO₂

M. Gartner, M. Crisan, L. Predoana, M. Zaharescu, A. Barau, S. Preda[a]

Institute of Physical Chemistry "Ilie Murgulescu", Romanian Academy, 202 Splaiul Independentei, 060021, Bucharest, Romania

[a] predas@icf.ro

Keywords: TiO₂ nanopowders, SiO₂ nanopowders, SiO₂-TiO₂ thin films.

Abstract. In this work, we report the sol-gel alkoxide route preparation of nanostructured SiO_2 and TiO_2 powders as well as TiO_2-SiO_2 thin films obtained by dip-coating. Thermal analysis, morphology and structure were characterized for powders and correlation between preparation method and optical properties of binary materials (SiO_2-TiO_2) for thin films was approached. Spectroscopic Ellipsometry (SE), Fourier Transform Infrared Spectroscopy (FTIR) and scanning electron microscopy (SEM) have been used for the physical characterization of the films.

Introduction

Sol-gel synthesis was extensively investigated starting in 1950s since films with controlled composition, thickness and multiple layers were obtained [1]. The most promising advantage of this method consists in the fact that materials with predetermined structure were obtained by varying the experimental conditions. It is well-known that alkoxides are the most common and well-established precursors used in the sol-gel route. Hydrolysis-condensation with small water content lead to linear polymers from which thin films can be deposited, before gelation of the sol, by common processes such as dip- and spin-coating. The process represents a powerful tool for making nanopowders and different types of transparent materials with interesting mechanical, electrical, optical and photonic properties. A particular emphasis on the SiO_2-TiO_2 system was given for the intended application as antireflection and protective coatings [2], active [3] and passive [4] optical planar waveguides, channel waveguides [5], interference filters [6] weather resistant optical memory discs [7] bone-bonding interfaces for load bearing implants [8], barriers to corrosion induced by high temperature, atmospheric exposure, H_2SO_4 attack or concentrated electrolyte immersion [9, 10].

In this work, we report the experiments performed for synthesizing monocomponent nanostructured oxides and coating thin films in the binary system by sol-gel method.

Experimental

Powder preparation. The composition of the starting solution and the experimental conditions used are as follows: a) for SiO_2 powder, the reagents were ethyl orthosilicate (TEOS), (Serva), absolute ethylic alcohol p.a. reagent (Reactivul) as solvent, ammonia water p.a. reagent (Reactivul) as catalyst and distilled water in the following molar ratios: $C_2H_5OH : H_2O : NH_4OH : TEOS = 54.6 : 30 : 2 : 1$, at a pH of 8.5 for 6 hours magnetic stirring; b) for TiO_2 powder, the reagents were tetraethyl orthotitatanate (TEOT), (Aldrich), absolute ethylic alcohol p.a. reagent (Reactivul) as solvent and distilled water in the following molar ratios: $C_2H_5OH : H_2O : TEOT = 85.5 : 5 : 1$ at a pH of 5.5 for 30 min. magnetic stirring. All experiments were made at room temperature.

Films preparation. In the binary systems (with molar composition: 70SiO_2-30TiO_2) a two steps method of preparation of the solutions was approached. In the first step, $Si(OC_2H_5)_4$ was prehydrolysed with under-stoichiometric amount of water. In the second step, alcoholic solutions of $Ti(OC_2H_5)_4$ was added to the prehydrolized $Si(OC_2H_5)_4$ and the mixture (pH = 4) was reacted under vigorous stirring for 2h at 50°C. The hydrolysis reaction was carried out in a closed system, flushed with purified nitrogen, via a dropping funnel.

Supported SiO_2-TiO_2 films on silicon wafer substrates and silicon wafer/SiO_2 substrates have been obtained by dipping, (with a withdrawal speed of 5 cm/min) and by spinning (2000 rot/min) using the solutions of composition given in [11]. Before deposition the solutions were left for ageing 24 hours. The densification of the films was mainly realised by thermal treatment for 1 hour at 300°C, with heating rate of 1°C/min (we tried also thermal treatment of monolayer films in 200-600°C temperature range). For multi-layered films after each deposition, the same thermal treatment was applied.

Methods of characterization

Powders were characterized by DTA/TG analysis, using a MOM OD 103 Derivatograph between 20-1400°C with a heating rate of 7.5°C/min., TEM with JEOL-FX-2000 equipment, XRD with MXP 18 (Mac Science Co Ltd) equipment, BET adsorption of N_2 gas at the temperature of liquid N_2 for specific surface area determination and particle size distribution on Zetasizer ZS Red Badge ZE 2600 in ethanol.

The optical properties of SiO_2-TiO_2 films were determined by spectroellipsometry (SE) measurements in the visible range (0.4-0.7 µm). Their morphology was realized by scanning electron microscopy (SEM). The species existing on the surface of the films were detected by Fourier Transform Infrared Spectroscopy (FTIR).

Results and discussion

Powders. Due to the wet chemical method of synthesis, sol-gel powders contain adsorbed and structural water, as well as organic residues, which need to be eliminated by thermal treatment at different temperature, that are determined according to the results of the DTA/TGA, as follows: TiO_2 and SiO_2 powders losses the water and organic residues up to 300°C. Both types of powders need a thermal treatment at 450°C for water and organic residues removal.

The TEM micrographs have shown significant differences in the morphology of the obtained powders, as observed in Fig. 1 a) for SiO_2 powder and b) for TiO_2, respectively. TiO_2 powders with 5 nm particles have a tendency of agglomeration in aggregates of ~50 nm. SiO_2 powders consist of spherical particles of a mean size of 200 nm with no aggregation tendency. The results from TEM were confirmed by granulometric analysis presented in Fig. 2. Mean size of SiO_2 particles was ~200 nm and TiO_2 particles were agglomerated in ~700 nm aggregates.

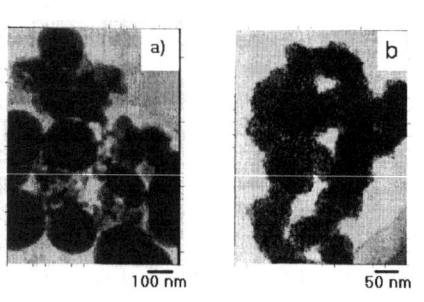

a) SiO_2 powder   b) TiO_2 powder

Figure 1 TEM micrographs of the sol-gel monocomponent powders

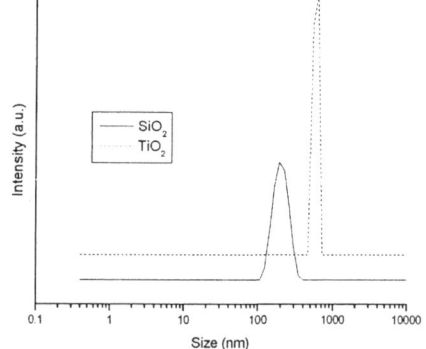

Figure 2 Particle size distribution of TiO_2 and SiO_2 powders in ethanol

From XRD patterns one observed a tendency to nano-crystallization of the TiO_2 powder to a crystalline anatase. SiO_2 powder remained into an amorphous structure.

The BET specific surface area data are in good agreement with TEM observations and the powders are of 13.38 m^2/g for SiO_2 and 154.18 m^2/g for TiO_2 respectively.

Films. In the conditions used, continuous and homogeneous films with a good adherence to the substrates were obtained.

SE measurements. The modelling of the experimental ellipsometric spectra was made with the multilayer and multicomponent Bruggemann's Effective Medium Approximation - BEMA model [12]. For the coatings deposited on glass or silicon substrate one layer model (SiO_2-TiO_2 film /glass or Si) was used. For the coatings deposited on SiO_2/silicon substrate two layers model (SiO_2-TiO_2 film/SiO_2/Si) was used. The parameters of the fit were: the thickness of the layers and the volume fraction of the components (SiO_2, TiO_2, air) From the best fit one obtained the refractive index (n), the thickness of the layers (d) and volume fractions of the components (Table 1). In the visible range, all films are transparent ($k<10^{-4}$).

The analysis of the SE results shows the influence of different parameters on the optical properties of the $70SiO_2$-$30TiO_2$ (%mol) films, such as:

Influence of the number of layers. The thickness of the SiO_2-TiO_2 films grows approximately linearly with the number of layers (Table 1).

Influence of the substrate. Our results are in agreement with the Yoldas' statement [13] that the topography of the substrate influences the densification/ crystallization of the films. The films deposited on the silicon substrate are denser and thinner then those deposited on SiO_2/Si or glass.

Table 1 The volume fraction of the components and the thickness of the layers obtained from SE measurements on mono and multilayer $70SiO_2$-$30TiO_2$(%mol) films

Support	DM	NL	NTT	SiO_2 [%]	TiO_2 [%]	Air [%]	D_{film} [Å]
Glass	dipping	1	0	72.57	17.58	9.85	1420
		1	1	76.33	23.65	0.02	1022
		2	2	70.00	30.00	0	1720
		3	3	70.00	30.00	0	3190
SiO_2/Si	dipping	1	0	69.42	30.58	0	1460
		1	1	70.00	30.00	0	930
		2	2	70.00	30.00	0	1797
	spinning	1	1	69.86	30.14	0	1570
Si	spinning	1	1	70.00	30.00	0	1395
		4	4	70.00	30.00	0	6792

*DM- deposition method; NL-number of layers; NTT-number of thermal treatments

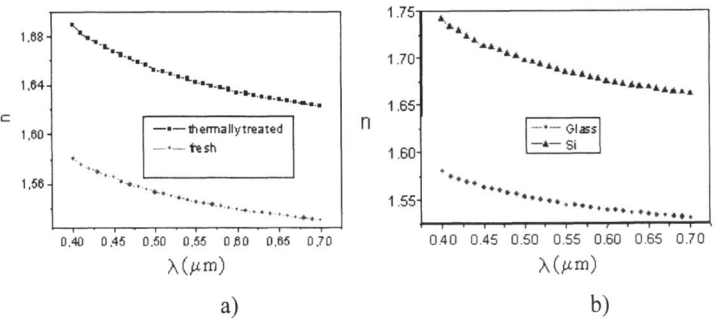

a)   b)

Figure 3 Dispersion of the refractive index of a monolayer of SiO_2-TiO_2 deposited by dipping: a) fresh and a thermally treated film deposited on glass, b) fresh film deposited on glass and silicon

Influence of the deposition method. The films deposited by spinning are more uniform and thicker than the films deposited by dipping (Table 1).

Influence of the thermal treatment. The lowest temperature that allows the formation of continuous, dense oxide films, free of organic traces, is 300°C. All films obtained below this temperature exhibit variation of optical constants with the thermal treatment. After a thermal treatment (1h at 300°C) the films are densified, decreasing in thickness with 20-30% (depending on the type of substrate used).

FTIR measurements. IR assignments of species existing on the surface are: 3400 cm^{-1} and 1600 cm^{-1} – bonding modes of the characteristic vibrations of structural OH and molecular water; 1160-1080 cm^{-1}–LO and TO modes of Si-O-Si; 950-800 cm^{-1} – stretching and bending vibrations of Si-O-Si; 920 cm^{-1} and 480 cm^{-1} – formation of the gel with tetracoordinated titania; 600 cm^{-1} - TiO$_2$.

IR assignments of species existing on the surface are: 3400 cm^{-1} and 1600 cm^{-1} – bonding modes of the characteristic vibrations of structural OH and molecular water; 1160-1080 cm^{-1}–LO and TO modes of Si-O-Si; 950-800 cm^{-1} – stretching and bending vibrations of Si-O-Si; 920 cm^{-1} and 480 cm^{-1} – formation of the gel with tetracoordinated titania; 600 cm^{-1} - TiO$_2$.

It was observed that IR spectra of films and gels are similar in the region of lattice vibrations. The abundant surface hydroxyl population diminishes with increasing temperature, this process being very important for films, the annealing at T<600°C is recommended.

Conclusions

Monodispersed, monomodal and spherical particles of SiO$_2$, and nanometric but agglomerated anatase TiO$_2$ particles were obtained by sol-gel method.

Continuous and homogeneous films with a good adherence on the substrate (glass, silicon and SiO$_2$/silicon) in the binary system (SiO$_2$-TiO$_2$) have been obtained by sol-gel (dipping or spinning) method, alkoxide route.

The influence of initial solutions composition and of the different experimental conditions (as type of support, method and number of depositions and thermal treatment) on the optical properties of the films has been investigated.

Acknowledgement: Programme of Excellence, Project Acronym COMPPPCNTVNPO, no. 29/2005

References

[1] C.J. Brinker and G.W. Scherer: *Sol-gel science, the physics and chemistry of sol-gel processing* (Academic Press, San Diego, CA, 1989)
[2] H. Schroeder, in: *Physics in thin films: Advances in research and development*, ed. G. Hass and R.E. Thun (Academic Press, New York, 1969) p. 87
[3] R.M. Almeida, X. Orignac and D. Barbier: J. Sol-Gel Sci. Technol. Vol. 2 (1994), p. 465
[4] M. Bahtat, J. Mugnier, C. Bovier, H. Roux and J. Serughetti: J. Non-Cryst. Solids Vols. 147&148 (1992), p. 23
[5] A.S. Holmes, R.R.A. Syms, M. Li and M. Green: Appl. Opt. Vol. 32 (1993), p. 4916
[6] B.D. Fabes, D.B. Birnie III and B.J.J. Zelinski: Thin Solids Films Vol. 254 (1995), p. 175
[7] A. Matsuda, Y. Matsuno, S. Katayama and T. Tsuno: J. Mater. Sci. Lett. Vol. 8 (1989), p. 902
[8] M. Shirkhanzadeh: J. Mater. Sci. Mater. Medicine Vol. 6 (1995), p. 206
[9] M. Atik, P. de Lima-Neto, L.A. Avaca, M.A. Aegerter and J. Zarzycki: J. Mater. Sci. Lett. Vol. 13 (1994), p. 1081
[10] M. Atik, P. de Lima-Neto, L.A. Avaca and M.A. Aegerter: Ceram. Int. Vol. 21 (1995), p. 403
[11] M. Zaharescu, C. Pirlog, M. Crisan, M. Gartner, M. Sahini and A. Vasilescu: Rev. Roum. Chim. Vol. 37 (1992), p. 173
[12] D.A.G. Bruggeman: Ann. Phys. Vol. 24 (1935), p. 636
[13] B. Yoldas: Appl.Opt. Vol. 21 (1982), p. 1982

Author Index

Keyword Index